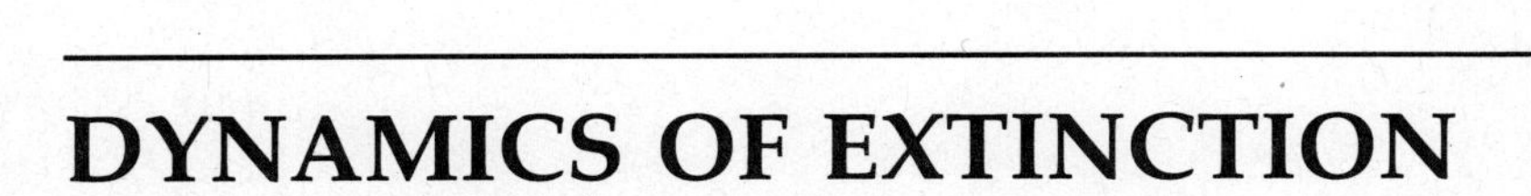

DYNAMICS OF EXTINCTION

DYNAMICS OF EXTINCTION

Edited by

DAVID K. ELLIOTT

Geology Department
Northern Arizona University
Flagstaff, Arizona

A Wiley-Interscience Publication

JOHN WILEY & SONS

New York / Chichester / Brisbane / Toronto / Singapore

Library of Congress Cataloging-in-Publication Data:

Main entry under title:

Dynamics of extinction.

"A Wiley-Interscience publication."
1. Extinction (Biology) I. Elliott, David K., 1947–
QE721.2.E97D96 1985 575′.7 85-17974
ISBN 0-471-81034-7

Printed in the United States of America

10 9 8 7 6 5 4 3 2

CONTRIBUTORS

WILLIAM J. BOECKLEN, Department of Biological Sciences and Department of Mathematics, Northern Arizona University, Flagstaff, Arizona

WILLIAM A. CLEMENS, Department of Paleontology, University of California, Berkeley, California

EDWIN H. COLBERT, Department of Geology, The Museum of Northern Arizona, Flagstaff, Arizona

PAUL R. EHRLICH, Department of Biological Sciences, Stanford University, Stanford, California

RUSSELL W. GRAHAM, Quaternary Studies Center, Illinois State Museum, Springfield, Illinois

DAVID JABLONSKI, Department of the Geophysical Sciences, University of Chicago, Chicago, Illinois

JERE H. LIPPS, Department of Geology, University of California, Davis, California

PAUL S. MARTIN, Department of Geosciences, Laboratory of Paleoenvironmental Studies, The University of Arizona, Tucson, Arizona

DIGBY J. MCLAREN, Department of Geology, University of Ottawa, Ottawa, Ontario, Canada

DAVID M. RAUP, Department of the Geophysical Sciences, University of Chicago, Chicago, Illinois

J. JOHN SEPKOSKI, JR., Department of the Geophysical Sciences, University of Chicago, Chicago, Illinois

DANIEL SIMBERLOFF, Department of Biological Science, Florida State University, Tallahassee, Florida

GEERAT J. VERMEIJ, Department of Zoology, University of Maryland, College Park, Maryland

PREFACE

Ideas concerning the causes of extinctions of organisms have been debated in paleontology for many years. However, in 1980, these discussions were focused by a paper authored by Walter Alvarez and others suggesting that the Cretaceous–Tertiary boundary was marked by the impact of a very large extraterrestrial body. High concentrations of iridium at the boundary were cited as the initial evidence for the impact, and it was postulated that this event might be related to the widespread extinctions of plants, invertebrates, and vertebrates that are seen in the fossil record at this level. Since then this contention seems to have been strengthened by the discovery of similar iridium anomalies occurring at more than 50 sites of similar age around the world and also at other levels at which extinction events have taken place.

The findings of the Alvarez group had a profound effect on thinking about the causes of extinction, and debate raged about the strengths and weaknesses of the proposed hypothesis. In particular, it became important to know in more detail exactly when groups of organisms became extinct and also if they were already waning before the final extinction event occurred. Detailed work in this area has now shown that a period of reduction in faunal diversity did in fact precede the Cretaceous–Tertiary extinctions. This has also resulted in the work by Raup and Sepkoski (1984, *Proceedings of the National Academy of Sciences*, **81,** 801–805) demonstrating a 26-m.y. periodicity to the pattern of extinctions through Phanerozoic time. Most recently, this periodicity has been related to a similar periodicity of meteor impacts (Whitmore et al. and Davis et al., *Nature*, April 19, 1984). They postulate that an as yet undiscovered solar companion, called Nemesis, may disrupt the orbits of comets and asteroids periodically, resulting in a much greater incidence of impact events.

Discussion on this topic has been interdisciplinary, bringing together evidence from astronomy, biology, geochemistry and paleontology, and other areas. It was felt therefore that a conference on this subject should have as its main aim the fostering of this interdisciplinary debate. Hence the contributors to the conference entitled "Dynamics of Extinction," held at Northern Arizona University, August 9–11, 1983, were selected for their ability to bring expertise from different disciplines to bear on a solution to the problems of extinction events. This volume includes contributions from most of the speakers at that conference and, therefore, reflects its wide-ranging treatment of the subject. However, the volume was not intended as simply a record of what transpired at the meeting and hence each author has updated and revised his contribution to reflect the most recent developments in what is a very rapidly evolving subject.

The conference was the second in an annual series sponsored by the Ralph M. Bilby Research Center of Northern Arizona University, and both the conference and this volume were made possible by the continued development of research fostered by Eugene Hughes, President, and Joseph Cox, Vice President for Academic Affairs. Considerable support was also given by Henry Hooper, Dean of Graduate Studies, who provided funds to cover the expenses of the conference, and Richard Faust, Director of the Bilby Research Center, who provided organizational expertise and other valuable help.

I would also like to thank those who helped with the organization of the conference or who reviewed manuscripts for this volume: Larry D. Agenbroad, Charles W. Barnes, Stanley S. Beus, Ronald C. Blakey, William J. Boecklen, Augustus C. Cotera, Richard Estes, George E. Goslow, Russell W. Graham, David Jablonski, Paul S. Martin, Larry T. Middleton, J. Dale Nations, John H. Ostrom, J. John Sepkoski, Jr., and Timothy D. Walker.

DAVID K. ELLIOTT

Flagstaff, Arizona
February 1986

CONTENTS

PART IV MODERN EXTINCTIONS

PART V MODELING EXTINCTION EVENTS

PART I

EXTINCTIONS AND THE FOSSIL RECORD

1

PERIODICITY IN MARINE EXTINCTION EVENTS

J. JOHN SEPKOSKI, JR.

DAVID M. RAUP

Department of the Geophysical Sciences
University of Chicago
Chicago, Illinois

INTRODUCTION

Mass extinction is perhaps the most enigmatic of paleontological phenomena. Although much has been written about extinction and many causal hypotheses offered, paleontology still lacks a coherent theory for why numerous lineages among diverse taxa in disparate regions of the globe should disappear simultaneously at certain horizons in the geologic record. Traditionally it has been assumed that each event of mass extinction is essentially unique and can be explained independently of other events. Explanations generally have invoked changes in factors that might affect the biotic environment, such as climate, global habitat area, continental configuration, orogenic activity, oceanic or atmospheric chemistry, and extraterrestrial phenomena (see Flessa, 1979). Because these factors vary irregularly or even randomly through time and because each extinction is viewed as unique, it has usually been assumed, at least implicitly, that mass extinctions are irregularly distributed in time with the timing of each event independent of that of the others.

Fischer and Arthur (1977) presented a radical alternative to this implicit viewpoint. On the basis of large-scale data on diversity in Mesozoic and Cenozoic pelagic ecosystems, they argued that extinction events are not independent but rather have recurred at regular intervals of approximately 32 m.y. duration (see also Fischer, 1981). In 1984, we (Raup and Sepkoski) offered statistical support for Fischer and Arthur's hypothesis. We showed that "peaks" (i.e., local maxima) in the percentages of marine animal families becoming extinct in each of 39 stratigraphic stages between the Permian and Tertiary displayed a statistically significant tendency to occur at regularly spaced intervals, although we found the mean waiting time to be closer to 26 m.y. (evidently reflecting mostly differences between the time scales used in our and Fischer and Arthur's analyses). We interpreted this pattern as indicating that extinction events have occurred periodically through time and to imply a single, ultimate forcing agent rather than a plethora of independent causes.

In this chapter we wish to examine the periodicity of extinction events in more detail. In particular, we wish to analyze the temporal distribution of specific, identifiable extinction events as opposed to the continuous fluctuations in extinction intensity examined in our previous work. Below, we first discuss the nature and limitations of the data base on the global fossil record in order to establish limits of resolution in statistical analyses. We then consider which peaks in extinction intensity appear to differ significantly from background levels and present new analyses of the temporal distribution of these peaks. Finally, we consider some possible causes of periodicity and of interdependence among extinction events over the last quarter billion years of Earth history.

DATA ON GLOBAL EXTINCTIONS

The primary data base for the analyses presented here is the same as used by Raup and Sepkoski (1984). The data are derived from the *Compendium of Fossil Marine Families* (Sepkoski, 1982a) with various corrections and additions. This compendium consists of a compilation of the stratigraphic ranges of some 3500 marine families of protozoans, invertebrates, and vertebrates listed within a uniform stratigraphic framework of 82 international stages. The families represent the entire preserved fauna of a single, well-defined major ecosystem, the world ocean, which has a far more complete and better documented fossil record than its counterparts in nonmarine ecosystems. Data have been compiled (but not published) for nonmarine animals, but the derived patterns of diversity and extinction appear strongly influenced by the stratigraphic distribution of fossiliferous rocks (see Padian and Clemens, 1985; Sepkoski and Hulver, 1985; Benton, 1985) and therefore are not as amenable to statistical analysis as are patterns for marine taxa.

We have concentrated our analyses on data from the mid-Permian (Leonardian Stage) to the Pleistocene (which is a slightly longer interval than analyzed in Raup and Sepkoski, 1984). Data from earlier portions of the Paleozoic have lower taxonomic and especially stratigraphic resolution. There are 43 stages between the mid-Permian and Pleistocene providing an average resolution of approximately 6 m.y. on familial extinctions. This contrasts with the 9.5 m.y. resolution afforded by the 34 stages of the pre-Leonardian Paleozoic. Approximately 1800 marine animal families (exclusive of soft-bodied taxa) have been described from mid-Permian to Pleistocene strata. Of these, 970 families are extinct. The largest taxonomic group represented among these extinct families is the Cephalopoda (chiefly

Ammonoidea), which contributes 25% of the extinctions. Other important groups include the echinoderms (12% of extinct families), vertebrates (12%), brachiopods (9%), sponges (8%), gastropods (7%), and bivalves (7%).

Although these data provide a large base for statistical analysis, they are not without limitations. There are a number of problems that affect taxonomic, stratigraphic, and chronometric resolution, as outlined below.

1. *Taxonomic Resolution.* Families are rather large and arbitrary taxonomic units that on average contain about 10^2 species. Thus, some variation in evolutionary rates among species is lost at the family level (Valentine, 1974; Raup, 1975, 1979). As McLaren (1983, 1984) has pointed out, the simultaneous extinction of a few very abundant species might constitute an important event but one that would be unrecognized at the family level unless accompanied by extinctions of species within smaller families. However, as discussed below, even the smaller mass extinctions first documented in detailed, local biostratigraphic studies are recognizable in the Mesozoic–Cenozoic familial data.

2. *Stratigraphic Resolution.* Stages are the smallest stratigraphic units that can be consistently recognized for all fossil groups on a worldwide basis. However, because stages have durations of several million years each, calculated extinction rates represent averages for intervals that may be considerably longer than any given extinction event. Thus, these rates may not accurately reflect the magnitude of an event and cannot be used to determine the abruptness or fine-scale patterns of extinction within it (McLaren, 1984; Flessa et al., 1986; Sepkoski, 1986). In our analyses, we are further limited in that we cannot recognize distinct extinction events separated by less than one stage. This problem does not appear critical for the Mesozoic and Cenozoic based on what is known about extinction events in these eras (Sepkoski, 1982b). It may be more important, though, for the Paleozoic where extinction events are not as well documented and occasionally may be clumped in time (Sepkoski; 1986). McGhee (1982) and House (1985), for example, have described several pulses of extinction of brachiopods and ammonoids, respectively, separated by only a few million years in the Devonian; with our data base, we can see only one peak in global extinction in the contiguous stages (cf. Raup and Sepkoski, 1982).

3. *Chronometric Resolution.* A third problem in determining magnitudes and patterns of extinction in global data is the uncertainty in the absolute time scale (cf Hallam, 1984). This uncertainty arises from experimental, geologic, and stratigraphic errors in aging calibration points and then the use of relatively small numbers of calibration points to interpolate ages of most stage boundaries. In the analyses presented here, we have used the geologic time scale of Harland et al. (1982). This time scale is similar to that of Armstrong (1978) and Palmer (1983) and seems generally preferable to that of Odin (1982) (see Dalrymple, 1983). Harland et al. (1982) provide estimates of uncertainties in the listed ages of stage boundaries, which range up to 34 m.y. for parts of the Jurassic.

Despite these limitations, the familial data still provide opportunities for studying extinction that are unparalleled in other data sets:

1. The familial data are global and therefore avoid problems of local hiatuses, facies changes, and so on, that affect higher-resolution data from individual stratigraphic sections and cores.
2. The data are comprehensive, covering all marine animal groups that have a fossil record; no data set with comparable stratigraphic resolution has yet been compiled for all animals at lower taxonomic levels.
3. The data for the most part are uniformly sampled over the entire fossil record, permitting unbiased estimates of long-term extinction patterns and allowing comparisons of patterns from one geologic interval to another.

LATE PHANEROZOIC EXTINCTION PATTERNS

Measures of Extinction Intensity

Given the interval nature of the compilation of stratigraphic ranges, four basic metrics of the intensity of familial extinction can be computed for each stage:

1. Simple number of extinctions.
2. Percent (or proportion of) extinctions (=number of extinctions relative to standing familial diversity).
3. Total rate of extinction (=number of extinctions relative to stage duration).
4. Per-family (or per-capita) rate of extinction (=total rate relative to standing diversity).

Each of these metrics has limitations, and none is clearly superior to the others. Simple number of extinctions is the least assuming, but it does not scale for either the number of families at risk or the time interval over which extinctions occur. Percent extinction scales for families at risk but in so doing adds statistical error since standing diversity is not precisely known. Total extinction rate scales for time interval but introduces even more error (perhaps more than 100% in some instances) since some stage durations may be poorly estimated. Finally, per-family extinction rate scales for both families at risk and time interval but is probably subject to the greatest amount of statistical error.

Rather than selecting a single metric for analysis, as was done in our previous analysis, we have chosen here to use all four measures of extinction intensity. The data for each of the metrics over all 43 stages from the Leondardian to Pleistocene are listed in Table 1 and illustrated in Figure 1. These data (and particularly that for percent extinction) differ from those analyzed by Raup and Sepkoski (1984) in that all fossil families, rather than just those that are now extinct, have been included. We have also used families with times of extinction known only to the series or system level, distributing these low-resolution data evenly among the stages within the coarser stratigraphic units. The only families we have excluded are the soft-bodied and lightly sclerotized taxa (e.g., medusoid coelenterates and

Table 1. Data for Metrics of Familial Extinction Analyzed in This Study[a]

Geologic Stage		Raup and Sepkoski (1984)	Extinction metrics: Simple Number	Percent	Total Rate	Per-Family Rate
Q:	Pleistocene	–	1	0.1	0.5	0.06
T:	Pliocene	–	6	0.8	2.0	0.25
	Upper Miocene	–	11	1.3	1.7	0.21
	Middle Miocene	25.0	12	1.3	3.7	0.43
	Lower Miocene	12.5	8	1.0	0.7	0.10
	Upper Oligocene	20.0	8	1.1	1.0	0.13
	Lower Oligocene	29.4	9	1.2	1.8	0.23
	Upper Eocene	45.8	15	2.0	3.8	0.50
	Middle Eocene	36.7	16	2.2	1.9	0.26
	Lower Eocene	11.1	7	1.1	1.7	0.25
	Paleocene	21.9	} 16	} 2.7	} 1.6	} 0.26
	Danian	22.2				
K:	Maestrichtian	66.3	90	13.9	11.2	1.74
	Campanian	14.7	28	4.4	2.8	0.44
	Santonian	9.2	20	3.3	4.4	0.72
	Coniacian	5.8	17	2.8	16.8	2.82
	Turonian	9.9	15	2.5	5.9	1.01
	Cenomanian	18.9	36	6.1	5.5	0.93
	Albian	12.0	18	3.4	1.2	0.22
	Aptian	10.2	15	3.1	2.5	0.52
	Barremian	6.0	10	2.1	1.7	0.36
	Hauterivian	9.5	12	2.7	2.1	0.45
	Valangian	3.6	8	1.9	1.2	0.26
	Berriasian	3.8	8	1.9	1.4	0.31
J:	Tithonian	19.5	30	6.5	5.1	1.09
	Kimmeridgian	9.8	19	4.4	3.2	0.73
	Oxfordian	6.0	14	3.4	2.1	0.48
	Callovian	7.6	11	2.8	1.9	0.47
	Bathonian	3.5	6	1.6	1.1	0.27
	Bajocian	11.6	13	3.6	1.0	0.27
	Toarcian	7.1	7	2.1	1.2	0.35
	Pliensbachian	15.2	17	5.1	2.8	0.86
	Sinemurian	10.1	9	2.9	1.5	0.49
	Hettangian	2.6	3	1.1	0.4	0.15
Tr:	Rhaetian	15.9	36	12.2	6.0	2.03
	Norian	38.6	36	11.6	6.0	1.94
	Carnian	26.4	36	11.3	6.0	1.88
	Ladinian	15.9	18	6.0	2.6	0.86
	Anisian	12.8	12	4.7	2.5	0.94
	Olenekian	24.0	22	9.1	8.8	3.64
	Induan	21.0	15	6.8	6.0	2.71
P:	Dzhulfian	52.5	81	28.1	16.3	5.61
	Guadalupian	–	154	35.6	30.9	7.12
	Leonardian	–	48	10.8	4.8	1.08

[a]Data on percent extinction from Raup and Sepkoski, computed from a highly culled data set, are included for comparison. Rate metrics are in units of m.y.$^{-1}$.

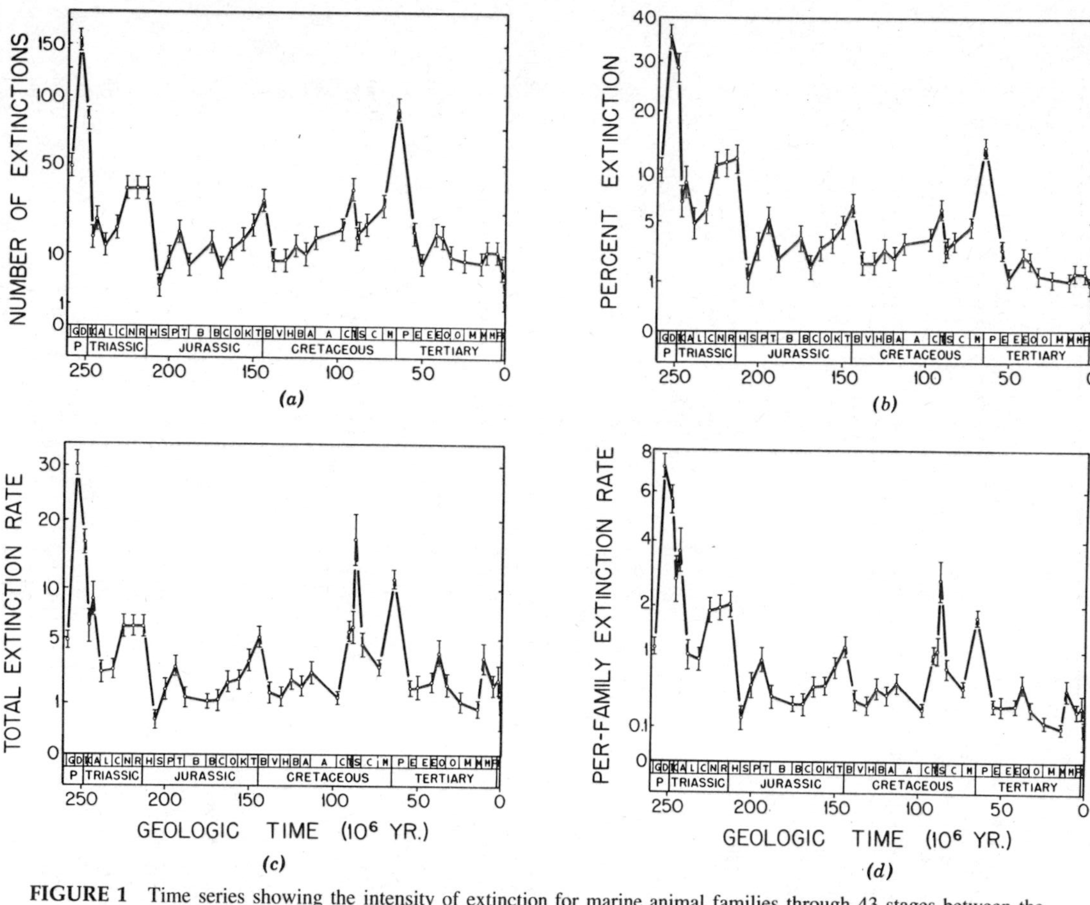

FIGURE 1 Time series showing the intensity of extinction for marine animal families through 43 stages between the mid-Permian and Pleistocene, as measured by (*a*) simple numbers of extinctions per stage, (*b*) percent extinction per stage, (*c*) total rate of extinction per stage, and (*d*) per-family rate of extinction per stage. Error bars indicate one estimated standard error on either side of the observation (see text). A square-root scale is used on the ordinates in order to reduce the variation at higher extinction intensities and to reflect the presumed Poisson character of the extinction metrics.

many small arthropods) whose stratigraphic ranges reflect only the distribution of exceptional fossil deposits (cf. Raup and Sepkoski, 1982).

We have also attempted in this analysis to estimate uncertainties in the extinction metrics, as shown by the "error bars" in Figure 1. The logic employed in estimating standard errors is the same as that used by Raup et al. (1983) and essentially reflects the "counting error" associated with multiple discrete events. We assume that extinction approximates a Poisson process, with the probability of precisely k extinctions occurring during a given interval of time Δt being

$$P(E_i = k) = \frac{e^{-\lambda}\lambda^k}{k!}$$

where E_i is the number of extinctions observed during the ith time interval and λ is the expectation for that interval. Given this model, E_i can be characterized as drawn from a Poisson distribution with a mean of λ and a variance of λ. If the square root of E_i is considered, its frequency distribution should approximate a Gaussian distribution with a mean of $\sqrt{\lambda}$ and a variance of ¼ (see Johnson and Kotz, 1969, p. 99; Bishop et al., 1975, p. 492). Thus, one standard error about the observed number of extinctions can be computed as

$$\left[\sqrt{E_i} \pm \sqrt{\frac{1}{4}}\right]^2$$

This is the error interval illustrated in Figure 1.

Analogous standard errors can be computed for the other extinction metrics if it is assumed that standing diversities (D_i) and stage durations (Δt_i) are known precisely. For percent extinction, E_i/D_i, one standard error about the observation is

$$\left[\sqrt{\frac{E_i}{D_i}} \pm \sqrt{\frac{1}{4D_i}}\right]^2$$

A similar estimate results if E_i/D_i ($=p_i$) is assumed to be a binomial variate and the error is computed as

$$p_i \pm \sqrt{\frac{p_i(1 - p_i)}{D_i}}$$

One standard error about the total extinction rate, $E_i/\Delta t_i$, using the Poisson model, can be estimated as

$$\left[\sqrt{\frac{E_i}{\Delta t_i}} \pm \sqrt{\frac{1}{4\Delta t_i}}\right]^2$$

and about the per-family extinction rate, $E_i/D_i\Delta t_i$, as

$$\left[\sqrt{\frac{E_i}{D_i\ \Delta t_i}} \pm \sqrt{\frac{1}{4D_i\ \Delta t_i}}\right]^2$$

These last two standard errors must be considered minimum estimates (and therefore mostly of heuristic value) since considerable additional error may exist in the value for Δt_i.

Extinction Events in the Permian to Pleistocene Interval

Since fluctuations in measured extinction intensities can reflect sampling noise as well as evolutionary information, it is of interest to know which fluctuations stand out as significantly higher than the normal extinction of most time intervals. One procedure for determining this is to fit a specified function to the data and search for statistical outliers, as done by Raup and Sepkoski (1982). However, because neither the statistical distribution of all extinction intensities nor the exact function for background extinction is known a priori, this procedure may be rather insensitive to outliers (cf. Quinn, 1983; Raup et al., 1983). An alternative procedure is to examine fluctuations only within their local neighborhood. A significantly high fluctuation can be considered as any local maximum in measured extinction intensity that exceeds the two nearest local minima by more than 1.96 standard errors (for 95% confidence). Local minima rather than adjacent stages should be used because the observed effects of even an abrupt extinction events may not be confined to a single stage. As Signor and Lipps (1982) have argued, failure to sample families within the stratigraphic units containing their actual extinctions tends to smear extinction records backward in time (see also Jablonski, this volume); this artifact of sampling can cause observed extinction intensities to increase over one or more stages prior to an extinction event. Conversely, extinction intensities may remain high for one or more stages following an extinction event as a result of high turnover rates in the wake of the extinction event. This poorly understood evolutionary phenomenon has been documented for taxa in the first part of the Triassic following the Late Permian mass extinction (Van Valen, 1984), for foraminifers in the early Danian following the Maestrichtian mass extinction (Smit and ten Kate, 1982), and for trilobites in several intervals of the Late Cambrian following biomere events (Stitt, 1971; Palmer, 1979, 1982).

Using the procedure outlined above, eight stages between the mid-Permian and Pleistocene can be identified as having positive fluctuations in extinction intensity that are significantly above adjacent local minima for at least some of the extinction metrics, as documented in Table 2. These stages constitute only two-thirds of the "extinction peaks" identified by Raup and Sepkoski (1984) but, very importantly, represent the subset of stages that previously have been recognized as containing extinction events on the basis of local biostratigraphic data or global diversity data for species or genera (which further indicates that the familial data are reflecting real evolutionary fluctuations). The eight stages and the nature of their contained extinctions are listed below.

1. *Guadalupian.* This stage and the succeeding "Dzhulfian" (=Tatarian) span the Late Permian mass extinction, the most severe event of the Phanerozoic marine record (Newell, 1967; Sepkoski, 1982b). During these two stages, more than 50% of marine families and possibly more than 95% of marine species became extinct (Raup, 1979; Sepkoski, 1986). Nearly all animal taxa were affected, with brachiopods, crinoids, corals, and cephalopods each losing 20 or more families; among classes, rostroconchs, trilobites, and blastoids disappeared entirely. The extinction metrics in Figure 1 show the Guadalupian as containing a higher intensity of extinction than the "Dzhulfian". It is not clear, however, whether this is real or if it reflects the worldwide dearth of latest Permian marine sections (Kummel and Teichert, 1970). Higher-resolution substage data on ammonoid families and subfamilies compiled from House and Senior (1981) indicate that these well-studied cephalopods suffered their greatest extinction in the latest "Dzhulfian", and detailed biostratigraphic studies of latest Permian sections in China (Sheng et al., 1984) indicate a major pulse of extinction very close to the Permo-Triassic Boundary; both analyses suggest that the Guadalupian maximum for all families may reflect sampling error (see also Jablonski, 1986, this volume).

2. *Rhaetian.* The Late Triassic contains one of the five largest mass extinctions of the Phanerozoic marine record (Hallam, 1981; Raup and Sepkoski, 1982; Sepkoski, 1982b). Cephalopods, brachiopods, bivalves, gastropods, and marine reptiles all suffered major declines in diversity, and the last conodonts and conulariids disappeared at this time. Traditionally, this extinction event has been referred to as the Norian mass extinction. However, the extinction metrics in Figure 1 show high extinction intensities throughout the Late Triassic, with the maximum generally in the terminal Rhaetian "Stage". This pattern is consistent with the substage data from House and Senior (1981), which show the major pulse of ammonoid extinction to be in the Rhaetian (=upper Norian, sensu Tozer, 1979). The high levels of familial extinction in the Norian in Figure 1 may be indicative of a protracted extinction event or, quite possibly, simply confusion over the definition and correlation of the Rhaetian (see, e.g., Hallam, 1981; Isozaki and Matsuda, 1983). The high level of extinction in the earlier Carnian (contributed in part by sponges) is probably spurious, reflecting incomplete paleontologic sampling of Upper Triassic sequences.

3. *Pliensbachian.* The familial data in Figure 1 exhibit a low but persistent maximum in extinction intensity in the Pliensbachian Stage of the Early Jurassic contributed by numerous extinctions among cephalopods as well as scattered extinctions among a variety of other groups. This maximum appears to reflect the event that Hallam (1976, 1977) identified as an extinction event among marine bivalves in the early Toarcian (see also Sepkoski, 1982b). Hallam's data on bivalve genera do show a local minimum in diversity in the Toarcian but a maximum in numbers of extinctions in the preceding Pliensbachian, as do data for bivalve species. Data from House and Senior (1981) indicate that the greatest drop in ammonoid diversity occurred between the early and late Pliensbachian, leaving the precise timing and duration of this event uncertain.

4. *Tithonian.* This stage at the end of the Jurassic contains a pronounced peak in all four extinction metrics in Figure 1, contributed largely by extinctions among cephalopod and bivalve families. This event is also reflected in large numbers of extinctions among bivalve (Hallam, 1976, 1977) and other genera (Sepkoski, 1986) and in a severe drop in species diversity of planktonic dinoflagellates (Tappan and Loeblich, 1971). Substage data from House and Senior (1981) indicate high numbers of ammonoid extinctions throughout the Tithonian with the greatest number occurring in the last half of the stage.

5. *Cenomanian:* The data on simple numbers and percentages of familial extinctions in Figure 1 show a strong peak in the Cenomanian Stage of the Cretaceous, contributed by scattered extinctions among cephalopods, echinoids, osteichthyan fishes, sponges, and other taxa. Among the plankton, this event is reflected in pronounced drops in diversities of dinoflagellate and globigerinid species (Lipps, 1970; Tappan and Loeblich, 1971; Fischer and Arthur, 1977). Kauffman (1979, 1983) indicates that the majority of these extinctions occurred within a short interval (<1 m.y.) just below the Cenomanian–Turonian Boundary. In Figure 1, the continued high total and per-family rates of extinction, culminating in a very high maximum in the Coniacian, are most likely spurious, reflecting the short estimated duration (1 m.y.) of the Coniacian Stage; the fact that the magnitudes of these peaks exceed that of the Maestrichtian suggests that the duration of the Coniacian is underestimated in the time scale of Harland et al. (1982) (cf. Kauffman, 1979; Kennedy and Odin, 1982).

6. *Maestrichtian.* This well-documented extinction event at the end of the Cretaceous stands out as a major peak in all of the extinction metrics in Figure 1. The peak is induced by numerous familial extinctions among cephalopods (including all remaining ammonoids), bivalves (including all remaining rudistids), gastropods, bryozoans, echinoids, sponges (including the last stromatoporoids), osteichthyan fishes, and marine reptiles (including the last plesiosaurs). Among microplanktonic groups, the Maestrichtian mass extinction is reflected in major declines in species or generic diversities of acritarchs, dinoflagellates, coccolithoporids, and foraminifers (Lipps, 1970; Tappan and Loeblich, 1971, 1972; Fischer and Arthur, 1977; Thierstein, 1982); detailed local biostratigraphic studies of latest Cretaceous microplankton indicate most of these extinctions occurred abruptly at the Maestrichtian–Danian Boundary (e.g., Smit, 1982; Smit and Romein, 1985). Available data for the macroplankton and benthos, however, suggest that some of these extinctions may have been somewhat graded (House and Senior, 1981; Ward, 1983; Alvarez et al., 1984).

7. *Late Eocene (Priabonian).* A Late Eocene extinction event is best documented for marine microplankton. Dinoflagellates, coccolithoporids, ebridians, silicoflagellates, and planktonic foraminifers all show major declines in species or generic diversity around the Eocene–Oligocene Boundary (Lipps, 1970; Tappan and Loeblich, 1971, 1972; Fischer and Arthur, 1977). The data on familial extinctions in Figure 1 indicate that the Eocene event also affected the marine fauna as a whole, although no specific taxonomic group appears to have suffered an inordinate

number of familial extinctions; lower resolution data on marine genera (Sepkoski, 1986) also indicate widespread extinction among the macrobenthos. Middle Eocene, as opposed to Late Eocene, maxima in the time series for raw numbers and percentages of extinctions (Table 2) probably reflect the lack of normalization for interval duration in the two metrics as well as the graded nature of Late Eocene extinctions; Keller (1986) has shown that the Eocene extinctions occurred in a steplike sequence of events over the last 3.4 m.y. of that epoch and extended back into the late Middle Eocene.

Table 2. Results of Tests of Significance on Extinction Maxima in the Leonardian-to-Pleistocene Interval.[a]

Geologic Stage	Simple Number	Extinction metrics: Percent	Extinction metrics: Total Rate	Per-Family Rate
Q: Pleistocene				
T: Pliocene			*x*/	*x*/
Upper Miocene				
Middle Miocene	*x*/	*/	*x*/*	*x*/*
Lower Miocene				
Upper Oligocene				
Lower Oligocene				
Upper Eocene			*/*x*	*/*x*
Middle Eocene	*x*/*x*	*x*/*x*		
Lower Eocene				
Paleocene[b]				
K: Maestrichtian	*/*	*/*	*/*	*/*
Campanian				
Santonian				
Coniacian			*/*	*/*
Turonian				
Cenomanian	*/*	*/*		
Albian				
Aptian			*x*/	*x*/
Barremian				
Hauterivian				
Valangian				
Berriasian				
J: Tithonian	*/*	*/*	*/*	*/*
Kimmeridgian				
Oxfordian				
Callovian				
Bathonian				
Bajocian			*x*/	*x*/
Toarcian				

Table 2. ***(continued)***

Geologic Stage	Simple Number	Extinction metrics: Percent	Extinction metrics: Total Rate	Per-Family Rate
Pliensbachian	*x*/*	*/*	*x*/*	*/*
Sinemurian				
Hettangian				
Tr: Rhaetian		*/*		*/*
Norian	*/*		*/*	
Carnian				
Ladinian				
Anisian				
Olenekian	*x*/	*/	*x*/	*/
Induan				
P: Dzhulfian				
Guadalupian	*/*	*/*	*/*	*/*
Leonardian				

[a]Symbols indicate significance of the maxima relative to adjacent minima, arranged as *younger minimum/older minimum*, where * = significant at the 99% confidence level and *x* = significant at the 95% level.

[b]"Paleocene" and Danian from Sepkoski's (1982a) compendium were lumped together because of ambiguities concerning the use of "Paleocene" by many authors [i.e., post-Danian Paleocene (≈ Thanetian) versus the entire Paleocene series].

8. *Middle Miocene (Langhian–Serravallian).* This event is the most poorly documented of the mass extinctions suggested by the familial data. The maxima in the extinction metrics in Figure 1 are produced by a relatively small number of families: six families known to become extinct in the Middle Miocene plus another six families resulting from spreading of series-level data through the Miocene. Still, this number of extinctions appears significantly higher than the neighboring very low background levels of extinction, which reflect the strong secular decline in extinction rates through the Cenozoic (Raup and Sepkoski, 1982; Van Valen, 1984) as well as sampling biases termed the "Pull of the Recent" (Raup, 1978, 1979). Some kind of extinction event around this time is also suggested by high rates of species or generic extinction computed from series-level data for dinoflagellates, ebridians, and silicoflagellates (Tappan and Loeblich, 1971, 1972) and from stage-level data for coccolithoporids, discoasters, and foraminifers (Haq, 1973; Fischer and Arthur, 1977; Hoffman and Kitchell, 1984; Ujiié, 1984). Many of these planktonic groups continue to exhibit high rates of extinction and consequent declines in diversity into the Pliocene and Pleistocene. Little is known about detailed patterns of extinction for the macrobenthos in the Miocene, although high rates of extinction have been documented for bivalve and gastropod species in the Plio-Pleistocene (Stanley, 1979, 1984a,b; Stanley et al., 1980; Stanley and Campbell, 1981; Raffi et al., 1985). The familial data in Figure 1 do show a low maximum in extinction

intensity in the Pliocene, but this peak appears lower than that in the Middle Miocene.

The four other "extinction peaks" (Olenekian, Bajocian, Callovian, and Hauterivian) identified by Raup and Sepkoski (1984) do not appear clearly significant in this analysis. All of the extinction metrics in Figure 1 do show a small peak in the Olenekian (=Smithian + Spathian) Stage of the Early Triassic, which is significantly higher than the local minimum in the Middle Triassic (Table 2). However, this peak reflects only extinctions among ammonoid cephalopods (which are accompanied by equal numbers of originations throughout the Olenekian; see House and Senior, 1981) and should not be considered an extinction event (cf. Jablonski, this volume; Flessa et al., 1986). Similarly, the small peak in raw numbers and percentages of extinctions in the "Bajocian" (=Aalenian + Bajocian, sensu stricto) in the Middle Jurassic also probably does not reflect an extinction event but rather a relatively large number of cumulated extinctions over this comparatively long interval of time; this peak disappears when the extinction metrics are normalized for time.

Neither the Callovian (Middle Jurassic) nor the Hauterivian (Early Cretaceous) contain significant peaks in any of the extinction metrics in Figure 1. However, the Callovian does contain a strong peak in familial extinctions for cephalopods alone, which is significantly higher than the minima on either side. The Aptian Stage of the mid-Cretaceous also shows a significant peak in cephalopod extinctions as well as a hint of significance for all families in Table 2. [Data on marine genera as a whole also exhibit a low peak of extinction over the Aptian (Sepkoski, 1986).] Families of cephalopods, and especially ammonoid cephalopods, display very high

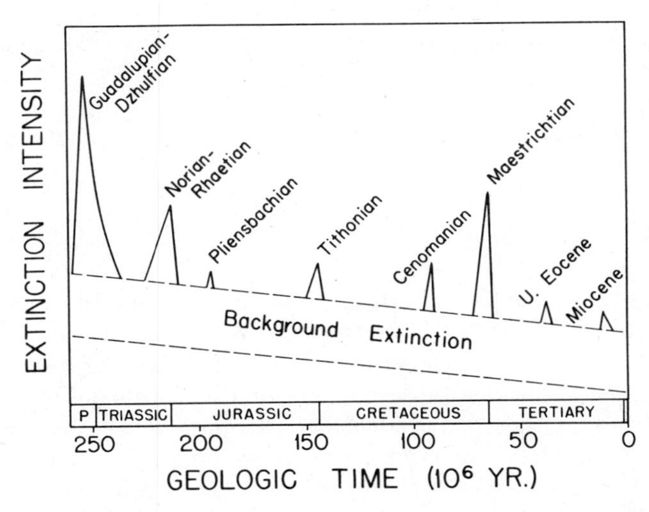

FIGURE 2. A model of extinction intensity in the marine realm over the last 270 m.y. of geologic time, based on Raup and Sepkoski (1982). Background extinction, the intensity of normal intervals of time, is shown as a decreasing function of time. Superimposed on the background are eight peaks of extinction indicated by the familial data in Figure 1. The heights of these peaks (which are left unscaled) reflect the relative magnitudes of the events at the familial and stage levels.

rates of evolutionary turnover in the fossil record (Stanley, 1979; Ward and Signor, 1983) and may be more sensitive to perturbations of the biosphere than other animal groups. However, in the absence of corroborative support from other taxonomic groups or from local biostratigraphic studies, neither the Callovian nor the Aptian can be treated as containing significant extinction events for the analysis here.

Figure 2 summarizes the analysis of familial extinction intensities presented above. This figure is essentially a qualitative refinement of Figure 1 in Raup and Sepkoski (1982). Background extinction, the normal intensity of most intervals of geologic time, is represented as being confined to a relatively narrow band that declines in magnitude toward the Recent. Superimposed upon this band are the eight extinction events discussed above. The absolute magnitudes of these extinction peaks are left unscaled in Figure 2 but are drawn to reflect the relative magnitudes indicated by the stage-level familial data. Below, we shall analyze the temporal distribution of these eight events.

ANALYSIS OF PERIODICITY

The new measures of extinction intensity and definitions of significant extinction events permit us to test our previous finding of periodicity in mass extinctions and refine our estimate of the length of the purported period. There are two basic ways of doing this:

1. Treat the measures of extinction intensity as continuous time series and search for temporal regularities in the fluctuations.
2. Treat the identified extinction peaks as discrete events and test for regularity in their timings.

Below, we use standard Fourier analysis for the first test and a nonparametric randomization procedure for the second.

Fourier Analysis

Fourier analysis is a simple but effective procedure for investigating whether a given time series contains periodic components. Within the context of the familial extinction data, it has the advantage that all data enter into the computations but has the disadvantages that it assumes that background extinction is continuous with extinction events and that this continuous relationship with extinction events of highly varying magnitude can be approximated by simple sine–cosine functions.

For the analyses presented here, we have used the Fourier program in Davis (1973, pp. 266–267). For input, the extinction intensity computed for a given stage was assigned to each million-year increment within that stage. Thus, the initial data appeared as a step function with a total length of 268 m.y. (beginning of Leonardian to end of Pleistocene). These data were then transformed to square roots to reflect

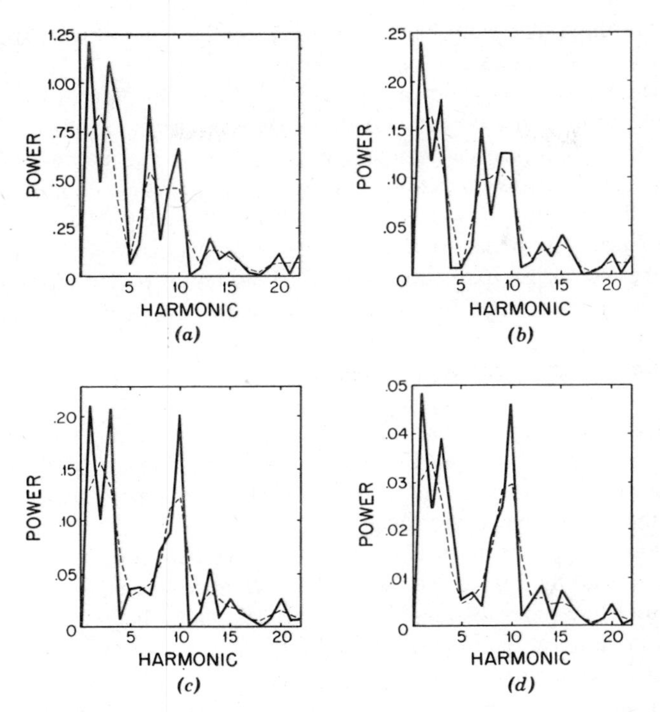

FIGURE 3. Fourier power spectra for the corresponding four extinction time series in Figure 1. The power spectra are displayed for the zeroth harmonic (which has zero power due to the detrending of the data) to the 22nd harmonic (which is the Nyquist limit for the 268-m.y.-long time series). Heavy, solid lines are the raw power spectra, and lighter, dashed lines are smoothed spectra (averaged by the algorithm $\frac{1}{4}P_{h-1} + \frac{1}{2}P_h + \frac{1}{4}P_{h+1}$, where P = power and h = harmonic number).

the presumed Poisson nature of extinctions and were detrended by computing residuals from a linear regression on geologic time.

Figure 3 displays power spectra resulting from Fourier analyses of each of the extinction metrics. All four spectra show a similar pattern that is somewhat easier to interpret than the single power spectrum presented in our previous analysis (see Figure 2 in Raup and Sepkoski, 1984). The spectra in Figure 3 all have strong peaks at low harmonics, corresponding to wavelengths of 268 and 89 m.y., which blend together in the smoothed power spectra. These peaks result from long-term trends in the data and appear basically uninteresting. Considerably more interesting is the persistent peak at the tenth harmonic, which corresponds to a period around (but not necessarily precisely at) 26.8 m.y. (=268/10). This peak in power is attributable to the maxima in the extinction time series and thus corroborates our previous finding of a 26-m.y. periodicity in mass extinctions. The peak at the tenth harmonic is most pronounced in the power spectra for the rate metrics in Figures 3*c*; *d* and is least pronounced in the spectrum for percent extinction in Figure 3*b*. In the latter, the ninth and tenth harmonics have nearly equal power, suggesting an intermediate period around 28.3 m.y.

The spectra for raw numbers and percent extinction in Figures 3*a*, *b* display a fourth strong peak at the seventh harmonic, which corresponds to a wavelength around 38 m.y. This peak is attributable not to extinction maxima but rather to minima between the mass extinctions. A. G. Fischer (personal communication) has also observed a tendency toward a regular 38-m.y. spacing of extinction minima in our time series for percent extinction. However, because this tendency is not evident in the metrics for extinction rate (which do not display peaks at the seventh harmonic), we are uncertain how to interpret the spacing of the minima.

Nonparametric Randomization Test

In our previous study, we presented a randomization test (similar, but not identical, to bootstrap procedures; see Diaconis and Efron, 1983; Connor, 1986) to refine the results of our Fourier analysis and to test their statistical significance. This test was applied to the entire time series for percent extinction by (1) identifying peaks in the extinction metric; (2) measuring the differences between these peaks and the positions predicted by a fitted, perfectly periodic function (see Stothers [1979] for a similar procedure); and (3) randomizing the data and repeating steps 1 and 2 in order to generate a test distribution for nonperiodic time series. If the goodness of fit (measured as the standard deviation of differences between observed and predicted timings of peaks) for the observed time series was less than those for 99% of the randomized versions of the data for a given period, we could conclude that the observed time series displayed a significant periodicity. We found that the time series for percent extinction computed from the highly culled familial data set did indeed show a highly significant periodicity at 26 m.y.

One problem in this previous analysis is that we used all peaks in percent extinction whether or not they were significantly higher than the surrounding background. However, it is not essential that all extinction peaks be used in the randomization procedure. Instead, a time series can be reduced to a step function that has unit height over each stage (or some portion thereof) with a significant extinction event and is zero elsewhere. This function can then be randomized with the constraint that each stage contain only one extinction event and that no two events fall in contiguous stages (such pairs would be impossible to distinguish in the actual data).

Figure 4 summarizes the results of the randomization test applied to the eight extinction events illustrated in Figure 2. Each event was treated as if it occurred at the end of its stage (see Table 3). The test was run for period lengths from 12 m.y. (the Nyquist limit given an average stage duration of approximately 6 m.y.) to an arbitrary upper limit of 60 m.y. Five hundred randomizations of the data were performed to generate the test distribution.

As evident in Figure 4, the randomization test of the eight significant extinction events produces results very similar to those presented by Raup and Sepkoski (1984), despite 33% fewer "extinction peaks." The solid curve, which indicates the standard deviations for the actual data, dips below the 99% limit for the randomized versions

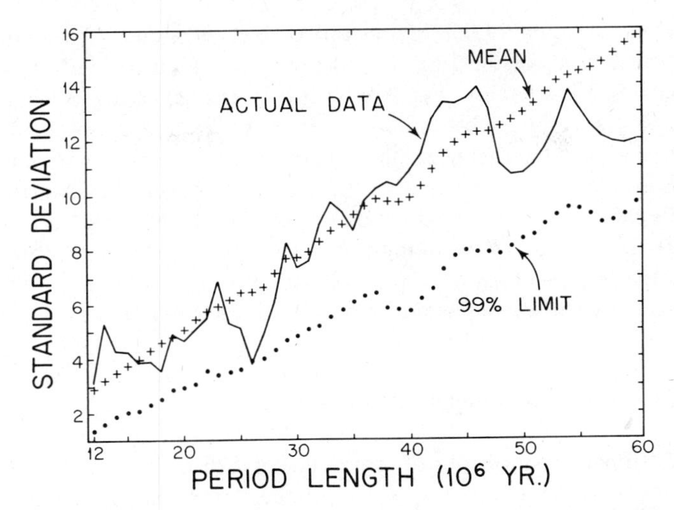

FIGURE 4. Results of the randomization test for periodicity applied to the temporal distribution of the eight significant extinction events listed in Table 3. The solid curve shows the goodness of fit for the actual data, measured as the standard deviation of differences between observed and predicted timings of extinction events (ordinate) for periods of 12–60 m.y. (abscissa). Crosses show the mean standard deviation for randomized versions of the data, and dots show their lower 99% limit.

of the data (dotted line) only once, at a period length of 26 m.y., which is considerably less than the average spacing of 33.5 m.y. (=268 m.y./8) between the events. At 26 m.y., the standard deviation between observed and predicted timings of extinction events is 3.8 m.y., which is about the same as that found previously for the 12 extinction peaks. Beyond 26 m.y., the standard deviation for observed events does not differ significantly from that for the randomized data, and this test shows no hint of significance around 30 m.y., as found previously (see also Rampino and Stothers, 1984a,b).

This result is independent of possible periodic components in the time scale, which was not randomized in this test. Some tendencies toward periodicity in the Harland et al. (1982) time scale are evident in Figure 4 in the scallops in the curve for the 99% limit, centered on periods of 40, 48, and 57 m.y. The trend of the curve for the standard deviations for the actual data parallel these scallops to some extent, as might be expected. However, no such irregularities occur at lower period lengths, suggesting that the significance of the results at 26 m.y. are robust to periodic tendencies in the time scale.

Actually, the test results presented in Figure 4 are rather conservative. The designated stages for the extinction events were taken directly from the positions of the maxima in Figure 1, although there is uncertainty in the precise timing of some of these events, especially those in the Late Permian and Late Triassic, as discussed above. If the Dzhulfian is substituted for the Guadalupian in the randomization test, the standard deviation of differences between observed and predicted timings of extinction events at 26 m.y. falls to 2.9 m.y., which is less than the standard deviations for 99.9% of the randomized versions. If, additionally, the Norian is substituted for the Rhaetian, the standard deviation falls to 1.1 m.y. at 26 m.y. and is considerably less than 100% of the randomized versions at that period length. These two substitutions are entirely plausible given the uncertainties in the data, and the consequent results suggest that the 26 m.y. periodicity in mass extinctions is indeed a strong element in the data. This conclusion is further strengthened if only the four youngest extinction events (Middle Miocene to Cenomanian), which have the most accurate stratigraphic and chronometric ages, are tested; the randomization test results in a standard deviation of only 0.46 m.y. at the 26 m.y. period length, which was equalled by only 1 of 5000 randomized versions of the data over the 100 m.y. interval tested.

Uncertainty in Period Length and Regularity

Given the estimated periodicity of approximately 26 m.y., each extinction event in the analysis can be assigned a "cycle number" reflecting which multiple of 26 m.y. falls closest to it. These cycle numbers are plotted against the best estimates of the ages of the extinction events in Figure 5, based on the time scale of Harland et al. (1982). The error bars about these points represent uncertainties in these ages stemming from two sources (Table 3):

1. Stratigraphic uncertainty reflecting the fact that the stratigraphic ages of some of the events are known only within broad bounds.

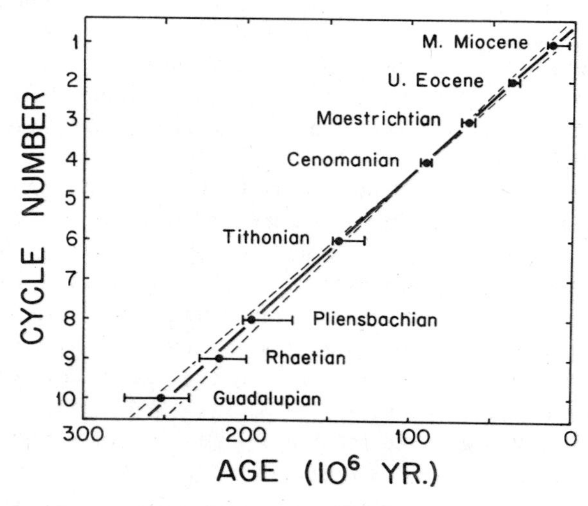

FIGURE 5. A plot of cycle number versus age for the eight significant extinction events listed in Table 3. Points indicate the median estimated ages of the events, and error bars indicate the maximum uncertainty in these ages (see Table 3). The heavy line through the points represents a linear least-squares fit and has a slope of 26.2 m.y. (= best-fit period) relative to the ordinate. The lighter dashed lines represent fits of minimum and maximum slopes that still fall within the error bars about all points.

Table 3. Ages of Significant Mass Extinctions in the Leonardian-to-Pleistocene Interval[a]

Cycle Number	Extinction Event	Age of Top of Stage	Age of Stratigraphic Bounds	Total Uncertainty in Age
1	Middle Miocene	11.3	2[b]–14.4	1.6–15.1
2	Upper Eocene	38	38–40	33–40
3	Maestrichtian	65	64–65	60–69
4	Cenomanian	91	90–91	87.5–93.5
6	Tithonian	144	144–147	127–147
8	Pliensbachian	194	194–200	171–202
9	Rhaetian	213	213–225[c]	200–228
10	Guadalupian	253	248[d]–258	235–275

[a]The ages of the tops of stages in which extinction events occur (or probably occur) were used in the randomization test for periodicity. The ages of the stratigraphic bounds indicate the maximum stratigraphic intervals within which the extinction events may have occurred (based on the time scale of Harland et al., 1982). The total uncertainty in ages represents the maximum possible spread of ages of the stratigraphic bounds, based on the uncertainties in stage ages listed by Harland et al. (1982), Odin (1982), and Palmer (1983).

[b]Includes the Pliocene, an interval of higher than normal extinction intensities.

[c]Includes the Norian.

[d]Includes the Dzhulfian.

2. Chronometric uncertainty reflecting possible error in the absolute ages of these bounds.

In determining the total errors, we tried to be conservative, using not only the possible errors in boundary ages listed by Harland et al. (1982) but also those listed by Odin (1982) and Palmer (1983). We selected the youngest and oldest possible ages permitted by any of the three time scales. Thus, the error bars represent the maximum uncertainties assignable to the ages of the extinction events given the current time scales.

The heavy line through the points in Figure 5 represents a linear least-squares fit of the ages of the extinction events (x-axis) to the cycle numbers (y-axis). Although such a regression line lacks statistical meaning, its slope does permit an unbiased estimate of period length that is more precise than that provided by the Fourier analysis or randomization test. The slope of the line illustrated in Figure 5 is 26.2 m.y., which is thus our best estimate of the extinction periodicity. A rough estimate of possible uncertainty about this value can be obtained from the slopes of other lines that fall within the error bars about the points. The two dashed lines in Figure 5 represent the extremes of these permissible lines; they are drawn such that their slopes are maximally different from 26.2 but still fall within the limits of error for all points. The slopes of these lines are 24.4 and 27.9 m.y., which can be considered to be near the maximum and minimum possible period lengths. If these values are treated as bounds of a 95% confidence interval, then the standard error about 26.2 m.y. is 0.9, or roughly 1, m.y.

The regression line in Figure 5 also provides a visual impression of the fit of the extinction events to the estimated periodicity. As is evident, the first four, best-dated extinction events (Middle Miocene to Cenomanian) fall remarkably close to the line. There is then a gap where a fifth event should occur. The succeeding Tithonian event again falls very close to the line, despite the rather broad range of possible ages. Beyond the Tithonian, there is a second gap where a seventh event should occur, followed by the last three events, which show more scatter about the line (which is hardly suprising given the greater stratigraphic and chronometric uncertainty in the ages of these older mass extinctions).

The occurrence of two gaps in the record of extinction events is bothersome. Assuming the basic observation of periodicity is correct, there are several possible explanations for these gaps. It could be that the forcing agent of extinction events is in fact not perfectly periodic and that it occasionally misses a "beat." Alternatively, it could be that our knowledge of extinction events over the last 268 m.y. is still quite imperfect. Identified extinction events vary greatly in magnitude, and it is very possible that there are smaller events that are completely damped in the familial data. Such damping could be exacerbated by the somewhat longer average stage durations in the mid-Jurassic to mid-Cretaceous interval: 7.3 m.y./stage from the Bajocian to Albian versus 5.6 m.y./stage for the remainder of the Leonardian to Pleistocene interval. However, there are hints that extinction events might occur within the two gaps. As noted previously, the familial data for ammonoids display significant extinction peaks in the Callovian and Aptian Stages. These fall very close to the predicted times for the missing extinction events: 170 m.y., which is within the latest Bathonian very near the Callovian Boundary, for the seventh event, and 118 m.y., which is within the early Aptian, for the fifth event. However, until there is more definitive biostratigraphic documentation of these possible events, they must remain implicit questions marks in Figure 5.

PERIODICITY IN THE PALEOZOIC?

The question of whether the periodicity in extinction events observed over the Mesozoic–Cenozoic interval extends back into the Paleozoic is difficult, if not impossible, to answer, given the present state of our knowledge of the global fossil record. As mentioned previously, the average duration of stages in the Paleozoic is nearly 10 m.y., providing fewer than three sampling intervals per 26-m.y. period; in the 75-m.y.-long Carboniferous, the situation is even worse, with the average stage duration in the Sepkoski (1982a) compendium being almost 12.5 m.y., which is very close to the Nyquist limit for detecting periodicities as short as 26 m.y. These longer stage lengths also lessen our ability to distinguish extinction events. Since the extinction metrics are averages for the sampling intervals, longer intervals mean that high intensities associated with discrete events are being averaged with more background extinction, decreasing the observed differences between stages with and without extinction events. Thus, a greater proportion of smaller events may be unrecognizable in the Paleozoic familial data. This problem is further exacerbated by the greater uncertainty in estimated stage durations and ages; more than half of the ages for Paleozoic stage boundaries in the time scale of Harland et al. (1982) have listed uncertainties of greater than 13 m.y., which is half a period length.

If an analysis similar to that conducted for the Leonardian-to-Pleistocene interval is performed on the familial data for the pre-Leonardian Paleozoic, seven extinction maxima can be identified as significantly higher than neighboring background in the majority of the extinction metrics. These maxima include the two largest mass extinctions of the pre-Permian Paleozoic (Raup and Sepkoski, 1982; Sepkoski, 1982b): the terminal Ordovician event (in the Ashgillian Stage) and the Late Devonian event (appearing as an extinction maximum in the Famennian Stage preceded by high intensities in the Frasnian and Givetian). The other five significant maxima are in the Botomian and Trempealeauan Stages of the Cambrian (previously identified as containing biomere events; see Palmer, 1982, 1984), the Late Silurian (Ludlovian–Pridolian), and the Namurian and Stephanian Stages of the Carboniferous. Hints of possible significance are also seen in the mid–Middle Cambrian (containing another possible biomere event) and in the Llandvirnian Stage of the Ordovician.

Taken together, these 7 to 9 events are fewer than the 12 that would be predicted for the Paleozoic, given a 26-m.y. period. Indeed, Fourier analyses of the Paleozoic extinction metrics give only very weak peaks in the power spectra around wavelengths of 26 m.y. Instead, they display much stronger peaks centered on 37 m.y. This same result is obtained from the randomization test performed either on all data from the Paleozoic time series (cf. Raup and Sepkoski, 1984) or on just the distribution of significant extinction peaks, as done in the Leonardian-to-Pleistocene analyses above. Both tests show that the standard deviation of differences between observed and predicted timings of extinction peaks falls below 99% of standard deviations for randomized versions of the data at period lengths of 37 and 38 m.y. and below 95% at period lengths from 36 to 39 m.y.

These results should not be interpreted as indicating a significant but longer periodicity of extinction events in the Paleozoic, especially since not all older events can be identified in the familial data (see Sepkoski, 1986). As we found in our previous analyses (Raup and Sepkoski, 1984), there is a substantial probability ($p > .05$) that a random sequence of events will display clusters of "significant" periodicities at period lengths greater than 30 m.y. Thus, both the Fourier and the randomization results could reflect simply chance regularity, induced perhaps by the relatively wide spacing (i.e., more than two stages) of the events. (Indeed, we note that the average spacing of events, which is $322/9 = 36$ m.y. where 322 is the length of the Paleozoic time series, is very close to the period identified in the statistical tests.) Thus, we conclude that better data sampled from finer time intervals is essential before the question of periodicity in Paleozoic extinction events can be resolved.

POSSIBLE CAUSES OF PERIODICITY IN EXTINCTION EVENTS

The observation that extinction events in the marine realm are periodically arrayed over at least the last quarter billion years has important consequences for causal hypotheses. Perhaps most importantly, it indicates that mass extinctions are not independent events but rather are dependent on some single ultimate cause (sensu McLaren, 1983) that must recur at regular intervals. However, the nature of this ultimate cause, or forcing agent, remains enigmatic. Most well-known periodic processes that influence the Earth's surface and atmosphere (including Milankovitch cycles) have characteristic time scales of 10^0–10^5 yr (McCrea, 1981). Deep-Earth processes with cyclic behaviors that may affect near-surface tectonics have characteristic time scales of 10^8–10^9 yr (Richter, 1984) (except perhaps for the forcing agent responsible for the still unproven periodicity in magnetic reversals; see Negi and Tiwari, 1983; Lutz, 1985; Raup, 1985a,b). This leaves no obvious terrestrial or solar process operating cyclically on time scales of 10^6–10^7 yr, which encompasses the observed periodicity in extinction events. Thus, by an argument of elimination, we speculated previously that the ultimate forcing agent must be astronomical. This argument is supported by two independent sets of observations:

1. The discovery of impact debris and geochemical anomalies with extraterrestrial signatures at two of the eight post-Leonardian extinction horizons: the terminal Maestrichtian (Alvarez et al., 1980, 1982a; Alvarez, 1983; Luck and Turekian, 1983; Montanari et al., 1983; Bohor et al., 1984; Pillmore et al., 1984; Smit and Kyte, 1984) and the Late Eocene (Glass and Zwart, 1977; Glass and Crosbie, 1982; Alvarez et al., 1982b; Asaro et al., 1982; Keller, 1986; Ganapathy, 1982); a possible iridium anomaly has also been reported at the Permo-Triassic Boundary (Sun Yi-Ying et al., 1984; Xu Dao-Yi et al., 1985), and meteoritic debris and geochemical anomalies have been found at the top of the Callovian, within 7 m.y. of where the missing seventh extinction event should occur (Brochwicz-Lewiński et al., 1984a,b).
2. The observation that terrestrial cratering may display a periodicity that is in phase with that of extinction events (Alvarez and Muller, 1984; Rampino and Stothers, 1984a,b).

We review this second observation in more detail below.

Periodicity in Terrestrial Cratering

Alvarez and Muller (1984) analyzed the ages of impact craters in a highly culled subset of Grieve's (1982) compilation of 88 well-documented terrestrial craters. Their primary data set consisted of 11 craters, all greater than 10 km in diameter and with uncertainties in ages of less than ±20 m.y.; these craters were selected from the age interval of 5–250 m.y., approximately the same as the familial data analyzed here. The ages of the craters had been determined largely by direct radiometric dating of associated melt products and not by stratigraphic procedures, making the cratering data entirely independent of the extinction records. Through a series of Fourier analyses and Monte Carlo simulations, Alvarez and Muller demonstrated a significant ($p < .01$) periodicity in the ages of the 11 craters, with a mean period length of 28.4 ± 1 m.y. and a phase relationship congruent with the extinction data over the younger half of the analyzed time interval. (Actually, the "cycle slippage" that Alvarez and Muller observed over the 150–200-m.y. interval need not exist; given the uncertainties in current geologic time scales, we see no statistically significant difference between 28.4 ± 1 m.y. and 26.2 ± 1 m.y., and the periodicities might ultimately appear completely in phase given revisions of the crater and stage ages.)

Because of the small sample size of Alvarez and Muller's data set, we have reanalyzed Grieve's original cratering data using the randomization test developed for the extinction data. The test procedure is the same as that described above except that we did not use a template of stratigraphic stages when assigning ages to observed or randomized events. Since the estimates of crater ages are mostly independent of the stratigraphic time scale, any million-year increment is permissible when randomly distributing ages. However, ages listed in Grieve's compilation do show strong digit preference, with 0 and 5 appearing frequently as a result of rounding (see also Rampino and Stothers, 1984a). Therefore, in the randomization procedure, we rounded some ages to generate the same frequency of 0's and 5's observed in the actual data.

Application of the randomization test to Alvarez and Muller's data set (either the 11 selected craters or the full 16 craters listed in their Table 1) produced results identical to theirs. At 28 m.y. (and at only this period length), the standard deviation of differences between observed and predicted ages for craters fell below 99% of those for the randomized versions. However, when larger subsets of Grieve's data were selected, somewhat more ambiguous results were obtained. Figure 6 illustrates results from the bootstrap test applied to 22 crater ages drawn from the time interval of 0–268 m.y. This set of ages was derived from the 41 craters analyzed by Rampino and Stothers (1984a) after craters with equal or contiguous ages (e.g., 1, 2, 3, and 4 m.y.) were lumped together. As shown in Figure 6, no period produced a standard deviation for the actual data that fell below the 99% limit for randomized versions. However, at four period lengths, equal to 21, 27, 30, and 31 m.y., standard deviations for the real data dropped below the 95% limit. The 27-m.y. period gave the strongest signal, with the standard deviation for the real data falling below 98.3% of standard deviations for randomized versions. The 21-m.y. period was the weakest and corresponds to the secondary peak found by Alvarez and Muller, who argued on the basis of simulations that this result was probably spurious, a result of noise in the data.

The marginally significant results obtained for period lengths of 30 and 31 m.y. are consistent with the 31-m.y. period in cratering ages found by Rampino and Stothers (1984a) (see also Kerr, 1985). However, the significance of this period length appears dependent on just three craters older than 210 m.y. (craters 29, 47, and 77 in Grieve's list). Two of these craters have listed age uncertainties of ± 40 m.y. or more, and crater 29 has no error margin assigned to it. Without these three craters of rather uncertain age, the significance of the 30–31-m.y. period disappears

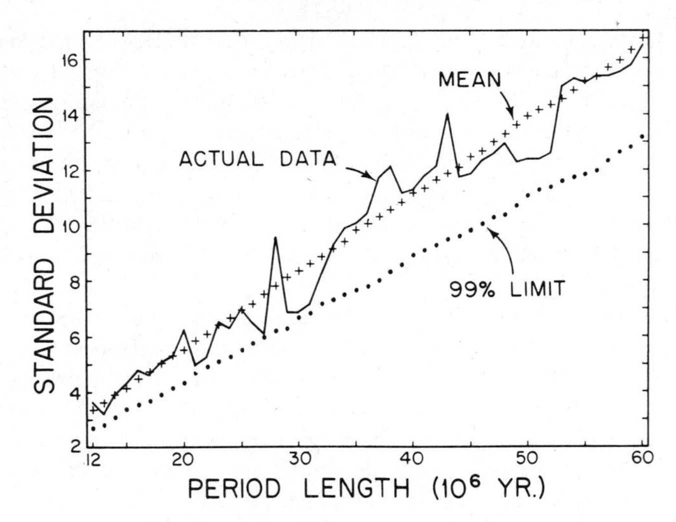

FIGURE 6. Results of the randomization test for periodicity applied to 22 ages of terrestrial impact craters derived from Grieve's (1982) compilation (see text). Plotting conventions are the same as in Figure 4. Because a template of stratigraphic stages was not used in this analysis, the curves for the mean (crosses) and 99% limit (dots) for the standard deviations of the randomized data are more regular than those in Figure 4.

in the randomization test, indicating a lack of robustness. Thus, a length in the neighborhood of 27–28 m.y. appears to be the best estimate for cratering periodicity, but clearly more and better data are needed to demonstrate that this is indeed statistically significant.

If the possible periodicity in terrestrial cratering does stand up, then impacts of extraterrestrial bodies (asteroids or comets) may be sufficient to explain the observed periodicity in extinction events. Mechanisms may involve either catastrophic changes in atmospheric conditions, as hypothesized for a single large impact at the Cretaceous–Tertiary Boundary (Alvarez et al., 1980; Toon et al., 1982; Milne and McKay, 1982; Pollack et al., 1983; Wolbach et al., 1985), or perhaps less catastrophic, cumulative deleterious effects on global climate caused by multiple smaller impacts over an interval of 1–2 m.y. [See Stanley (1984a,b) for discussions of the association of mass extinctions with climatic coolings; cumulative climatic effects might also be the forcing agent of excursions in stable isotope ratios (Kauffman, personal communication; see also Kerr, 1984) and widespread black shale horizons (Fischer and Arthur, 1977; Ager, 1981; Hallam, 1984) associated with several of the periodic extinction events as well as with the "missing" event in the Aptian.]

The nature of the driving mechanism of periodic impacts and extinctions is still an open question. Since we presented our results at the "Dynamics of Extinction" Symposium, four possible astronomical mechanisms have been suggested:

1. Transit of the solar system through the spiral arms of the Galaxy.
2. Vertical oscillation of the solar system about the plane of the Galaxy.

3. Precession of an undetected tenth planet.
4. Orbital dynamics of an unobserved solar companion.

These four hypotheses are briefly summarized below.

Transit Through Galactic Arms

As the solar system orbits the Milky Way Galaxy, it passes through spiral arms, or density waves, approximately every 50 m.y. Napier and Clube (1979) and Clube and Napier (1982, 1984) have argued that interstellar "planetismals" are captured by the sun during these transits and that this capture could increase "asteroid" bombardment of the Earth during ensuing 20–30-m.y. intervals. It has also been suggested (e.g., Shoemaker, 1984) that tidal forces resulting from the higher density of the spiral arms might perturb the existing cometary cloud surrounding the solar system, increasing the flux of comets into the inner solar system and thus increasing the frequency of terrestrial impacts.

A major problem with these hypotheses is that they predict a periodicity in extinction events of about twice the observed length. This might be permissible if the observed record of mass extinction consisted of roughly equal numbers of periodic and aperiodic (i.e., random) events. Under such circumstances, the statistical tests employed here would generate estimated period lengths of exactly half the actual period. However, the observed record of mass extinctions, as summarized in Figure 5, appears too regular to reflect a mixture of periodic and random events, despite the two gaps in the record; in particular, it seems unlikely that the last four extinction events would display such a low variance in timings if they included two random events. Thus, transit through the galactic arms probably can be rejected as a plausible ultimate cause of periodic extinction.

Oscillation About the Galactic Plane

As the solar system orbits the Galaxy, it oscillates vertically through the galactic plane. Current estimates of the period of this oscillation are around 67 m.y. (range = 52 − 74 m.y.; see Innanen et al., 1978; Bahcall and Bahcall, 1985), meaning that approximately every 33 m.y. the solar system passes through the plane and also every 33 m.y. it reaches its oscillatory extremes, which are about 80 to 100 pc above or below the galactic plane. Schwartz and James (1984) speculated that at the extremes the Earth might be subjected to greater fluxes of soft x-rays and hard UV radiation emanating from the galactic center and that these greater fluxes might disturb the ionization balance of the upper atmosphere, inducing climatic changes and thus causing mass extinctions (see also Hatfield and Camp, 1970). Rampino and Stothers (1984a,b), on the other hand, argued that tidal forces from intermediate-sized molecular clouds of 10^3–10^4 solar masses, concentrated near the galactic plane, might perturb the Oort Cloud and inner cometary reservoir as the solar system approached the plane. Using the model developed by Hills (1981), they suggested that such perturbations could produce comet showers of up to several million years duration, during which one large and several smaller comets might impact the Earth.

An obvious problem with these hypotheses is the disparity between the 33-m.y. half-period for solar oscillation and the 26-m.y. period for extinction events. Rampino and Stothers resolved this problem by arguing that (1) the data of Raup and Sepkoski (1984) fit a 30-m.y. period equally well and (2) because molecular clouds are randomly distributed through a finite thickness about the galactic plane, there should be a fair amount of stochasticity in extinction events so that the period estimated from a small sample of events may not correspond precisely to the oscillatory half-period. However, in view of the new data and analyses presented here, we feel that a 30-m.y. period is untenable as an estimate of the extinction periodicity, leaving the disparity with the oscillatory half-period unresolved (see also Stigler, 1985). Furthermore, Thaddeus and Chanan (1985) have argued that matter near the galactic plane may be less concentrated than previously assumed, with the implication that no periodicity should be detectable over a 270 m.y. time series.

Tenth Planet

It has long been speculated that an undetected tenth planet, often called "Planet X," might lie beyond the orbit of Pluto within the solar system. Whitmire and Matese (1985) argued that such a planet might be able to produce periodic comet showers if its orbit were highly inclined and if the inner edge of the comet cloud extended as a thin disk almost to the orbit of Neptune. Under such conditions, the orbit of Planet X, hypothesized to have a semimajor axis of about 10^2 AU, would precess through the planetary plane. If the precession period were approximately 56 m.y., the planet, with an estimated one to five Earth masses, would sweep comets out of the inner disk as it passed through the plane every half period. Of the scattered comets, perhaps 10^5 would cross the Earth's orbit and become potential (although low-probability) impactors.

This hypothesis is largely ad hoc (although somewhat less so than the solar companion hypothesis discussed below) since the precession period is determined entirely from the periodicities in extinction and cratering that are being explained. Questions have also been raised as to whether the approach to the comet disk would be sufficiently rapid to induce a brief, intense comet shower and whether the unobserved (and therefore presumably small) planet would have sufficient mass to scatter a large number of comets (see Kerr, 1985). Thus, without detection, this hypothesis must be considered very speculative.

Unobserved Binary Companion

Davis et al. (1984) and Whitmire and Jackson (1984) independently speculated that the Sun might be part of a binary star system with an undetected companion of low mass and luminosity (see also Muller, 1985). Beginning with the assumption of an orbital periodicity of 26 to 28 m.y., they calculated a major axis of 2×10^5

AU ($\approx$ 3 light years) for the companion's orbit and a perihelion distance of 2×10^4–3×10^4 AU ($\approx$0.3–0.5 light years) from the Sun. This would bring the companion, with an estimated size of 10^{-3}–10^{-1} solar mass, into the inner comet reservoir, producing a comet shower similar to that hypothesized by Rampino and Stothers. Again using the model of Hills (1981), Davis et al. estimated that at perihelion the companion, which they dubbed "Nemesis," would perturb up to 10^9 comets into the inner solar system, of which 10–200 ($\bar{x}$ = 25) might impact the Earth over an interval of 10^5–10^6 yr. Studies of the stability of the hypothesized companion (Hills, 1984; Hut, 1984; Torbett and Smoluchowski, 1984) indicate that the orbital periodicity should be somewhat irregular, varying over 10 to 20%, and that the expected half life for the orbital configuration should be on the order of 1000 m.y. (This last result has the interesting implication that sometime before 270 m.y. B.P. the companion, if it exists, may have been perturbed outward from a smaller, more stable orbit, and that prior to this event it caused more frequent, although perhaps less intense, comet showers and mass extinctions.)

The principal problem with this hypothesis is that it is entirely ad hoc, even if plausible. It does explain the periodicities in both extinction events and terrestrial cratering (given the unconstrained estimate of orbital periodicity), and it permits variation in the timing and magnitude of mass extinction events through the expected variation in the orbital periodicity and the number and mass distributions of impacting comets. However, unless the postulated companion is discovered (which may be difficult given the low radiation of an astronomical body of 10^{-3}–10^{-1} solar mass), this hypothesis also must be considered speculative, and the entire problem of what ultimately causes extinction events remains yet to be solved.

CONCLUSIONS

Our conclusion that extinction events in the marine realm have been periodic in their recurrence, with a period length of 26.2 ± 1 m.y., must still be considered a hypothesis that needs further testing. Both patterns of extinction and possible ultimate causes require much more investigation. We still need good synoptic data on extinction patterns in the Paleozoic and in the mid-Jurassic and mid-Cretaceous interval in order to determine the regularity and persistence of the observed periodicity. We also need detailed comparative information on fine-scale patterns within extinction events; with such information we can ask whether all extinction events are fundamentally similar, differing only in magnitude, or if there are several distinct classes of events, distinguishable by duration, selectivity, geographic domain, geologic setting, and so on. These questions have important ramifications for hypothesized ultimate causes.

If the hypothesis of periodicity survives its tests, it will have important implications for Earth history and macroevolution. It leads to the prediction that other geologic phenomena often associated with extinction events might also be periodic and therefore interrelated. These phenomena include sea level changes, large-scale climatic fluctuations, black shale events, and others. Because of the long time scale involved, demonstration of periodicity in these various geologic phenomena may again implicate extraterrestrial forcing agents.

The frequency of extinction events has important consequences for the history of life (Gould 1984, 1985; Raup, 1985c; Sepkoski, 1985). We can speculate that the biosphere has never had the opportunity to equilibrate fully. Instead, recurrent perturbations, coupled with possible indiscriminant elimination of evolutionary lineages, may have maintained the biosphere in a constant state of flux. Lineages may never have been permitted sufficient time to optimize their adaptations to the physical and biotic environment, and biotic interactions may never have ultimately determined evolutionary pathways in the resultant unsaturated ecosystems. This situation, which is essentially one of intermediate disturbance operating on macroevolutionary time scales, may have allowed novel and untuned adaptations to appear geologically frequently in the absence of intense biotic pressure. Thus, mass extinction, rather than being simply an agent of destruction, may have played a creative role in maintaining a diversity of morphologies and adaptations throughout the history of life.

ACKNOWLEDGMENTS

We thank L. W. Alvarez and S. M. Stigler for valuable suggestions concerning statistical analysis. Execution of this work received partial support from NASA Grants NAG 2-237 and NAG 2-282.

REFERENCES

Ager, D. V., 1981, Marine cycles in the Mesozoic, *J. Geol. Soc. Lond.*, **138**, 159–176.

Alvarez, L. W., 1983, Experimental evidence that an asteroid impact led to the extinction of many species 65 Myr ago. *Proc. Natl. Acad. Sci. U.S.A.*, **80**, 627–642.

Alvarez, W., and Muller, R. A., 1984, Evidence from crater ages for periodic impacts on the Earth, *Nature*, **308**, 718–720.

Alvarez, L. W., Alvarez, W., Asaro, F., and Michel, H. V., 1980, Extraterrestrial cause for the Cretaceous-Tertiary extinction, *Science*, **208**, 1095–1108.

Alvarez, W., Alvarez, L. W., Asaro, F., and Michel, H. V., 1982a, Current status of the impact theory for the terminal Cretaceous extinction, in Silver, L. T., and Schultz, P. H., eds., *Geological Implications of Impacts of Large Asteroids and Comets on the Earth*, Geological Society of America Special Paper 190, pp. 305–316.

Alvarez, W., Asaro, F., Michel, H. V., and Alvarez, L. W., 1982b, Iridium anomaly approximately synchronous with terminal Eocene extinctions, *Science*, **216**, 886–888.

Alvarez, W., Kauffman, E. G., Surlyk, F., Alvarez, L. W., Asaro, F., and Michel, H. V., 1984, Impact theory of mass extinctions and the invertebrate fossil record, *Science*, **223**, 1135–1141.

Armstrong, R. L., 1978, Pre-Cenozoic Phanerozoic time scale: Computer file of critical dates and consequences of new and in-progress decay-constant revisions, in Cohee, G. V., Glaessner, M. F., and Hedberg, H. D., eds., *The Geologic Time Scale*, American Association of Petroleum Geologists Studies in Geology no. 6, pp. 73–92.

Asaro, F., Alvarez, L. W., Alvarez, W., and Michel, H. V., 1982, Geochemical anomalies near the Eocene/Oligocene and Permian/Triassic Boundaries, in Silver, L. T., and Schultz, P. H., eds., *Geological Implications of Impacts of Large Asteroids and Comets on the Earth*, Geological Society of American Special Paper 190, pp. 517–528.

Bahcall, J. N., and Bahcall, S., 1985, The Sun's motion perpendicular to the galactic plane, *Nature*, **316**, 706–708.

Benton, M. J., 1985, Mass extinction among non-marine tetrapods, *Nature*, **316**, 811–814.

Bishop, Y. M. M., Fienberg, S. E., and Holland, P. W., 1975, *Discrete Multivariate Analysis: Theory and Practice*, Cambridge, Massachusetts, The MIT Press, 557 pp.

Bohor, B. F., Foord, E. E., Modreski, P. J., and Triplehorn, D. M., 1984, Mineralogic evidence for an impact event at the Cretaceous-Tertiary Boundary, *Science*, **224**, 867–869.

Brochwicz-Lewiński, W., Gasiewicz, A., Strzelecki, R., Suffczyński, S., Szatkowski, K., Tarkowski, R., and Żbik, M., 1984, Anomalia geochemiczna na pograniczu jury środkowej i górnej w południowej Polsce, *Przeglad Geol.*, **32**, 647–650.

Brochwicz-Lewiński, W., Gasiewicz, A., Suffczyński, S., Szatkowski, K., and Żbik, M., 1984, Lacunes et condensations à la limite Jurassique moyen-supérieur dans le Sud de la Pologne: Manifestation d'un phénomène mondial? *C.R. Acad. Sc. Paris*, Sér. II, **299**, 1359–1362.

Clube, S. V. M., and Napier, W. M., 1982, The role of episodic bombardment in geophysics, *Earth and Planet. Sci. Lett.* **57**, 251–262.

Clube, S. V. M., and Napier, W. M., 1984, Comet capture from molecular clouds: A dynamical constraint on star and planet formation, *Monthly Notices of the Royal Astronomical Society*, **208**, 575–588.

Connor, E. F., 1986, Time series analysis and the fossil record, in Raup, D. M., and Jablonski, D., eds., *Pattern and Process in the History of Life*, Berlin, Springer-Verlag (in press).

Dalrymple, G. B., 1983, Geologic time [review of Harland, W. B., et al., A geologic time scale], *Science*, **221**, 944–945.

Davis, J. C., 1973, *Statistics and Data Analysis in Geology*, New York, Wiley, 550 pp.

Davis, M., Hut, P., and Muller, R. A., 1984, Extinction of species by periodic comet showers, *Nature*, **308**, 715–717.

Diaconis, P., and Efron, B., 1983, Computer-intensive methods in statistics, *Sci. Am.*, **248**(5), 116–130.

Fischer, A. G., 1981, Climatic oscillations in the biosphere, in Nitecki, M. H., ed., *Biotic Crises in Ecological and Evolutionary Time*, New York, Academic Press, pp. 103–131.

Fischer, A. G., and Arthur, M. A., 1977, Secular variations in the pelagic realm, in Cook, H. E., and Enos, P., eds., *Deep-Water Carbonate Environments*, Society of Economic Paleontologists and Mineralogists Special Publication 25, pp. 19–50.

Flessa, K. W., 1979, Extinction, in Fairbridge, R. W., and Jablonski, D., eds., *The Encyclopedia of Paleontology*, Stroudsburg, Pennsylvania, Dowden, Hutchinson & Ross, pp. 300–305.

Flessa, K. W., Erben, H. K., Hallam, A., Hsü, K. J. Hüssner, H. M., Jablonski, D., Raup, D. M., Sepkoski, J. J., Jr., Soulé, M. E., Sousa, W., Stinnesbeck, W., and Vermeij, G. J., 1986, Causes and consequences of extinction, in Raup, D. M., and Jablonski, D., eds., *Pattern and Process in the History of Life*, Berlin, Springer-Verlag (in press).

Ganapathy, B., 1982, Evidence for a major meteorite impact on the Earth 34 million years ago: Implication for Eocene extinction, *Science*, **216**, 885–886.

Glass, B. P., and Crosbie, J. R., 1982, Age of Eocene/Oligocene Boundary based on extrapolation from North American microtektite layer, *Am. Assoc. Petrol. Geol. Bull.*, **66**, 471–476.

Glass, B. P., and Zwart, M. J., 1977, North American microtektites, radiolarian extinctions and the age of the Eocene-Oligocene Boundary, in Swain, F. M., ed., *Stratigraphic Micropaleontology of Atlantic Basin and Borderlands*, Amsterdam, Elsevier, pp. 553–568.

Gould, S. J., 1984, The cosmic dance of Siva, *Nat. Hist.*, **93(8),** 14–19.

Gould, S. J., 1985, The paradox of the first tier: An agenda for paleobiology, *Paleobiology*, **11,** 2–12.

Grieve, R. A. F., 1982, The record of impact on Earth: Implications for a major Cretaceous/Tertiary impact event, in Silver, L. T., and Schultz, P. H., eds., *Geological Implications of Impacts of Large Asteroids and Comets on the Earth*, Geological Society of America Special Paper 190, pp. 25–38.

Hallam, A., 1976, Stratigraphic distribution and ecology of European Jurassic bivalves, *Lethaia*, 9, 245–259.

Hallam, A., 1977, Jurassic bivalve biogeography, *Paleobiology*, **3**, 58–73.

Hallam, A., 1981, The end-Triassic bivalve extinction event, *Palaeogeogr. Palaeoclimatol. Palaeoecol.* **35**, 1–44.

Hallam, A., 1984. The causes of mass extinction, *Nature*, **308**, 686–687.

Haq, B. U., 1973, Transgressions, climatic change and the diversity of calcareous nannoplankton, *Marine Geol.* **15**, M25–M30.

Harland, W. B., Cox, A. V., Llwellyn, P. G., Pickton, C. A. G., Smith, A. G., and Walters, R., 1982, *A Geologic Time Scale*, Cambridge, Massachusetts, Cambridge University Press, 131 pp.

Hatfield, C. B., and Camp, M. J., 1970, Mass extinctions correlated with periodic galactic events, *Geol. Soc. Am. Bull.*, **81**, 911–914.

Hills, J. G., 1981, Comet showers and the steady-state infall of comets from the Oort Cloud, *Astron. J.*, **86**, 1730–1740.

Hills, J. G., 1984, Dynamical constraints on the mass and perihelion distance of Nemesis and the stability of its orbit, *Nature*, **311,** 636–638.

Hoffman, A., and Kitchell, J. A., 1984, Evolution in a pelagic planktic system: A paleobiologic test of models of multispecies evolution, *Paleobiology*, **10**, pp. 9–33.

House, M. R., 1985, Correlation of mid-Palaeozoic ammonoid evolutionary events with global sedimentary perturbations, *Nature*, **313,** 17–22.

House, M. R., and Senior, J. R., eds., 1981, *The Ammonoidea*, Systematic Association Special Volume 18, London, Academic Press, 593 pp.

Hut, P., 1984, How stable is an astronomical clock that can trigger mass extinctions on Earth?, *Nature*, **311,** 633–641.

Innanen, K. A., Patrick, A. T., and Duley, W. W., 1978, The interaction of the spiral density wave and the Sun's galactic orbit, *Astrophys. Space Sci.* **57**, 511–515.

Isozaki, Y., and Matsuda, T., 1983, Middle and Late Triassic conodonts from bedded chert sequences in the Mino-Tamba Belt, southwest Japan. Part 2: *Misikella* and *Parvigondolella*, *J. Geosci.* Osaka City Univ., **26**, 65–86.

Jablonski, D., 1986, Evolutionary consequences of mass extinction, in Raup, D. M., and Jablonski, D., eds., *Pattern and Process in the History of Life*, Berlin, Springer-Verlag (in press).

Johnson, N. L., and Kotz, S., 1969, *Discrete Distributions*, New York, Wiley, 328 pp.

Kauffman, E. G., 1979, Cretaceous, in Robison, R. A., and Teichert, C., eds., *Treatise on Invertebrate Paleontology, Part A*, Lawrence, Kansas, Geological Society of America and University of Kansas, pp. A418–A487.

Kauffman, E. G., 1983, Mass extinction within the Cretaceous: Earthbound events for calibration of extraterrestrial effects, *Geol. Soc. Am. Abstr. Progr.*, **15**, 608.

Keller, G., 1986, Stepwise mass extinctions and impact events: Late Eocene to Early Oligocene, in Pomeizol, L. H., ed., *Geological Events at the Eocene-Oligocene Boundary*, Amsterdam, Elsevier (in press).

Kennedy, W. J., and Odin, G. S., 1982, The Jurassic and Cretaceous time scale in 1981, in Odin, G. S., ed., *Numerical Dating in Stratigraphy, Part I*, Chichester, Wiley, pp. 557–592.

Kerr, R. A., 1984, Periodic impacts and extinctions reported, *Science*, **223**, 1277–1279.

Kerr, R. A., 1985, Periodic extinctions and impacts challenged, *Science*, **227**, 1451–1453.

Kummel, B., and Teichert, C., 1970, Stratigraphy and paleontology of the Permian-Triassic beds, Salt Range and Trans-Indus Ranges, West Pakistan, in Kummel, B., and Teichert, C., eds., *Stratigraphic Boundary Problems: Permian and Triassic of West Pakistan*, Lawrence, Kansas, University Press of Kansas, pp. 1–110.

Lipps, J. H., 1970, Plankton evolution, *Evolution*, **24**, 1–22.

Luck, J. M., and Turekian, K. K., 1983, Osmium-187/osmium-186 in manganese nodules and the Cretaceous-Tertiary Boundary, *Science*, **222**, 613–615.

Lutz, T. M., 1985, The magnetic reversal record is not periodic, *Nature*, **317**, 404–407.

McCrea, W. H., 1981, Long time-scale fluctuations in the evolution of the earth, *Proc. Roy. Soc. Lond., Series A*, **375**, 1–41.

McGhee, G. R., Jr., 1982, The Frasnian-Famennian extinction event: A preliminary analysis of Appalachian marine ecosystems, in Silver, L. T., and Schultz, P. H., eds., *Geological Implications of Impacts of Large Asteroids and Comets on the Earth*, Geological Society of America Special Paper 190, pp. 491–500.

McLaren, D. J., 1983, Bolides and biostratigraphy, *Geolo. Soc. Am. Bull.*, **94**, 313–324.

McLaren, D. J., 1984, Abrupt extinctions, *Terra Cognita*, **4**, 27–32.

Milne, D. H., and McKay, C. P., 1982, Response of marine plankton communities to a global atmospheric darkening, in Silver, L. T., and Schultz, P. H., eds., *Geological Implications of Impacts of Large Asteroids and Comets on the Earth*, Geological Society of America Special Paper 190, pp. 297–304.

Montanari, A., Hay, R. L., Alvarez, W., Asaro, F., Michel, H. V., Alvarez, L. W., and Smit, J., 1983, Spheroids at the Cretaceous-Tertiary Boundary are altered impact droplets of basaltic composition, *Geology*, **11**, 668–671.

Muller, R. A., 1985, Evidence for a solar companion star, in Papagiannis, M. D., ed., *The Search for Extraterrestrial Life: Recent Developments*, Dordrecht, Reidel, pp. 233–243.

Napier, W. M., and Clube, S. V. M., 1979, A theory of terrestrial catastrophism, *Nature*, **282**, 455–459.

Negi, J. G., and Tiwari, R. K., 1983, Matching long term periodicities of geomagnetic reversals and galactic motions of the solar system, *Geophys. Res. Lett.*, **10**, 713–716.

Newell, N. D., 1967, Revolution in the history of life, in Albritton, C. C., ed., *Uniformity and Simplicity: A Symposium on the Principle of the Uniformity of Nature*, Geological Society of America Special Paper 89, pp. 63–91.

Odin, G. S., ed., 1982, *Numerical Dating in Stratigraphy*, Somerset, New Jersey, Wiley, 968 pp.

Padian, K., and Clemens, W. A., 1985, Terrestrial vertebrate diversity: Episodes and insights, in Valentine, J. W., ed., *Phanerozoic Diversity Patterns: Profiles in Macroevolution*, Princeton, New Jersey, American Association for the Advancement of Science and Princeton University Press (in press).

Palmer, A. R., 1979, Biomere boundaries re-examined, *Alcheringa*, **3**, 33–41.

Palmer, A. R., 1982, Biomere boundaries: A possible test for extraterrestrial perturbation of the biosphere, in Silver, L. T., and Schultz, P. H., eds., *Geological Implications of Impacts of Large Asteroids and Comets on the Earth*, Geological Society of America Special Paper 190, pp. 469–476.

Palmer, A. R., 1983, The Decade of North American Geology 1983 geologic time scale, *Geology*, **11**, 503–504.

Palmer, A. R., 1984, The biomere problem: Evolution of an idea, *J. Paleontol.*, **58**, 599–611.

Pillmore, C. L., Tschudy, R. H., Orth, C. J., Gilmore, J. S., and Knight, J. D., 1984, Geologic framework of nonmarine Cretaceous-Tertiary Boundary sites, Raton Basin, New Mexico and Colorado, *Science*, **223**, 1180–1183.

Pollack, J. B., Toon, O. B., Ackerman, T. P., Turco, R. P., and McKay, C. P., 1983, Environmental effects of an impact-generated dust cloud: Implications for the Cretaceous-Tertiary extinctions, *Science*, **219**, 287–289.

Quinn, J. F., 1983, Mass extinctions in the fossil record [technical comment], *Science*, 219, 1239–1240.

Raffi, S., Stanley, S. M., and Marasti, R., 1985, Biogeographic patterns and Plio-Pleistocene extinction of Bivalvia in the Mediterranean and southern North Sea, *Paleobiology*, **11**, 368–388.

Rampino, M. R., and Stothers, R. B., 1984a, Terrestrial mass extinctions, cometary impacts and the sun's motion perpendicular to the galactic plane, *Nature*, **308**, 709–712.

Rampino, M. R., and Stothers, R. D., 1984b, Geological rhythms and cometary impacts, *Science*, **226**, 1427–1431.

Raup, D. M., 1975, Taxonomic diversity estimation using rarefaction, *Paleobiology*, **1**, 333–342.

Raup, D. M., 1978, Cohort analysis of generic survivorship, *Paleobiology*, **4**, 1–15.

Raup, D. M., 1979, Size of the Permo-Triassic bottleneck and its evolutionary implications, *Science*, **206**, 217–218.

Raup, D. M., 1985a, Magnetic reversals and mass extinctions, *Nature*, **314**, 341–343.

Raup, D. M., 1985b, Rise and fall of periodicity, *Nature*, **317**, 384–385.

Raup, D. M., 1985c, Life, terrestrial environments, and events in space, in Milne, D., Raup, D. M., Billingham, J., Niklas, K. J., and Padian, K., eds., *The Evolution of Complex and Higher Organisms*, NASA Special Publication 478, pp. 9–24.

Raup, D. M., and Sepkoski, J. J., Jr., 1982, Mass extinctions in the marine fossil record, *Science*, **215**, 1501–1503.

Raup, D. M., and Sepkoki, J. J., Jr., 1984, Periodicity of extinctions in the geologic past, *Proc. Natl. Acad. Sci. U.S.A.*, **81**, 801–805.

Raup, D. M., Sepkoski, J. J., Jr., and Stigler, S. M., 1983, Mass extinctions in the fossil record [technical comment], *Science*, **219**, 1240–1241.

Richter, 1984, Time and space scales of mantle convection, in Holland, H. D., and Trendall, A. F., eds., *Patterns of Change in Earth Evolution*, Berlin, Springer-Verlag, pp. 271–289.

Schwartz, R. D., and James, P. B., 1984, Periodic mass extinctions and the Sun's oscillation about the galactic plane, *Nature*, **308**, 712–713.

Sepkoski, J. J., Jr., 1982a, *A Compendium of Fossil Marine Families*, Milwaukee Public Museum Contributions in Biology and Geology no. 51, 125 pp.

Sepkoski, J. J., Jr., 1982b, Mass extinctions in the Phanerozoic oceans: A review, in Silver, L. T., and Schultz, P. H., eds., *Geological Implications of Impacts of Large Asteroids and Comets on the Earth*, Geological Society of America Special Paper 190, pp. 283–289.

Sepkoski, J. J., Jr., 1985, Some implications of mass extinction for the evolution of complex life, in Papagiannis, M. D., ed., *The Search for Extraterrestrial Life: Recent Developments*, Dordrecht, Reidel, pp. 223–232.

Sepkoski, J. J., Jr., 1986, Phanerozoic overview of mass extinction, in Raup, D. M., and Jablonski, D., eds., *Pattern and Process in the History of Life*, Berlin, Springer-Verlag (in press).

Sepkoski, J. J., Jr., and Hulver, M. L., 1985, An atlas of Phanerozoic clade diversity diagrams, in Valentine, J. W., ed., *Phanerozoic Diversity Patterns: Profiles in Macroevolution*, Princeton, New Jersey, American Association for the Advancement of Science and Princeton University Press (in press).

Sheng, J.-Z., Chen, C.-Z., Wang, Y.-G., Rui, L., Liao, Z.-T., Bando, Y., Ishi, K.-I., Nakazawa, K., and Nakamura, K., 1984, Permian-Triassic Boundary in middle and eastern Tethys, *Jour. Fac. Sci.*, Hokkaido Univ., Ser. IV, **21,** 133–181.

Shoemaker, E. M., 1984, Large body impacts through geologic time, in Holland, H. D., and Trendall, A. F., eds., *Patterns of Change in Earth Evolution*, Berlin, Springer-Verlag, pp. 15–40.

Signor, P. W., III, and Lipps, J. H., 1982, Sampling bias, gradual extinction patterns, and catastrophes in the fossil record, in Silver, L. T., and Schultz, P. H., eds., *Geological Implications of Impacts of Large Asteroids and Comets on the Earth*, Geological Society of America Special Paper 190, pp. 291–296.

Smit, J., 1982, Extinction and evolution of planktonic Foraminifera at the Cretaceous/Tertiary Boundary after a major impact, in Silver, L. T., and Schultz, P. H., eds., *Geological Implications of Impacts of Large Asteroids and Comets on the Earth*, Geological Society of America Special Paper 190, pp. 329–352.

Smit, J., and Kyte, F. T., 1984, Siderophile-rich magnetic spheroids from the Cretaceous-Tertiary Boundary in Umbria, Italy, *Nature,* **310,** 403–405.

Smit, J., and Romein, A. J. T., 1985, Sequence of events across the Cretaceous-Tertiary Boundary, *Earth and Planet. Sci. Lett.*, **74,** 155–170.

Smit J., and ten Kate, W. G. H. Z., 1982, Trace element patterns at the Cretaceous–Tertiary Boundary: Consequences of a large impact, *Cretac. Res.*, **3**, 307–332.

Stanley, S. M., 1979, *Macroevolution: Pattern and Process*, San Francisco, Freeman, 332 pp.

Stanley, S. M., 1984a, Marine mass extinctions: A dominant role for temperature, in Nitecki, N. H., ed., *Extinctions*, Chicago, University of Chicago Press, pp. 69–117.

Stanley, S. M., 1984b, Temperature and biotic crises in the marine realm, *Geology*, **12**, 205–208.

Stanley, S. M., and Campbell, L. D., 1981, Neogene mass extinction of western Atlantic molluscs, *Nature*, **293**, 457–459.

Stanley, S. M., Addicott, W. O., and Chinzei, K., 1980, Lyellian curves in paleontology: Possibilities and limitations, *Geology*, **8**, 422–426.

Stigler, S. M., 1985, Terrestrial mass extinctions and galactic plane crossings, *Nature*, **313,** 159.

Stitt, J. H., 1971, Repeating evolutionary pattern in Late Cambrian trilobite biomeres, *J. Paleontol.*, **45**, 178–181.

Stothers, R., 1979, Solar activity cycle during classical antiquity, *Astron. Astrophys.*, **77,** 121–127.

Sun, Y.-Y., Chai, Z.-F., Ma, S.-L., Mao, X.-Y., Xu, D.-Y., Zhang, Q.-W., Yang, Z.-Z., Sheng, J.-Z., Chen, C.-Z., Rui, L., Liang, X.-L., Zhao, J.-M., and He, J.-W., 1984, The discovery of iridium anomaly in the Permian-Triassic Boundary Clay in Changxing, Zhejiang, China, and its significance, in *Developments in Geoscience*, Beijing, Science Press, pp. 235–245.

Tappan, H., and Loeblich, A. R., Jr., 1971, Geobiologic implications of phytoplankton evolution and time–space distribution, *Geological Society of America Special Paper*, **127**,247–340.

Tappan, H., and Loeblich, A. R., Jr., 1972, Fluctuating rates of protistan evolution, diversification and extinction, *24th Int. Geol. Cong., Montreal*, **7**, 205–213.

Thaddeus, P., and Chanan, G. A., 1985, Cometary impacts, molecular clouds, and the motion of the Sun perpendicular to the galactic plane, *Nature*, **314,** 73–75.

Thierstein, H. R., 1982, Terminal Cretaceous plankton extinctions: A critical assessment, in Silver, L. T., and Schultz, P. H., eds., *Geological Implications of Impacts of Large Asteroids and Comets on the Earth*, Geological Society of America Special Paper 190, pp. 385–400.

Toon, O. B., Pollack, J. B., Ackerman, T. P., Turco, R. P., McKay, C. P., and Liu, M. S.,

1982, Evolution of an impact-generated dust cloud and its effects on the atmosphere, in Silver, L. T., and Schultz, P. H., eds., *Geological Implications of Impacts of Large Asteroids and Comets on the Earth*, Geological Society of America Special Paper 190, pp. 187–200.

Torbett, M. V., and Smoluchowski, R., 1984, Orbital stability of the unseen solar companion linked to periodic extinction events, *Nature*, **311**, 641–642.

Tozer, E. T., 1979, Latest Triassic ammonoid faunas and biochronology, western Canada, *Geological Survey of Canada Paper*, **79-1B**, 127–135.

Ujiié, H., 1984, A Middle Miocene hiatus in the Pacific region: Its stratigraphic and paleoceanographic significance, *Palaeogeogr., Palaeoclimatol., Palaeoecol.*, **46**, 143–164.

Valentine, J. W., 1974, Temporal bias in extinctions among taxonomic categories, *J. Paleontol.*, **48**, 549–552.

Van Valen, L. M., 1984, A resetting of Phanerozoic community evolution, *Nature*, **307**, 50–52.

Ward, P., 1983, The extinction of the ammonites, *Sci. Am.*, **249**, 136–147.

Ward, P. D., and Signor, P. W., III, 1983, Evolutionary tempo in Jurassic and Cretaceous ammonites, *Paleobiology*, **9**, 183–198.

Whitmire, D. P., and Jackson, A. A., IV, 1984, Are periodic mass extinctions driven by a distant solar companion?, *Nature*, **308**, 713–715.

Whitmire, D. P., and Matese, J. J., 1985, Periodic comet showers and Planet X, *Nature*, **313**, 36–38.

Wolbach, W. S., Lewis, R. S., and Anders, E., 1985, Cretaceous extinctions: Evidence for wildfires and search for meteoritic material, *Science*, **230**, 167–170.

Xu Dao-Yi, Ma Shu-Lan, Chai Zhi-Fang, Mao Xuo-Ying, Sun Yi-Ying, Zhang Qin-Wen, and Yang Zheng-Zhong, 1985, Abundance variation of iridium and trace elements at the Permian-Triassic Boundary at Shangsi in China, *Nature*, **314**, 154–156.

2

ABRUPT EXTINCTIONS

DIGBY J. McLAREN

Department of Geology
University of Ottawa
Ottawa, Ontario
K1N 6N5, Canada

Publication 16-83 of the
Ottawa-Carleton Centre
for Geoscience Studies.

INTRODUCTION

The stratigraphic column was put together in the early days of geology. The systems and their subdivisions were defined for the most part in Western Europe, with additions from Eastern Europe and Western North America. Boundaries were selected between the major stratigraphic units recognized and tended to be distinguished by some event, frequently a major faunal change. Some of these events represented time missing from the succession, but it was not clear whether this was always true. Stratigraphy developed as an empirical science based on local observations, and it was soon realized that it might be possible to construct a relative time scale for the succession of fossiliferous rocks, the Phanerozoic, with few or no major gaps in the sequence. Names were attached to particular sequences of rocks representing finite amounts of time that can now be approximately quantified, following the development of a chronology based on some form of isotope decay. Gradualism, therefore, became a necessary part of stratigraphic principles, and sudden breaks in the succession, whether lithologic or biologic, were suspect.

Parallel with the development of stratigraphic principles, but postdating the recognition that a relative time scale could be constructed, Darwin's explanation of evolution by natural selection gave a firmer theoretical framework to the use of fossils in constructing such time scales and correlating from place to place. Here, again, gradualism appeared to be the necessary premise in searching for a continuous evolutionary sequence of life forms through time.

In stratigraphy, however, certain horizons, or boundaries, continued to look as though they represented a sudden event. I distinguished such boundaries as horizons across which "something happened" as opposed to boundaries selected more or less arbitrarily for their convenience, which could be categorized as quiet boundaries (McLaren, 1970). There is nothing new in this, but it did draw attention to the fact that in spite of detailed studies in stratigraphy throughout the Phanerozoic, it appeared that sudden events really happened. For a recent discussion on continuity and discontinuity in stratigraphy, see Plaziat and Ellenberger (1982).

There has been increasing evidence and support for the idea of discontinuities in gradualistic evolution. Currently known as "punctuated equilibrium" (Eldredge and Gould, 1972), such ideas were implicit in Darwin's *Origin* and certainly in some of the writings on the modern synthesis of evolutionary theory in the 1940s and 1950s (Huxley, 1982). Recently, two relatively new concepts have, to some degree, overtaken discussions on event stratigraphy and punctuated equilibrium which demand integration into the resolution of current controversy. I refer to two widely different physical discoveries both of which impinge on the empirical structures of stratigraphy and evolutionary genetics. First is the concept of impacts of large asteroids on the earth; second is the advent of molecular biology and the realization that genetic adaptations can and do occur rapidly in response to ecologic perturbation. It is interesting to note, furthermore, that these two concepts may, in turn, interact and offer both a peculiarly accurate stratigraphic horizon marker and a powerful evolutionary driving force.

This brief chapter is principally concerned with arguing the effects of the first of these ideas because I was forced to examine the possibility of some such mechanism as a direct result of field observations at a particular horizon in the stratigraphic column. Lack of knowledge, however, precludes any useful observations on molecular biology. The reader is referred to the remarkable paper by Schopf (1981) for a stimulating introduction to the whole subject from the point of view of a paleontologist.

Current controversy concerning duration of taxa, differences between paleontologic and biologic species, punctuated equilibria, and the mechanisms of speciation in general will not be dealt with here. This chapter is concerned primarily with physical punctuation of the stratigraphic column that may be recongized by abrupt faunal changes. I refer to observable extinctions that take place at a bedding plane such as the end of the Frasnian or end of the Cretaceous and the difficulties of detecting other such abrupt events by means of current systems of charting the progression of life through time. The driving force of this discussion, as it were, rests with the recognition of at least one causative mechanism, namely large-body impact, together with the confidence engendered by assurance from molecular biology that, given adequate environmental stimulus, evolution might be very rapid indeed.

EXTINCTIONS AND TAXA PLOTS

How are extinctions to be recognized? Life is changing continuously, and Raup and Sepkoski (1982) have shown that the mean extinction rate of animal families through time is on the order of 3 or 4 per million years, which they describe as the background extinction rate. Superimposed on this generalization, there are brief periods of greatly accelerated extinction, with five particularly well marked at the end of the Ordovician, the Frasnian Stage of the Devonian, the Permian, the Trias, and the Cretaceous. These are properly called mass extinctions (see further discussion in Sepkoski, 1982).

Various ways have been adopted to demonstrate biological changes during geological time by some form of plotting taxa. For instance, Newell (1962) plotted percentages of total families of animals against time, showing, on two curves, first and last occurrences at the end of each series. This practice has continued with a shortening of the time interval. So far, families against stages seems to be the best we have achieved (Raup and Sepkoski, 1982). Other techniques plot biological diversity against time by counting the numbers of taxa, commonly families, extant at any defined moment or in a defined interval of time (e.g., Valentine and Moores, 1972, and many others). An interesting development plots numbers of families for each metazoan class present in each stage-level stratigraphic interval through time. Such spindle diagrams show the patterns of the waxing and waning of various groups through time. (e.g., Sepkoski, 1981).

SOME DIFFICULTIES

It is difficult to move from such statistical approaches to a consideration of what actually happened at any particular moment in earth history, in spite of the value of such laborious compilations and the elegance of the statistical procedures applied to them. There are unavoidable fallacies involved, and these fall into three groups:

1. The relationship between rank of taxa and number of individuals or biomass of a particular taxon is hugely variable. In talking of extinctions, reality would seem to dictate that one should attempt to examine how many animals or plants disappeared. It is claimed that number of taxa is a more objective measure than estimates of biomass, although this might be disputed when one considers that at a fossiliferous locality the numbers of individuals collected plotted against the rank order of species within a defined group results in the "hollow curve of distribution" (Schopf, 1982). The same applies in general to plots of any rank of taxon against a higher taxon within a fossil group. This means that the longer one collects from a particular locality, the more taxa will be found, without necessarily changing the ultimate biomass in any significant way. Here we must firmly face the problem of quantification in field observation, the measurable versus the anecdotal. I must refer to the late Frasnian extinction, taking the corals as an example (Pedder, 1982; McLaren, 1982, 1983). Virtually all shallow, warm-water corals (142 species)

disappeared at one bedding plane wherever the interval is preserved, and this represented a huge biomass covering vast regions of warm shallow seas from northwest North America east to Western Australia. It can be claimed that the biomass disappearance of shallow warm-water corals was close to 100%. No exceptions are known to this statement; it is deeply significant statistically. Some anecdote!

2. Plotting an interval of time against numbers of taxa introduces another difficulty. The interval selected, commonly a stage, with an average duration of some 7–8 m.y. is necessarily arbitrary. Thus, in plotting families against stages, and knowing that an average of about three, or four families disappear each 1 m.y., summing the total families that ended during the stage may not have much meaning. It should also be recognized that if an anomalous departure from background extinction levels is signaled at the end of an interval, it is still necessary to examine whether such a change is gradual throughout or a sudden event at some time during the interval or some variant of these possibilities. Furthermore, if a bedding plane extinction is in fact a synchronous horizon all over the world, as appears likely for the Frasnian and the Late Cretaceous, then the statistical significance of even a few extinctions at that point, is huge, as the time involved may be very short.

3. Although not necessarily stated, it would appear that the action of charting diversity or taxa ranges implicitly suggests that an extinction can only be demonstrated by disappearances of taxa. In fact, as has already been pointed out (Boucot, 1983; McLaren, 1983), there is no reason to suppose that a biomass extinction of considerable magnitude would necessarily include every individual of a particular taxon affected. There are many examples of major extinctions in which a relatively small number of individuals of some of the taxa affected continue. Although these may be trivial in numbers and importance, nevertheless, the taxa to which they belong will not show as extinctions on range charts at that horizon. Furthermore, diversity plots may show mass extinctions, but do so essentially accidentally. Each plot shows the diversity, that is, number of taxa extant, for each time unit and thus represents only the animals alive during that unit. If all taxa in a group became extinct at the end of a particular time unit and were replaced by an equal number during the next time unit, an extinction would not be apparent. The plot, in fact, shows only the net change in number of taxa between intervals, and there are many possible causes for such a change, for example, changes in provincialism due to plate movement, volume of sedimentary rock preserved, or a real evolutionary phenomenon (Raup, 1972; Sepkoski et al., 1981). The same comment applies to spindle diagrams. Again, referring to the late Frasnian extinction, Sepkoski shows about 28 families of corals in the Frasnian and Famennian Stages (1981). This may be correct, but what actually happened is that only 11 coral families were present in the latest Frasnian strata, and of these, six continued into the Famennian. Of these six, four are deep-water forms, and the other two families of shallow-water corals were markedly impoverished (Pedder, 1982). The event is much more dramatic when species and biomass are considered as discussed above. As a further example, on the same diagrams, Sepkoski shows about 55 families of Famennian brachiopods in juxtaposition to a larger number in the Frasnian. In fact, only 10 genera or genetic groups representing as many families have been interpreted as

having survived from the Frasnian into the Famennian (Johnson and Boucot, 1973). Many similar examples could be cited.

It appears, therefore, that however useful taxa counts plotted against intervals of time may be in charting the major evolutionary changes that take place during geological time, the technique is not particularly well suited for discovering or demonstrating sudden extinctions of whatever magnitude. Sudden extinction events can only be recognized by detailed stratigraphic and paleontological collecting across a suspected horizon. A count of individuals and species above and below may prove adequate indicators, but an attempt must be made on a regional or worldwide basis to assess the biomass changes or, in other words, the magnitude of the extinction.

RECOGNITION

Is there more than one kind of extinction? It is convenient to talk about background versus mass extinctions, but are they real (Raup, 1981; Raup and Sepkoski, 1982)? We are trying to do two different things: (1) chart major changes in life during time and (2) detect punctuations in the orderly development of life, which evidently do occur and appear to have happened fairly quickly. Techniques to achieve these objectives differ, and every attempt should be made to bring them together. The emphasis on how to recognize extinction horizons and to assess the duration of extinction events may well be the most important single problem for biostratigraphy in providing correlatable horizons and for evolutionary studies in understanding the driving force of evolution. If we can demonstrate that an extinction has taken place in a relatively short time, although possibly in a strictly defined environment, there may be an important message for overall studies in biotic diversity. Diversity plots tend to smooth out the fluctuations induced by sudden extinctions, especially when there is a rapid reestablishment of equilibrium after an event. Diversity studies must try to include such short-term fluctuations in their modeling if we are to arrive at a true understanding of the role of extinctions in the development of life.

DURATION OF EVENTS

Geology and paleontology deal with a very imperfect time scale whose limits of resolution are commonly undetermined or at best poorly known. The advantage of attempting correlation together with geochemical and sedimentological studies on a worldwide basis, however, allows us to overcome some of the difficulties to a certain extent. Thus, the worldwide extinction of planktonic foraminifera at the end of the Maastrichtian and their replacement largely by new forms in the basal Danian was probably a rapid event, certainly much less than half a million years and possibly almost instantaneous. Geochemical and sedimentological studies carried out at this horizon also suggest some very rapid event (e.g., many papers in Silver and Schultz,

1982; Smit and ten Kate, 1982). Similarly, McLaren (1982) has given reasons for supposing that an extinction of warm-water shallow benthos at the end of the Frasnian took place in less than one conodont zone, approximately half a million years, and probably very much less. In fact, the disappearance of biomass takes place at a bedding plane at every section affected in the world. It would appear, in both these instances, that the event was indeed sudden.

The record is not complete for most other suggested mass extinctions, and it is difficult to determine whether they took place rapidly or over a geologically measurable period of time. It would certainly appear that the great extinctions that occurred during and at the end of the Late Permian were spread over at least two stages. Nevertheless, the possibility of individual rapid extinctions within these intervals has not yet been resolved. It seems likely that the extinction at the recently defined limit between the Ordovician and Silurian Systems was rapid and may well represent a bedding plane extinction over a wide region of the earth (T. E. Bolton, 1983, personal communication). Further discussions on duration of extinctions and event stratigraphy may be found in Walliser (1983).

PATTERNS OF EXTINCTION

Extinctions appear to be selective, and evidence suggests that some rapid extinctions affected only the biota of certain major ecological groups, such as shallow-water benthos, filter-feeding organisms, terrestrial vertebrates, and shallow-water plankton. They may involve large numbers of individuals or biomass, but, as has been pointed out above, all individuals of a taxon affected may not disappear immediately. Frequently a few linger on to disappear in the next time interval, and occasionally survivors may continue for a long time (e.g., the coelacanth). A model of extinction patterns uncomplicated by large numbers of other organisms has been provided by Palmer with his study of the biomere boundaries in the Middle and Upper Cambrian, affecting the trilobite faunas (1965, 1982; Stitt, 1971). The pattern of extinction in each case is clear: massive disappearance of many taxa, continuation and brief proliferation of a few opportunistic taxa before they, too, become extinct; and the appearance, with the opportunists, of new forms without ancestors in the immediately preceding faunas that form the root stock for the succeeding radiation (Palmer, 1983, personal communication).

Similar patterns are found in extinctions at other horizons except that, commonly, one or a few lineages continue through the horizon as the precursors of the next major radiation, for example, the Frasnian–Famennian extinction, the Maastrichtian–Danian extinction (Silver and Schultz, 1982), ammonoid extinctions several times (House, 1963), mammal-like reptile extinctions (Kemp, 1982), and others. The pattern of the forms of life that develop immediately after a major extinction such as the Frasnian and Late Cretaceous is complex and cannot be dealt with in this chapter. Nevertheless, one must emphasize that in most known cases, the existence of an empty environment immediately after a major biomass extinction over a large region gives rise to what Fischer and Arthur have called an oligotaxic pulse (1977), and the early faunas are commonly rich in numbers and few in species. Furthermore, the ancestry of many of the forms that appear is not easily traced through an obvious lineage from the previous population. In other words, there appears to be a "macro-evolutionary" jump, even though the time may be, geologically speaking, brief.

CAUSES

During the last few years, it has been demonstrated that the earth, as well as all other observable bodies in the inner solar system, have been struck by asteroids of various sizes during their history. For the burgeoning literature demonstrating such events and assessing their effects, the reader is referred to the report of the Snowbird Conference held in October 1981 (Silver and Schultz, 1982), the catalogue of terrestrial craters (Grieve and Robertson, 1979), as well as a general summary of some Phanerozoic possibilities (McLaren, 1983).

Here I should digress briefly and discuss a situation that is somewhat unusual within our empirical science. We are familiar with stories of how ideas in geology have been held back by physicists denying effects because there were no possible causes. Yet there have always been some geologists who maintained that empirical evidence led to inductions that differed from the current physical model. For instance, Darwin and Chamberlin argued from geological evidence that the world was much older than Lord Kelvin allowed, while Du Toit adduced massive evidence for continental drift in spite of the fact that Jeffreys proved that it was impossible. Paradoxically, today we have astrophysicists offering us a cause that might well have some importance for geology and evolutionary theory, while some of us deny its effects. We are told that geological processes only take place as quickly as one's fingernails grow and that therefore it is pointless to look for empirical evidence for events that might have taken place more rapidly. Now I maintain that it is difficult to deny the reality of explosive asteroid impacts with energies as high as 10^{23} joules or more, with a frequency defined within limits arrived at by observation and by theoretical means. Apollo objects can be observed and counted, and terrestrial impact craters exist and can be examined and dated. The causes may begin in outer space, but the effects are very much down-to-earth and, in fact, represent a normal and continuing phenomenon in the evolution of the solar system, including this planet. In general, no claim is being made or has been made that impacts cause all extinctions or that any one extinction has been caused only by an impact. Like many other phenomena in geology, such causes may be multiple and resolved only with difficulty. But surely the burden of proof rests, now, firmly on the shoulders of those who would deny the existence of this violent phenomenon, and for them the proof must be in two phases: (1) evidence and argument leading to the denial of the demonstration of Earth-crossing asteroids and their periodic arrival on Earth and, if that cannot be achieved, (2) the demonstration that such energetic events have had no effect, detectable by geological means, on crustal evolution, including the development of life. If these two falsifications cannot be achieved, then let us get on with the job of finding out just how much effect a large-body impact may have under differing conditions and at different times in earth history.

From many lines of evidence I suggest that we must now accept that large-body impacts have occurred and continue to occur, and part of that evidence exists in the geological record. Such impacts must be considered as possible contributory causes to some extinctions in the stratigraphic record, and attempts must be made to assess the relative importance of such contributions. It is apparent that biotic extinctions may be caused by many immediate effects, and these may be generalized into three ultimate causes: internal changes in the Earth, solar and orbital changes within the solar system, and impacts or supernovas. It appears likely that certain immediate causes may result from more than one ultimate cause, and we may look for patterns by which these ultimate causes may be recognized (Figure 1). Plainly it is necessary to study the ecology of the organisms that were affected and their physical environment. In the marine realm, it would appear that the most likely immediate causes for abrupt, or bedding plane, extinctions include ocean anoxia, which might lead to black shale deposition (Berry and Wilde, 1978), turbidity, and, more controversially, rapid regression and temperature changes. Several of these, in turn, might be caused by a large-body impact. In the terrestrial realm, climatic stress would appear to be a major contender, followed by temperature changes and effects caused by large-body impacts or, possibly, major volcanic activity, including dust leading to darkness on the order of several months duration and high atmospheric concentrations of NOx. Many references might be given, but the reader is again referred to papers in Silver and Schultz (1982), Smit and ten Kate (1982), Johnson (1974), and Walliser (1983).

In conclusion, it must be stressed that abrupt extinctions exist but may be obscured by plotting taxa against time intervals; possibilities of accurate worldwide correlation appear real; the record of such sudden extinction events is probably far from complete; large-body impacts occur and may well be contributory causes to extinction

IMMEDIATE CAUSES		ULTIMATE CAUSES		
		MANTLE CONVECTION	SOLAR AND ORBITAL CHANGES	IMPACTS AND SUPERNOVAE
SEA-LEVEL CHANGES		X		
CLIMATIC CHANGES	LONG TERM	X	X	
	SHORT TERM		X	X
RADIATION INTENSITY CHANGES	LONG TERM		X	
	SHORT TERM	X	X	X
POISONING		X		X
SHOCK EFFECTS				X

Figure 1. Causes of extinctions: some possible immediate causes plotted against ultimate causes. An *X* indicates that an immediate cause could have been initiated by the one or more ultimate causes indicated (McLaren, 1983).

events; and large biomass extinction followed by rapid adaptive radiation of new forms may be a major factor in evolution. It would appear that the recognition of event stratigraphy together with the realization that genetic change can be driven by environmental stress give a new perspective to biochronology and macroevolution.

ACKNOWLEDGMENTS

This chapter was first presented at the second meeting of the European Union of Geosciences in Strasbourg, France, on March 30, 1983, and will appear in *Terra Cognita*, the journal of the EUG, the editors of which kindly gave permission for me to deliver the paper at the Flagstaff Conference and to publish a revised version in the proceedings of that conference. The manuscript benefited from critical reading by R. G. Blackadar and A. R. Palmer, who are gratefully acknowledged.

REFERENCES

Berry, W. B. N., and Wilde, P., 1978, Progressive ventilation of the oceans: An explanation for the distribution of the lower Paleozoic black shales, *Am. J. Sci.* **278**, 257–275.

Boucot, A. J., 1983, Does evolution take place in an ecological vacuum, *J. Paleontol.* **57**, 1–30.

Eldredge, N., and Gould, S. J., 1972, Punctuated equilibria: An alternative of phyletic gradualism, in Schopf T. J. M., ed., *Models in Paleobiology*, San Francisco, Freeman Cooper.

Fischer, A. G., and Arthur, M. A., 1977, *Secular Variations in the Pelagic Realm*, Society of Economic Paleontologists and Mineralogists Special Publication 25, pp. 19–50.

Grieve, R. A. F., and Robertson, P. B., 1979, The terrestrial cratering record, I. Current status of observations, *Icarus*, **38**, 212–229.

House, M. R., 1963, Bursts in evolution, *Advancement of Science*, March 1963, 499–507.

Huxley, A. F., 1982, Anniversary address of the President, *Proc. Roy. Soc. Lond.*, A**379**, v–xx.

Johnson, J. G., 1974, Extinction of perched faunas, *Geology*, **2**, 479–482.

Johnson, J. G., and Boucot, A. J., 1973, Devonian brachiopods, in Hallam, A., ed., *Atlas of Palaebiogeography*, Amsterdam, Elsevier Scientific Publishing Co., pp. 89–96.

Kemp, T. S., 1982, Mammal-like reptiles and the origins of mammals, Academic Press, New York 384 pp.

McLaren, D. J., 1970, Time, life and boundaries, *J. Paleontol.*, 44, 801–815.

McLaren, D. J., 1982, *Frasnian-Famennian Extinctions*, Geological Society of America Special Paper 190, pp. 477–484.

McLaren, D. J., 1983, *Bolides and Biostratigraphy*, Geological Society of America Bulletin 94, pp. 313–324.

Newell, N., 1962, Paleontological gaps and geochronology, *J. Paleontol.* **36**, 592–610.

Palmer, A. R., 1965, Biomere: A new kind of biostratigraphic unit, *J. Paleontol.*, **39**, 149–152.

Palmer, A. R., 1982, *Biomere Boundaries: A Possible Test for Extraterrestrial Perturbation of the Biosphere*, Geological Society of America Special Paper 190, pp. 469–476.

Pedder, A. E. H., 1982, *The Rugose Coral Record Across the Frasnian-Framennian Boundary*, Geological Society of America Special Paper 190, pp. 485–490.

Plaziat, J. C., and Ellenberger, F., 1982, A propos de la limite Crétacé-Tertiaire: La réconciliation moderne des conceptions continue et discontinue en stratigraphie et en tectonique, *Société Géologique de France Bulletin 7e série*, tome 24, 831–841.

Raup, D. M., 1972, Taxonomic diversity during the Phanerozoic, *Science*, **177**, 1065–1071.

Raup, D. M., 1981, Extinction: Bad genes or bad luck?, *Acta Geol. Hispan.*, **16**, 25–33.

Raup, D. M., and Sepkoski, J. J., 1982, Mass extinctions in the marine fossil record, *Science*, **215**, 1501–1503.

Schopf, T. J. M., 1981, Evidence from findings of molecular biology with regard to the rapidity of genomic change: Implications for species durations, in Niklas, K. J., ed., *Paleobotany, Paleoecology and Evolution*, New York: Praeger Publishers.

Schopf, T. J. M., 1982, A critical assessment of punctuated equilibria I. Duration of taxa; *Evolution*, **36**, 1144–1157.

Sepkoski, J. J., Jr., 1981, A factor analytic description of the Phanerozoic marine fossil record, *Paleobiology*, **7**, 36–53.

Sepkoski, J. J., Jr., 1982, *Mass Extinctions in the Phanerozoic Oceans: A Review*, Geological Society of America Special Paper 190, pp. 283–290.

Sepkoski, J. J., Jr., Bambach, R. K., Raup, D. M., Valentine, J. W., 1981, Phanerozoic marine diversity and the fossil record, *Nature*, **293**, 435–437.

Silver, L. T., and Schultz, P. H., eds., 1982, *Geological Implications of Impacts of Large Asteroids and Comets on the Earth*, Geological Society of America Special Paper 190, 528 pp.

Smit, J., and ten Kate, W. G. H. Z., 1982, Trace-element patterns at the Cretaceous-Tertiary boundary: Consequences of a large impact, *Cretac. Res.*, **3**, 307–332.

Stitt, J. H., 1971, Repeating evolutionary pattterns in Late Cambrian trilobite biomeres, *J. Paleontol.*, **45**, 178–181.

Valentine, J. W., and Moores, E. M., 1972, Global tectonics and the fossil record, *J. Geol.* 80, 167–184.

Walliser, O. H., 1983, Influence of geologic processes on evolution, ecology and stratigraphy, *Terra Cognita*, 3, 214 (abstract).

PART II

EXTINCTIONS IN THE GEOLOGICAL RECORD: MESOZOIC

3

MESOZOIC TETRAPOD EXTINCTIONS: A REVIEW

EDWIN H. COLBERT

Curator Emeritus
The American Museum of Natural History
New York, New York

Professor Emeritus
Columbia University
New York, New York

Curator of Vertebrate Paleontology
The Museum of Northern Arizona
Flagstaff, Arizona

INTRODUCTION

The natural extinctions of organisms through time has been a continuing, creative evolutionary process, a counterbalance to the origins of life forms. Without extinction there could be no progressive evolutionary development; the world would long ago have become a vast charnel house of ancient plants and animals in which such organisms as may have existed would carry on their life histories at an exceedingly low level of organization. And this above all else emphasizes the creative function of extinction. Yet important and crucial as has been the phenomenon of extinction, it has perhaps received much less attention from students of evolution than has the origin of living things. This may be owing to the fact that extinctions probably are more subtle and more difficult to study and interpret than are origins.

In any case, it is gratifying to see the renewed interest in extinctions as expressed by studies and publications that have characterized the past few years of research in the paleontological and biological sciences. Certainly some of this interest has been inspired by the amazing attention that has been devoted to the disappearance of the dinosaurs at the end of Cretaceous time, an interest characterized not only by the consideration given to the subject by scientists in various fields of geological, paleontological, and biological research, but also that has been generated among the general public, evidenced by the appearance of numerous books and periodical articles devoted to the subject.

Significant and spectacular as the extinction of the dinosaurs may have been, it is important to remember that this was only one fraction of the extinctions of Mesozoic tetrapods. As stated at the beginning of this chapter, since extinction has been a continuing process, it is not surprising to find extinctions occurring at various levels of intensity throughout the extent of Mesozoic time. It is to the problem of Mesozoic tetrapod extinctions that the present contribution will be devoted.

THE FACTS OF MESOZOIC EXTINCTION

There were 35 orders of tetrapods living during Mesozoic times, 4 of amphibians, 15 of reptiles, 7 of birds, and 9 of mammals. Of these, 21 became extinct during the Mesozoic, the extinctions encompassing 2 orders of amphibians, 10 of reptiles, 3 of birds, and 6 of mammals.*

The tetrapod orders here under consideration may be listed as follows.

Amphibia

Temnospondyli—labyrinthodont amphibians

Proanura—ancestral to the anurans—frogs and toads

Anura—frogs and toads

Urodela—salamanders

Reptilia

Cotylosauria—stem reptiles

Eosuchia—the first and most primitive diapsids, at present a rather heterogeneous group

*The problem of numbers is related to the classification used. In the present study, the classification of Mesozoic tetrapods is based in large part on Romer's classification of 1966, published in the third edition of his *Vertebrate Paleontology*. As for mammals, however, the classifications published in *Mesozoic Mammals* (Lillegraven, Kielan-Jaworowska, and Clemens, 1979) have been used. Except for the Eutheria, the classification by Clemens, Lillegraven, Lindsay, and Simpson in *Mesozoic Mammals* is used (p. 9). For the Eutheria, the classification by Kielan-Jaworowska, Brown, and Lillegraven is used (p. 223). In this present compendium, certain tetrapod orders are something less than satisfactory, notably the Eosuchia and the Protorosauria among the reptiles, both of which are in the nature of catchall assemblages. As for the eutherian mammals, Clemens et al. (Lillegraven et al., 1979) state that the "orders [are] uncertain or disputed in the Mesozoic." Since the Mesozoic orders are defined in the same volume by Kielan-Jaworowska et al., their classification is followed

Rhynchocephalia—"beaked reptiles," represented today by the tuatara, *Sphenodon**
Chelonia—turtles
Squamata—lizards and snakes
Thecodontia—Triassic archosaurs
Pterosauria—archosaurs, flying reptiles
Crocodilia—archosaurs, the crocodilians
Saurischia—archosaurs, saurischian dinosaurs
Ornithischia—archosaurs, ornithischian dinosaurs
Protorosauria—at present a "wastebasket group" of Triassic reptiles
Sauropterygia—marine nothosaurs and plesiosaurs
Placodontia—marine, mollusk-eating reptiles of Triassic age
Ichthyosauria—ichthyosaurs, of fishlike form
Therapsida—mammallike reptiles

Aves

Archaeopterygiformes—*Archaeopteryx*, the first bird, Jurassic
Hesperornithoformes—loonlike toothed birds, Cretaceous
Ichthyornithoformes—ternlike birds, Cretaceous
Gaviiformes—the divers, loons and grebes
Colymbiformes—doves and pigeons
Circoniiformes—waders, storks, and herons
Charadriformes—gulls and terns and their relatives

Mammalia

Multituberculata—earliest herbivores, with specialized teeth
Triconodonta—small carnivores with sharp-cusped teeth
Docodonta—Jurassic mammals with expanded tooth crowns
Symmetrodonta—ancient mammals with triangular-shaped cheek teeth
Eupantotheria—possible ancestors of later mammals
Marsupialia—pouched mammals
Proteutheria—very primitive eutherian mammals
Primates—today the lemurs, monkeys, apes, and man
Condylarthra—primitive hoofed mammals

Although the extinctions of 21 tetrapod orders extended throughout the span of Mesozoic time, it is interesting to see that 5 of the orders became extinct at the end of the Triassic period† and 9 at the close of Cretaceous history. Furthermore,

*Some authors regard the Triassic rhynchosaurs as a separate order rather than as members of the Rhynchocephilia.
†Six, if the rhynchosaurs are recognized as a separate order.

the orders Temnospondyli and Therapsida, although lingering well into the Jurassic, were largely exterminated at the close of Triassic time. So we see 15 or 16 Mesozoic tetrapod orders disappearing in two waves of extinction, with only 4 orders dying out at times other than the end of the Triassic (or near it) and the end of the Cretaceous. These were the Proanura, represented by a single Lower Triassic genus, the Archaeopterygiformes, again represented by a single genus, *Archaeopteryx* of Late Jurassic age, and 2 orders of mammals. Of the latter, the Docodonta is represented by a single family containing a few genera of Jurassic age. The other mammalian order, the Eupantotheria, was a group of some size and considerable consequence, since within it were mammals that may be closely identified with the origins of the modern mammals. This order does not extend beyond the end of Jurassic time.

PATTERNS OF MESOZOIC TETRAPOD EXTINCTIONS

Several patterns are evident in the extinction of Mesozoic tetrapod orders (Figure 1). One such pattern is the dwindling of an order through time toward what would seem to be its inevitable extinction. An outstanding example of this is to be seen in the therapsid or mammallike reptiles, an order of tetrapods that dominated Late

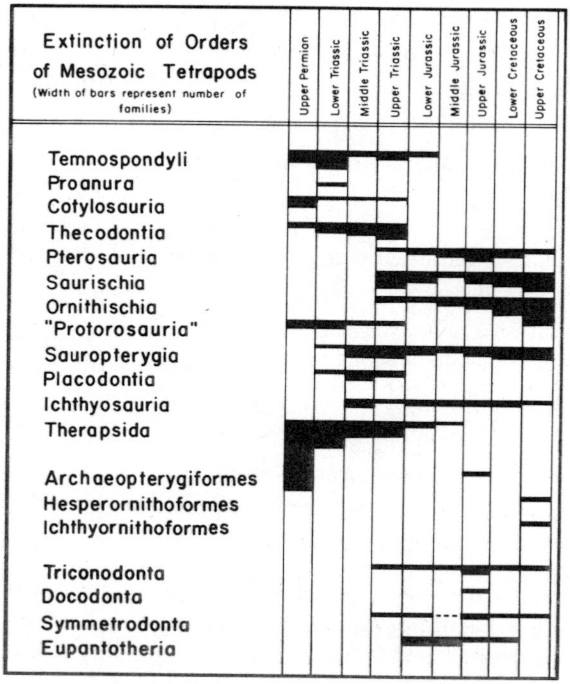

Figure 1. Orders of tetrapods that became extinct during Mesozoic time. The widths of the bars are proportional to the numbers of families in each order.

Permian history. But with the transition from Paleozoic to Mesozoic times, the therapsids suffered a drastic reduction in numbers and variety for reasons that may have been various and subtle. The beginning of Triassic time was marked by the rise of the archosaurs, the reptiles that were to dominate the Mesozoic, and perhaps this origin of a new and vigorous group of reptiles—having its beginning during the transition from the Permian into the Triassic—had something to do with the Mesozoic decline of the therapsids—already an attenuated order of reptiles. With the advent of the archosaurs, there was introduced a new and very successful combination of reptilian adaptations, notably the bringing of the feet beneath the body to form a narrow footfall trackway, generally an emphasis on the hind limbs for locomotion, reduction, and compaction of the bones in the ankle to make a strong joint directed fore and aft, generally a narrowing and deepening of the skull with the loss not only of various skull bones but also all except the marginal teeth, and the frequent development of a kinetic skull. All of these adaptations may be compared with the permanent quadrupedal stance of the therapsids, generally with primitive limbs and feet, the retention of virtually all of the bones seen in the primitive tetrapod skull, as well as the common retention of palatal teeth. Although the therapsids are notable for evolutionary advances within certain familial groups in the direction of the mammalian grade of organization, advances such as the form of the skull and jaws, the differentiation of the teeth, as well as differentiation of the vertebral column and ribs (indicating that in some of the advanced therapsids there may have been a diaphragm), these improvements evidently were not enough to prevail against the structural developments that led to the success of the archosaurs. Certainly the therapsids show a rapid decline through the Triassic period.

Of course, some of the Triassic decline of the therapsid reptiles may be the result of evolution; in other words, certain therapsid lines evolved themselves out of existence by becoming mammals.

In this respect the thecodont reptiles so characteristic of the Triassic may represent an example of the extinction of ancestors through competition from their descendants. The thecodonts were the direct ancestors of the crocodilians, the pterosaurs or flying reptiles and the two orders of dinosaurs, all of which appeared during Late Triassic time. With the rise of these progressive and sophisticated archosaurs the days of the thecodonts would seem to have been numbered. The disappearance of thecodonts at the end of the Triassic period was unequivocal at the time when the other archosaur were expanding prodigiously along various lines of adaptive radiation.

To go back to the phenomenon of a dwindling line, perhaps this is to be seen in the Triassic history of the temnospondyl amphibians, the Triassic members of which were almost the last labyrinthodonts. The labyrinthodonts were the first of the tetrapods, and for a time during late Paleozoic history they dominated the land. During Permiam times their dominance was challenged by the early reptiles, so that during the final years of Permian history the reptiles had prevailed over the labyrinthodonts. But the labyrinthodonts continued into and through the Triassic period, always as a diminishing group occupying ecological niches that had not been fully exploited by the reptiles. Thus, the last of the labyrinthodonts were large, fish-eating lake and swamp dwellers that managed to hold on in the face of competition from the aquatic archosaurs, namely the thecodontian phytosaurs and the earliest of the crocodilians. These rather fearsome-looking amphibians, seemingly so well established in many Late Triassic faunas, were destined soon to disappear without issue.*

The same can be said for the cotylosaurs, represented in Triassic times by small, lizardlike genera known as procolophonids. These, too, were the last of a long evolutionary line—not quite, but almost, as old as that of the labyrinthodont amphibians. The procolophonids were in effect ecologic "lizards," and it may be that the appearance of true lizard ancestors and related forms during Triassic time brought about the disappearance of the procolophonids.

Other orders that became extinct at the end of the Triassic period were the so-called Protorosauria, a restricted assemblage—whatever they may have been, the Placodontia, the highly specialized and to our eyes bizarre, shallow-water marine reptiles, adapted for feeding on mollusks, and the equally bizarre Rhynchosauria (if these constitute a separate order), bulky reptiles with strange teeth, perhaps adapted for feeding on roots and tubers. The extinction of the protorosaurs possibly represents the suppression of a not very successful assemblage of reptiles that "held on" from Permian into and through Triassic times in the face of competition from other, more aggressive reptiles that were then becoming established. The extinction of the placodonts is not so readily explained. It would seem logical that these mollusk-eating reptiles might have continued through the years of the Mesozoic, living after a fashion the type of life that is lived today by the walruses, yet they died out at the end of Triassic history. Likewise, the extinction of the rhynchosaurs is not readily explained.

So we see that the extinctions of tetrapod orders taking place at or near the end of Triassic time seemingly were of several kinds. There were the disappearances of long-established groups, such as the labyrinthodont amphibians and the therapsid reptiles. There were the extinctions resulting in part from the displacement of ancestors by their descendants, such as certain of the therapsids by the first mammals and thecodonts by the various archosaurs descended from them. And there were the extinctions of small "holdovers," such as the cotylosaurian procolophonids and the protorosaurs, by newly arisen competitors.†

*Labyrinthodonts have recently been found in Lower Jurassic rocks in Australia and Asia.

†Olsen and Galton (1977) have shown that the transition from Triassic into Jurassic time was not marked by a sudden extinction of various major groups, but rather by "gradual faunal replacement spread over the Late Triassic and at least the Early Jurassic." This conclusion is based in part by revisions of correlations of certain strata, whereby a part of the Newark Supergroup of eastern North America, all of the Glen Canyon Group of the southwestern states, and the Upper Stormberg Group of South Africa are classified as of Early Jurassic age rather than of Late Triassic age, as formerly has been the case. Thus, the supposed extinctions that marked the end of Triassic time are based in part on the definitions of ages, which are still open to some question. However that may be, the fact is that on the ordinal level there was a marked extinction at the end of the Triassic period as recorded by the decline of the temnospondyl amphibians (see footnote this page) and the extinction of the cotylosaurs, thecodonts, protorosaurs, placodonts, and rhynchosaurs. Moreover, although three families of therapsids persisted into the Jurassic, in the form of the tritylodonts, haramyids, and morganucodonts, this great reptilian

When we come to view extinctions of tetrapod orders at the close of the Cretaceous period, it is possible to distinguish two additional general patterns; the first that of the extinction of vigorous, highly varied orders, seemingly at the height of their evolutionary development, the second that of the extinction of small "experimental" orders evolving along lines parallel to those that were to be successfully established their relatives. The first pattern is, of course, seen in the extinction of the dinosaurs, pterosaurs, and Cretaceous marine reptiles, the second of certain early birds and of various lines of primitive mammals.

The extinction of the dinosaurs is one of the most celebrated and widely debated examples of extinction. Both orders of these reptiles were at the zenith of their separate but related evolutionary histories in Late Cretaceous times, being present in greater variety and numbers than at any previous stage in their development. Yet at the end of the Cretaceous, they succumbed—to what we do not know and perhaps shall never know. It would appear that a great taxonomic wave of extinction engulfed the dinosaurs suddenly, in geological terms, and completely. A similar wave of extinction also affected the marine reptiles—the plesiosaurs and ichthyosaurs (and mosasaurs), as likewise the flying reptiles or pterosaurs. In the case of the marine reptiles competition from the burgeoning bony fishes may have been a large factor in their extinction, although this is hard to visualize. One can think that these reptiles might have continued at least into the early years of Cenozoic time, performing ecologic functions now carried out by the small cetaceans and the pinniped carnivores. Why did they not continue, subsequently to be replaced by the marine mammals? As for the pterosaurs, it can be thought they they actually did succumb to competition from the early birds that were well established in Late Cretaceous time. (It should be noted here that the extinctions along strict taxonomic lines so evident in the abrupt disappearances of dinosaurs, plesiosaurs, and ichthyosaurs do not hold with respect to the mosasaurs. These were giant varanoid lizards—members of a superfamily that continues unabated today.)

The second pattern of Cretaceous extinction can be thought of as "attrition" during the early phases of avian and mammalian evolution. The hesperornithids were probably "experimental prototypes" that gave way to the presumably better adapted loons and grebes that had their beginnings in Cretaceous times. The same may have been true for the ichthyornids, which were displaced by the gulls and terns.

The various orders of Mesozoic mammals also have for the most part the appearance of experimental prototypes that in the long run were unsuccessful. Such were the multituberculates (holding over from the Mesozoic into early Cenozoic time), seemingly an attempt at being "rodents," and the triconodonts, docodonts (restricted to the Upper Jurassic), and symmetrodonts, evolutionary attempts along lines that later were followed by certain marsupials and insectivores. As opposed to these

---

order had essentially reached the end of its evolutionary history during the Triassic–Jurassic transition. Perhaps although the Late Triassic–Early Jurassic extinctions were not as sudden in a geological sense as those that occurred at the end of Cretaceous times, they nevertheless marked a profound change in tetrapod faunas during this crucial span of geologic history.

evolutionary experiments, there were the eupantotheres of Jurassic and Early Cretaceous age, representing a group from which the marsupials and the eutherian or placental mammals had their origins. Finally during Late Cretaceous times there were true marsupials, especially opossumlike animals, and true placentals, among which were the earliest of the insectivores, the primates, and the hoofed mammals known as condylarths.

When we turn to the extinctions of Mesozoic tetrapod families (Figure 2), it would appear that these emphasize the extinctions of orders. Here again there are two distinct phases of extinction, one at or near the end of the Triassic period and one at the end of the Cretaceous period. (In addition, there was a drastic extinction of tetrapod families at the end of the Permian period as well.) In percentages the

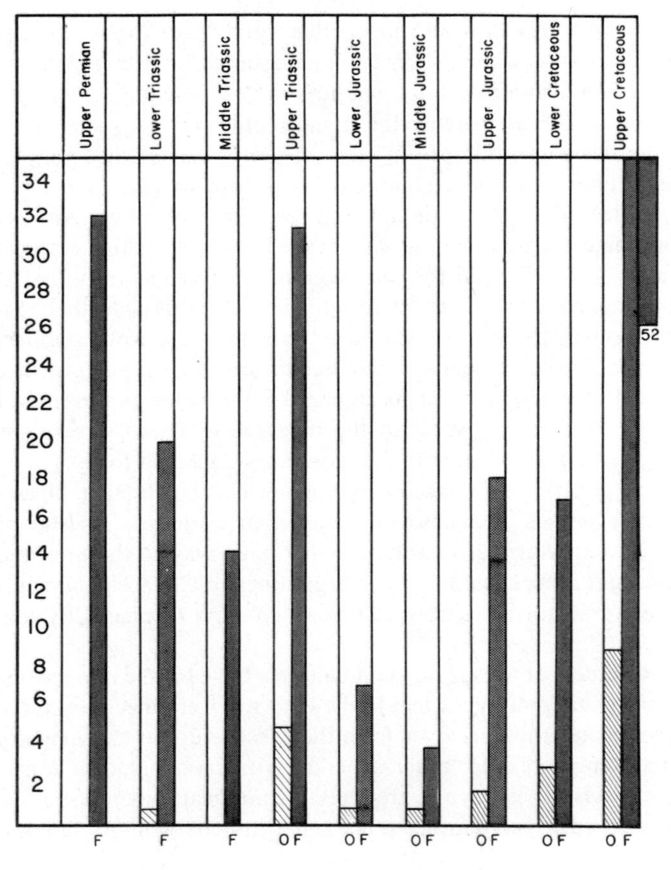

Figure 2. Extinctions of orders and families of Mesozoic tetrapods and extinctions of families of tetrapods at the end of the Permian Period.

figures are 71% extinction of families at the end of the Permian period, 80% at the end of the Triassic, and 61% at the end of the Cretaceous. Intermediate extinctions were for the most part also on a rather large scale; 67 and 50% at the end of the Lower and Middle Triassic, respectively; 30% at the end of the Lower Jurassic; 17 and 36% at the end of the Middle and Upper Jurassic; and 32% at the end of the Lower Cretaceous. These intermediate extinctions of families—much more striking than the intermediate extinctions of orders—reflect the disappearance of families within orders that continued—for example, numerous lesser taxa that might be regarded as experimental prototypes that failed. Figures on genera would further emphasize what the figures on families have shown, but this will not be attempted in this chapter. Nor will there be an attempt to analyze the figures on species, especially since fossil species are prone to be subjective and therefore not particularly useful in a discussion such as this. Here it may be pertinent to repeat the statement made at the beginning of this chapter, namely, that extinction is a continuing, creative process—a fact that is particularly revealed by the episodes of extinction of lesser taxa occurring throughout the extent of Mesozoic time, not merely at the end of the Triassic and Cretaceous periods.

COMPARISONS OF MESOZOIC TETRAPOD EXTINCTIONS

Even though there were tetrapod extinctions of consequence throughout the Mesozoic era, it may be interesting to compare the Triassic and Cretaceous peaks of Mesozoic extinction (Figure 3).

So far as the extinctions at the end of the Triassic period are concerned—extinctions that percentagewise at the family level exceed those at the end of the Cretaceous—it would appear that two of the major patterns involve the diminution of long persisting lines, combined in some cases with extinction through the evolution of certain members of a diminishing line into "higher" forms. Here one sees the decline of the labyrinthodont amphibians, the so-called protorosaurs and the therapsid reptiles, as well as the origins of mammals from certain therapsids, and of advanced archosaurs, rulers of the Mesozoic, from Triassic archosaurian ancestors.

In contrast, the extinctions at the end of the Cretaceous period would appear to fit two other patterns, namely, attrition through evolution and the almost inexplicable phenomenon of a terminal "wave" of extinction engulfing seemingly vigorous and successful lines of evolutionary development. As we have seen, the first of these patterns is exemplified by the elimination of various early, tentative lines of mammalian development, the second by the dramatic and sudden terminations of the dinosaurs and the marine reptiles.

Thus, there would seem to be some qualitative as well as quantitative differences between tetrapod extinctions that took place at the end of Triassic time as compared with those that marked the end of Cretaceous history. Why should there be these apparent differences?

Perhaps they may reflect the nature of the tetrapods that became extinct and of those that followed. The reptiles and amphibians that disappeared during the transition from the Triassic into Jurassic periods of Mesozoic history were replaced

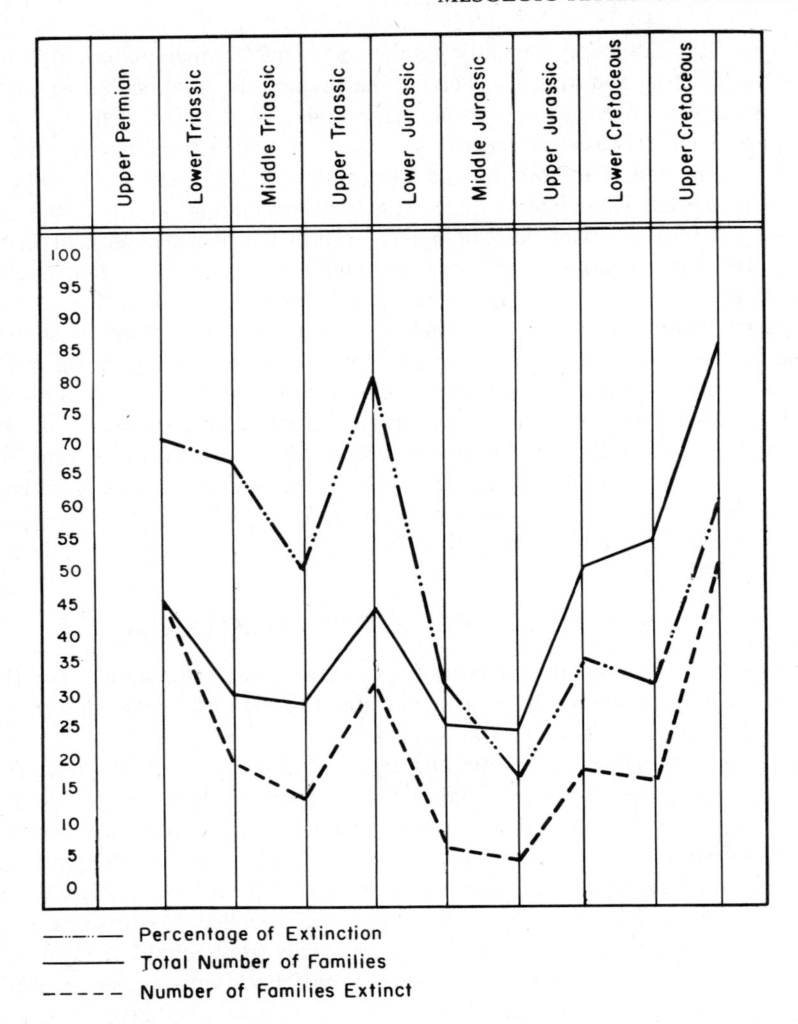

Figure 3. Exinction of families of Mesozoic tetrapods as compared with total numbers of families, and percentages of extinctions.

largely by reptiles and, to a lesser extent, by very primitive mammals. It was a case of the continuation of tetrapod evolutionary history within a reptillian framework. But the tetrapods that disappeared at the end of the Cretaceous period were to a considerable degree not immediately replaced. For the pterosaurs and dinosaurs, the marine reptiles and the very primitive mammals, the replacements, when they occurred, were made by new and unrelated tetrapods. So there was a break in the taxonomic continuity of life. Pterosaurs were replaced by birds, already well established in Late Cretaceous time, and by bats. Dinosaurs were replaced by a variety of mammals, large and small, ultimately descended from seemingly insignificant Cretaceous forebears. Marine reptiles were replaced after a certain hiatus in time by the whales and pinniped carnivores. The small host of Mesozoic mammals were largely replaced by marsupials and early placentals, descended from one Mesozoic order, to which other Mesozoic orders were sterile side branches.

EXTINCTIONS AND REPLACEMENTS

The possible replacements (using this word in a very general way) of ecological types within the orders that became extinct at the end of the Triassic and Cretaceous periods may be indicated in the following manner.

Extinct at or Near the End of the Triassic Period	Replacements
Temnospondyli	Large frogs (Jurassic)
Cotylosauria (Procolophonids)	Lizards (Jurassic)
Thecodontia	
Primitive thecondonts	Small carnivorous dinosaurs (Jurassic)
Armored thecodonts	Armored dinosaurs (Cretaceous)
Phytosaurs	Crocodilians (Jurassic)
Protorosauria	Lizards
Placodontia	Walruses (Miocene)
Therapsida	
Carnivorous theriodonts	Mesozoic mammals (Jurassic)
Dicynodonts	Herbivorous dinosaurs (Jurassic)
Tritylodonts	Multituberculates (Cretaceous and Paleocene)

Note: Multituberculates were ancient mammals that, with complex cheek teeth, in some respects paralleled the tritylodonts and were not unlike rodents.

Extinct at the End of the Cretaceous Period	Replacements
Pterosauria	Birds, perhaps bats
Saurischia	
Coelurids	Small creodonts (Paleocene)
Ostrich dinosaurs	Ratites (Eocene)
Carnosaurs	Large creodonts (Paleocene)
Sauropods	Large amblypods (Eocene)

Extinct at the End of the Cretaceous Period	Replacements
Ornithischia	
Iguanodontids	Coryphodonts (Paleocene)
Hadrosaurs	Amynodonts (Eocene)
Pachycephalosaurs	Amblypods (Eocene)
Ankylosaurs	Glyptodonts? (Pliocene)
Ceratopsians	Uintatheres (Eocene)
Plesiosaurs	Toxochelyid turtles (Paleocene)
Ichthyosaurs	Toothed cetaceans (Eocene)
Hesperornithoformes	Loons, grebes (Paleocene)
Ichthyornithoformes	Terns, gulls (Paleocene)
Triconodonta	Small carnivorous mammals (Paleocene)
Symmetrodonta	Small carnivorous mammals (Paleocene)

Note: Creodonts were primitive carnivorous mammals. Coryphodonts and amblypods were medium to large primitive hoofed mammals. Uintatheres were gigantic primitive hoofed mammals with horns. Glyptodonts were heavily armored giant edentates, related to armadillos.

This attempt to indicate possible ecological substitutions for groups that became extinct at the close of Triassic and of Cretaceous times is indeed very subjective. It does try to speculate as to the tetrapods that may have filled niches becoming vacant as a result of various extinctions. Some of the suggested replacements perhaps are not particularly realistic; for example, the substitution of Pliocene glyptodonts for Cretaceous ankylosaurs, a "replacement" involving a hiatus of millions of years. Yet perhaps the niche for large or gigantic armored herbivores did remain vacant during an interval that involved most of Cenozoic time. The same is true for the suggested replacement of placodonts by walruses—involving an even greater time gap. Therefore, the tabulation here presented should be looked at for what it is, a very tentative listing that can be challenged on almost every count. Only in a few cases, such as the substitution of crocodilians for phytosaurs in two successive periods, Triassic and Jurassic, is there reason for a large feeling of confidence.

CONCLUSIONS

What conclusions are to be drawn from this examination of Mesozoic tetrapod extinctions?

In the first place, it is evident that extinctions continued throughout the extent of Mesozoic time; the disappearance of various tetrapod groups at all taxonomic levels was taking place, as they had always taken place, as part of the evolutionary process. Old types of animals were constantly fading away, generally, but not always, to be replaced by ecological types adapted for the niche that was being or had been vacated.

Although the extinctions continued without abatement, there were two episodes of numerous extinctions; namely, first, at or near the end of the Triassic period and again at the end of the Cretaceous period. Although the great Cretaceous extinctions are commonly cited as one of the remarkable events in the history of tetrapod life, the Triassic extinctions were nonetheless also of large proportions. Indeed, at the family level the percentage of extinctions at the end of the Triassic was considerably greater than at the end of the Cretaceous.

The Triassic extinctions are largely to be accounted for by the diminution of long-established evolutionary lines, perhaps in part as a result of the appearance and rapid development of new lines better adapted to the changing environments of the Mesozoic scene. Also, some of the Triassic extinctions were the result of evolution; for example, of some lines of therapsid reptiles into early mammals and of some thecodont reptiles into more advanced archosaurian reptiles.

The Cretaceous extinctions were to some degree the result of the causes cited above but, in addition, were in part the result of evolutionary attrition—the disappearance of "experimental" groups during the early stages of major evolutionary deployments. Such would have been the extinction of certain groups of Mesozoic mammals. However, the Cretaceous extinctions were marked largely by the rather sudden disappearance of many well-established and seemingly highly successful groups that were remarkably adapted to the conditions under which they were prospering. Such were the extinctions of the dinosaurs and pterosaurs and of the marine reptiles. For many years numerous students have been trying to account for such extinctions with little success.

Qualitatively, there would appear to have been some difference between the Triassic and Cretaceous extinctions and the following replacements. In the former, there were replacements in kind of amphibians by amphibians and reptiles by reptiles. In the latter there were the most part replacements of reptiles by mammals, and these involved animals on significantly different levels of evolutionary development.

REFERENCES

Archibald, J. D., and Clemens, W. A., 1982, Late Cretaceous extinctions; *Am. Sci.*, **70**, 377–385.

Alvarez, L.W., Alvarez, W., Asaro, F. A., and Michel, H. V., 1980, Extraterrestrial cause for the Cretaceous-Tertiary extinction, *Science*, **208**, 1095–1108.

Benton, M. J., 1982, The Diapsida; revolution in reptile relationships, *Nature*, **296**, 306–307.

Benton, M. J., 1983, Large-scale repacements in the history of life, *Nature*, **302**, 16–17.

Colbert, E. H., 1958, Tetrapod extinctions at the end of the Triassic period, *Proc. Natl. Acad. Sci.*, **44**, pp. 973–977.

Ehrlich, P., and Ehrlich, A., 1983, *Extinction*, New York, Ballantine Books, 384 pp.

Lillegraven, J. A., Kielan-Jaworowska, Z., and Clemens, W. A., eds. 1979, *Mesozoic Mammals*, Berkeley, University of California Press, 311 pp.

McLean, D. M., 1981, A test of terminal Mesozoic "catastrophe," *Earth Planet. Sci. Lett.*, **53**, 103–108.

Newell, N. D., 1967, *Revolutions in the History of Life*, The Geological Society of America Special Paper 89, pp. 63–91.

Officer, C. B., and Drake, C. L., 1983, The Cretaceous-Tertiary transition, *Science*, **219**, 1383–1390.

Olsen, P. E., and Galton, P. M., 1977, Triassic-Jurassic tetrapod extinctions: Are they real?, *Science*, **197**, 983–986.

Romer, A. S., 1966, *Vertebrate Paleontology*, Chicago, Ill., The University of Chicago Press, 468 pp.

Russell, D. A., 1979, The Cretaceous-Tertiary boundary problem, *Episodes*, 1979, 21–24.

Russell, D. A., 1982a, *A Paleontological Consensus on the Extinction of the Dinosaurs?*, Geological Society of America Special Paper 190, 401–405.

Russell, D. A., 1982b, The mass extinctions of the late Mesozoic, Sci. Amer., **246**, 58–65.

Russell, D., and Tucker, W., 1971, Supernovae and the extinction of the dinosaurs, *Nature*, **29**, 553–554.

Simpson, G. G., 1953, *The Major Features of Evolution*, New York, Columbia University Press, 434 pp.

4

EVOLUTION OF THE TERRESTRIAL VERTEBRATE FAUNA DURING THE CRETACEOUS—TERTIARY TRANSITION

WILLIAM A. CLEMENS

Department of Paleontology
University of California
Berkeley, California

INTRODUCTION

"And on land the dinosaurs became extinct." Too frequently this is the complete summary of events and sole justification for the view that a "mass extinction" of terrestrial vertebrates occurred at the end of the Cretaceous. Other authors are somewhat more detailed in discriminating between the extinction of the large dinosaurs, which were exposed to one or another hypothesized horror, and the survival of small mammals, which hid under rocks and thereby escaped annihilation. In part these superficial accounts of change in the terrestrial biota appear to reflect the bias of most recent studies of patterns of extinction that have either emphasized or dealt exclusively with the record of marine organisms. For example, the widely cited study by Raup and Sepkoski (1982) is based on the evolutionary histories of families of marine invertebrates and vertebrates. It indicates that the rate of extinction of families of marine organisms increased at the end of the Cretaceous. Change in diversity of terrestrial organisms has yet to be analyzed at comparable levels of biological and temporal resolution (Padian and Clemens, 1985). However, the demise of the dinosaurs often is simply hypothesized as having constituted an extinction event of equivalent magnitude that resulted from the same factors that caused the change in the marine biota. These hypotheses concerning the pattern of evolution of the terrestrial biota must be tested with paleontological data.

The goal of this article is to review what currently is known of the patterns of change in the terrestrial vertebrate fauna during the Cretaceous–Tertiary transition. With what degrees of temporal and geographic resolution does the available fossil record document these changes? Is there evidence of an extraordinary increase in the rate of extinction of groups of terrestrial vertebrates at the end of the Cretaceous? Responses to these and related questions require a summation and evaluation of paleontological data, not the imposition of a hypothesis that prescribes the nature and causes of change. The status report presented here will necessarily be ephemeral because several extensive studies of pertinent faunas and floras are still underway. However, it is already apparent that the available record of changes in the terrestrial vertebrate fauna across the Cretaceous–Tertiary boundary, although limited, is not rudimentary and clearly documents a complex history of survival and extinction.

TEMPORAL AND PALEOBIOGEOGRAPHIC SCALES

Because analyses of patterns of extinction and origination involve comparisons of rates (evolutionary events per unit time), the temporal duration of the interval of earth history under study must be assessed. Obviously, as temporal correlations are made over larger and larger geographic areas, their degrees of resolution tend to decrease. Extinction "events" of great magnitude can be mistakenly generated if the extinctions of lineages that occurred over millions of years are simply summed and treated as having occurred simultaneously. Also, analyses of changes in faunal diversity, particularly on continental or global scales, are susceptible to bias from unequal paleobiogeographic sampling. For example, during a large part of the Late Cretaceous, North America was bisected by the Western Interior Sea. The composition of the terrestrial fauna of the western subcontinent varied latitudinally (Lehman, 1981). Additionally, the much smaller samples of the Late Cretaceous faunas of the eastern subcontinent include dinosaurian genera unknown in the faunas of the western subcontinent (Russell, 1984b). Studies of change in Mesozoic diversity of terrestrial vertebrates based on summaries of records from modern continents without attention to differences in paleobiogeographic scope of sampling (e.g., Russell, 1984a) cannot distinguish between evolutionary patterns and sampling biases.

In order to minimize the effects of poor temporal resolution in correlations made over long distances and inequality in sampling paleobiogeographic patterns of distribution, the primary data base for this study is drawn from analyses of fossiliferous deposits in the northern Western Interior of North America. As discussed below, the region has yielded the most complete fossil record of the evolution of a terrestrial fauna during the Cretaceous–Tertiary transition yet discovered. Specifically, the study area includes sites in Montana and Wyoming, the province of Alberta, Canada,

Table 1. Correlation of North American Mammalian or Vertebrate Ages [a]

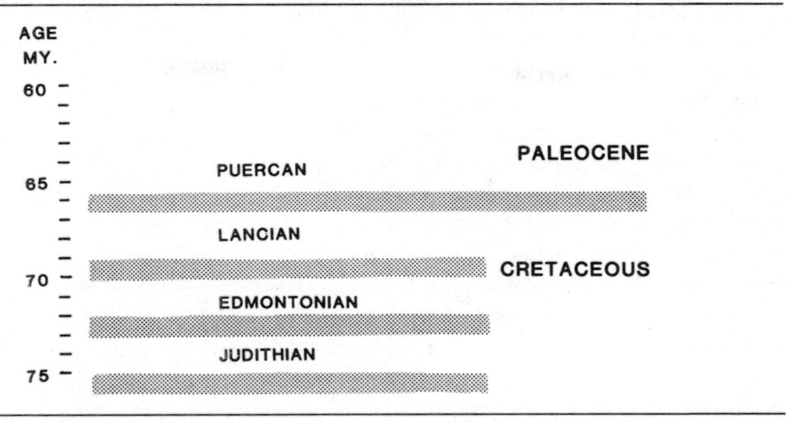

[a]Temporal scale calibrated in millions of years (left) and global chronostratigraphic units (right).

and some parts of immediately adjacent states and provinces. Within this area the extinction of dinosaurs and some other terrestrial vertebrates and evolution of early Paleocene (Puercan) faunas are most thoroughly documented in northeastern Montana. Local faunas from these sites record the changing terrestrial biota that inhabited the lowland areas to the west of the Cretaceous Western Interior Sea and much smaller Paleocene Cannonball Sea (note Appendix 1). Stratigraphers have ordered these local faunas in a series of North American terrestrial vertebrate (or mammal) "ages." The sequence of these "ages," best estimates of their durations, and correlations with global chronostratigraphic units are shown in Table 1.

Currently, a combination of magnetostratigraphic and biostratigraphic methods of correlation provides the most precise method of comparing the ages of the sites within the study area to those in the San Juan Basin, New Mexico, and further to the marine magnetic polarity time scale. The magnetostratigraphic record of the San Juan Basin is long, and the interval of reversed polarity in which the extinction of dinosaurs and some other terrestrial vertebrates occurred in New Mexico now can be correlated with the marine magnetic polarity chron 29R (Butler and Lindsay, 1983; Butler, 1984). In Montana and Alberta, these extinctions also occurred in an interval of reversed polarity. (The magnetostratigraphy of the Denver Basin sites has yet to be documented.) The magnetostratigraphic records of the northern areas are short and cannot be directly matched with the New Mexican. However, using *biostratigraphic* criteria, the interval of reversed polarity containing these northern records of last dinosaurian faunas can be correlated with chron 29R as now recognized in the San Juan Basin (Archibald et al., 1982).

Estimates of the duration of magnetic polarity chron 29R vary but approximate 500,000 yr (Dingus, 1983a). If the correlations adopted here are correct, the magnetostratigraphic time scale cannot reveal whether the extinctions of the dinosaurs occurred in Montana and New Mexico precisely at the same time or were discordant by as much as 10,000, 100,000 or 500,000 yr.

Alvarez and co-workers (1980) advanced the hypothesis that concentrations of iridium found at or near paleontologically defined Cretaceous–Tertiary boundaries were the product of an asteroid's impact. If so, these geochemical markers could provide a very precise temporal marker, even on a biologically significant time scale. However, their stratigraphic utility is open to question. Luck and Turekian (1983) indicate that iridium concentrations of extraterrestrial origin might not be geochemically distinguishable from those derived from the mantle, and Zoller and co-workers (1983) identified a modern volcanic source of iridium. If the concentrations of iridium found at or near the paleontologically defined Cretaceous–Tertiary boundary at Stevns Klint, Denmark, and Raton Basin, New Mexico, are both of extraterrestrial origin, Luck and Turekian (1983) suggest that they were derived from the impacts of bolides of different compositions. If two bolides produced these concentrations, what independent evidence establishes the contemporaneity of their impacts? The so-called Nemesis hypothesis (Davis, Hut, and Muller, 1984) adds to the uncertainty of correlations based on concentrations of iridium in its assertion that at various times the earth was bombarded by showers of comets that might also have swept in this element from extraterrestrial sources (Hsü et al., 1982). Until operational definitions for recognizing geochemical markers as signatures of the impacts of specific extraterrestrial objects are developed, the use of iridium concentrations in stratigraphic correlations is best regarded as a hypothesis to be tested (Padian et al., 1984).

In both marine (Officer and Drake, 1983) and nonmarine (Archibald and Clemens, 1982) deposits, paleontological records of extinctions of characteristic Cretaceous taxa are not always precisely at the same stratigraphic level as the iridium enrichments. If these concentrations of iridium found near the paleontologically defined Cretaceous–Tertiary boundary prove to be globally contemporaneous, then the paleontological records of extinction at different stratigraphic levels challenge hypotheses invoking an instantaneous causal factor of their demise.

At present, magnetostratigraphy provides the most precise method of global correlation and the only means to independently test correlations based on biostratigraphic or geochemical data. Where the magnetic polarity chron of the rocks containing the paleontologically defined Cretaceous–Tertiary boundary can be determined, the chron is or appears to be correlative with 29R, which had an estimated duration of about 500,000 yr. Thus, to the extent that they fall within a period of about a half million years' duration, the extinctions used to define the Cretaceous–Tertiary boundary and the time of formation of the concentrations of iridium are contemporaneous.

THE AVAILABLE FOSSIL RECORD

In order to recognize significant changes in evolutionary rates (for example, the occurrence of "mass" extinctions), data on taxonomic diversity within the same paleobiogeographic area before, during, and after the specific interval under study must be available. Figure 1 contains two maps illustrating the distribution of localities where pertinent concentrations of vertebrate fossils have been found. Figure

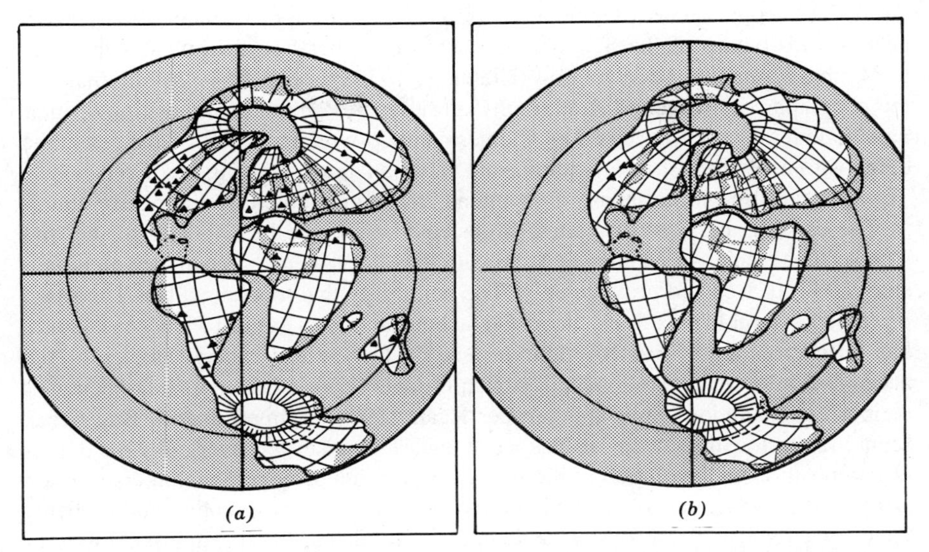

Figure 1. (*a*) Distribution of Late Cretaceous sites that have yielded scientifically significant collections of ornithischians and saurischians. Some triangles designate areas rich in dinosaurian fossils; others represent sites of discovery of a few fossils (data primarily from Charig, 1979). (*b*) Sites where fossiliferous terrestrial deposits of latest Cretaceous (i.e., within the last two or three million years of the end of the Cretaceous, Lancian age) are overlain by fossiliferous deposits of earliest Paleocene (Puercan) age. Maps redrawn from Lillegraven, et al., 1979.

1*a* documents the distribution of Late Cretaceous sites that have yielded the scientifically significant collections of dinosaurs (data primarily from Charig, 1979). Some of the triangles denote areas rich in dinosaurian fossils; others represent the sites of discovery of a single or a few specimens. Figure 1*b* records the sites where fossiliferous terrestrial deposits of latest Cretaceous age (i.e., within the last 2 or 3 m.y. of the Cretaceous; Lancian age in the Western Interior) have been discovered and are overlain by fossiliferous deposits of earliest Paleocene age (Puercan age, ca. 2 m.y. duration, note Table 1).

The data presented on the maps (Figure 1) document a major bias in discovery and collecting. Only in the Western Interior study area have stratigraphically controlled fossiliferous deposits of latest Cretaceous and earliest Paleocene age been found and investigated. From this geographically limited sample (along with snippets of information from Cretaceous, but not necessarily latest Cretaceous, and Paleocene, but not necessarily earliest Paleocene, faunas in other areas), many have tried to paint a global picture of the changes in the terrestrial fauna during the transition from the Cretaceous to the Tertiary.

Limiting studies to a North American perspective does not remarkably improve their resolution, for pertinent fossiliferous deposits have yet to be found in vast areas of the continent. The most detailed record available has been collected in the Western Interior study area. The core of this area is in northeastern Montana, where well-sampled local faunas of the Hell Creek Formation reveal the record of the latest Cretaceous faunas (Archibald, 1982). These are approximately contemporaneous with faunas of the Scollard Formation of Alberta (Lillegraven, 1969) and the Lance Formation of Wyoming (Clemens, 1973; Estes, 1964). A summary of the geographic position and stratigraphy of these sites and the composition of their mammalian faunas is presented by Clemens and co-workers (1979). Changes in the composition of the fish, amphibian, and reptilian faunas during the Cretaceous–Tertiary transition in northeastern Montana are being studied by Laurie Bryant.

The usual assignment of a latest Cretaceous age to the local faunas of the Bug Creek faunal facies of Montana (Archibald, 1982, 1984) and Saskatchewan (Johnston, 1980) is accepted provisionally. The possibility that this faunal facies is based on mixed assemblages of *very* latest Cretaceous and *very* earliest Paleocene faunas has been mooted. In light of Behrensmeyer's (1983) and others' studies of the accumulation of vertebrate fossils in stream channel deposits, the possibility demands serious consideration. Smit and van der Kaars (1984) claimed that all Bug Creek faunal facies localities were of Paleocene age. This assertion overlooks the details of composition and evolution of the fauna and its wide paleogeographic distribution (Archibald, 1984). Also, detailed sedimentologic and stratigraphic studies being conducted by David Fastovsky challenge the lithostratigraphic interpretations of Smit and van der Kaars (1984). Until these studies and a thorough taphonomic analysis are completed, assignment of a Late Cretaceous age to the Bug Creek faunal facies sites in Montana must remain equivocal. However, the data presently available support the view that these sites record the gradual transition from a dinosaurian-dominated, Cretaceous fauna to one characteristic of the Paleocene.

The Tullock Formation, which overlies the Hell Creek Formation in northeastern Montana, is yielding terrestrial vertebrate faunas of Puercan (earliest Paleocene) age that are currently being analyzed. The only other samples of northern Puercan faunas are provided by a limited collection derived from the Mantua Lentil, Bighorn Basin, Wyoming, and material from the Ravenscrag Formation, Saskatchewan, being analyzed by P. A. Johnston and R. C. Fox (Fox, 1984). Much larger Puercan local faunas are known from the San Juan Basin, New Mexico (see Savage and Russell, 1983, for a review of Paleocene mammalian faunas).

SURVIVAL AND EXTINCTION OF AQUATIC (NONMARINE) AND TERRESTRIAL VERTEBRATES

In general, the following accounts are divided into ordinal or higher taxonomic groupings; the classification usually follows Romer (1966). Except when indicated, the analysis is based only on faunas of the Western Interior study area.

Fish

Five genera of chondrichthyans (sharks and rays) represent four families in Cretaceous faunas. Three of the families and 2 genera are known from Tertiary deposits. The absence or decrease in abundance of the surviving forms in Paleocene deposits of the area could reflect waning influence of marine or brackish water conditions resulting from the regression of the Cretaceous Western Interior Sea (Estes, 1970) and, possibly, lower salinity of the Paleocene Cannonball Sea (Fox and Olsson, 1968).

The osteichthyan (bony fish) fauna includes a minimum of 13 genera representing 10 or more families. Chondrosteans and holosteans are most abundant; teleosteans are rare (Estes, 1970). The common genera of Late Cretaceous fishes include a sturgeon (*Acipenser*), a bowfin (*Amia*), and a gar (*Lepisosteus*). Almost all the families have representatives in the Recent fauna; only the extinction of *Belonostomus* (Aspidorhynchidae) appears to terminate a Mesozoic family and order. As the collections of Paleocene osteichthyans are studied in greater detail, probably some of the five apparent generic extinctions will prove to be artifacts of our limited knowledge of the available fossil record.

Amphibia

Within the last few years knowledge of the salamanders and frogs of the Cretaceous and Paleocene has been greatly augmented and refined (Estes, 1981; Estes and Sanchiz, 1982). Late Cretaceous faunas of the Western Interior study area include representatives of at least eight families of amphibians. All but one of these families survive into the early Tertiary. The rate of survival of genera is also high; only 4 of the 12 recognized genera lack Tertiary records, and one of these appears to be at least structurally ancestral to a Tertiary genus.

Reptilia

Chelonia. Hutchison (1982) and Archibald (1977) have found that approximately 18 genera of turtles representing at least 4 families were present in the Late Cretaceous faunas of Montana. Representatives of all these families are known in the Tertiary. Two genera of large turtles, *Basilemys* and *Helopanoplia*, became extinct at the end of the Cretaceous. The remaining lineages, which included other large turtles, *Trionyx* and *Adocus*, for example, appear to have survived. However, the Early Paleocene chelonian fauna of Montana is less diverse than that of the latest Cretaceous, for the ranges of some of the taxa shifted southward.

Eosuchia. *Champsosaurus*, a large, probably amphibious, crocodilianlike reptile is abundantly represented in both latest Cretaceous and early Paleocene faunas.

Eolacertilia. On the basis of his recent systematic analysis, Richard Estes (personal communication) regards the Eolacertilia as a nonsquamate group, thus requiring separate recognition in this tabulation. Most eolacertilian species are known from Late Permian, Triassic, and Jurassic faunas. Primarily on the basis of dental characters, the latest Cretaceous *Litakis* is recognized as the last member of the group (Estes, 1983b). It is a rare member of the herpetofauna of the Lance Formation, Wyoming, and as yet is unknown in other latest Cretaceous faunas of the northern Western Interior.

Lacertilia. Estes (1983a, 1983b) presented authoritative and comprehensive reviews of the fossil record of lizards that make it possible to assess their history during the Cretaceous–Tertiary transition at different levels of geographic and taxonomic resolution. On a global basis, all 10 families of terrestrial lacertilians currently known in Late Cretaceous faunas have Paleocene representatives. Estes's analysis (1983a, Fig. 1) suggests that globally the maximum number of lacertilian families ranging from the Cretaceous into the Tertiary could have been as high as 16. The only extinctions recorded are those of two families (Romer, 1966) of marine varanoids.

Narrowing the perspective again to the Late Cretaceous records from the Western Interior study area, 15 genera representing 7 families are currently recorded. All the families continued into the Cenozoic. However, at the beginning of the Paleocene, generic diversity of lizards in this area was greatly reduced. Estes (e.g., 1970) argued that the decrease in diversity is not the result of extensive extinction but of a southward shift in biogeographic ranges of the teiids and other lizards thought to have required humid subtropical or warm temperate climates. When these taxa are excluded from consideration, only 4 of the 15 Cretaceous genera appear to have lacked Paleocene representatives.

Serpentes. Two genera of snakes were rare members of latest Cretaceous faunas of Montana and Wyoming. The aniliid *Coniophus* survived into the Eocene. The second species of snake is a poorly known boid, a member of an extant family (Rage, 1984).

Crocodilia. Four genera have been identified in faunas of the northern Western Interior. The two abundant latest Cretaceous crocodilians, *Leidyosuchus* and *Brachychampsa*, also were members of the earliest Paleocene faunas of the area. Of the two other latest Cretaceous genera, *Thoracosaurus* is represented by species in the European Paleocene (Steel, 1975) and *Prodiplocynodon*, possibly, was closely related to Paleocene forms.

Dinosauria, the Saurischia and Ornithischia. The available fossil record is not adequate for a detailed assessment of the global patterns of saurischian and ornithischian evolution just before their extinctions, which are used to recognize the end of the Cretaceous in nonmarine deposits. However, the fossil record of the Western Interior study areas provides an opportunity to make a first assessment of these patterns in a limited geographic region. The richly fossiliferous strata of the Judith River (Montana) and Oldman (Alberta) formations are providing increasingly larger samples of terrestrial faunas that lived approximately 5–10 m.y. prior to the end of the Cretaceous. Records of dinosaurs from these two formations were merged to provide an estimate of diversity during the Judithian age (Table 2 and Appendix 1). Faunas of intermediate, Edmontonian age are known from sites within the study area, but the samples are smaller and not included in the compilations. It should be noted, however, that the Edmontonian faunas contain many ornithischians and saurischians known in Judithian faunas; a striking change in composition of the dinosaurian faunas of Alberta occurred between the Edmontonian and Lancian (Russell, 1983). The composite Lancian fauna is a summary of records from the Lance, Hell Creek, and Scollard formations. A summary of the comparison of

Table 2. Familial and Generic Diversity of Judithian and Lancian Dinosaurian Faunas of the Western Interior Study Area

	Judithian		Lancian	
	Families	Genera	Families	Genera
Saurischia	6	12	5	8
Ornithischia	7	20	8	14
Total	13	32	13	22

generic diversity of Judithian and Lancian saurischians and ornithischians is given in Table 2, and the data base is described in Appendix 1.

At the family level, the taxonomic diversity of Judithian and Lancian faunas of the northern Western Interior differs only slightly. But, on close inspection, saurischian family diversity might have remained constant for the validity of the Caenognathidae is questionable. The basic compilation (Appendix 1) shows that ornithischian diversity increased through addition of the Protoceratopsidae. Again, the numerical change requires qualification. Protoceratopsians were members of western North American, pre-Lancian faunas outside the study area, and the Lancian records of the family include only rare occurrences of *Leptoceratops* in the Scollard Formation, Alberta, and the "Lance" Formation of the Bighorn Basin, Wyoming (Ostrom, 1978). Thus, the general pattern is one of stability in representation of families from the Judithian to the Lancian.

In contrast, at the generic level there was a major change in the dinosaurian fauna with a reduction of Lancian diversity to about two-thirds of the Judithian level. The loss of diversity among the saurischians might be illusory. Current systematic analyses of Lancian faunas that, for the first time, emphasize samples of isolated skeletal elements have revealed several genera of small Lancian saurischian dinosaurs that await description (D. A. Russell and R. A. Long, personal communication; Russell, 1984b). However, similarly detailed studies of samples of Judithian faunas, requisite for a valid comparison, have yet to be undertaken.

The generic diversity of ornithischian dinosaurs decreased from 20 in the Judithian to 14 in the Lancian through the loss of many hadrosaurids and ceratopsids. In part, this results from a southward shift of biogeographic ranges. The hadrosaurids *Hadrosaurus* (including *Kritosaurus*) and *Parasaurolophus*, known from Judithian deposits in Montana or Alberta, are not present in Lancian faunas of the Western Interior study area but survived during the Lancian in New Mexico (Lehman, 1981).

Finally, the tabulation in Appendix 1 illustrates another aspect of the pattern of change in the composition of the dinosaurian faunas. Of the 30 formally recognized (i.e., named) Judithian genera, only 9 are recognized in the Lancian fauna. The remaining 10 formally recognized Lancian genera were either the products of evolution of local lineages or immigrants. During the 5–10 m.y. prior to the extinction of saurischians and ornithischians, the evolution of the dinosaurian fauna of the northern Western Interior involved extinction or southward shifts of ranges of genera that were partially balanced by evolutionary origination or immigration.

Pterosauria and Aves. Both birds and pterosaurs were members of the Lancian faunas of the Western Interior. Taphonomic biases against preservation of the delicate remains of these aerial vertebrates have drastically limited our knowledge of their history during the Cretaceous–Tertiary transition. Latest Cretaceous records of birds include a few fragmentary records of shorebirds (Brodkorb, 1963); the contemporary pterosaurs were apparently all pterodactyloids. No meaningful statements can be made about the pattern of pterosaur extinction nor the evolution of birds during the Cretaceous–Tertiary transition.

Mammalia

Representatives of three major (ordinal or superordinal) groups of mammals—multituberculates, marsupials, and eutherians (placentals)—dominated the Lancian mammalian fauna. One apparent relict of an earlier radiation of therian mammals was also present but dropped from the record before the Paleocene. (It is not included in the tabulation in Appendix 2). Each group has a different history of survival or extinction across the Cretaceous–Tertiary transition (Clemens, 1982).

Multituberculata. During the transition from the Cretaceous to the Tertiary, 4 of the 11 genera and 2 of the 8 families of these rodentlike mammals became extinct. However, in the Puercan the surviving lineages differentiated and the total diversity of the group increased in North American early and middle Paleocene faunas.

Marsupialia. This group almost shared the fate of the dinosaurs. Extinction claimed two of the three Cretaceous families and three of the four genera (Clemens, 1984). During the Cenozoic, marsupials never regained their Mesozoic relative abundance and diversity in Holarctic faunas.

Eutheria. Only one of the nine genera and one of the four families of Cretaceous eutherian (placental) mammals died out at the end of the period. The total diversity of eutherian mammals increased spectacularly, much more than the multituberculates, during the transition from the Cretaceous to the Tertiary and during the Paleocene. In fact, only one of the dozen or more Cretaceous lineages apparently lacks Paleocene descendants.

DISCUSSION

On biologically significant scales, the available sample of the fossil record of the Cretaceous–Tertiary transition provides a geographically limited and temporally coarse data base. Earliest Paleocene terrestrial faunas have been discovered only in North America, Europe, and Africa. The last records of Late Cretaceous terrestrial faunas on many continents are not certainly of latest Cretaceous age. Thus, the impression of abrupt and precisely contemporaneous global extinctions of dinosaurs or other groups of Cretaceous organisms could be the product of gaps in the known fossil record (Dingus and Sadler, 1982; Dingus, 1983b; Newell, 1982). Moreover, currently available methods of correlation are not precise enough to reveal biologically significant temporal differences in records of last or first occurrences of groups of organisms in widely separated areas. Demonstration that the extinctions used to mark the end of the Cretaceous occurred within the interval of reversed magnetic polarity 29R only indicates that they occurred within a period of approximately 500,000 yr. For example, we cannot answer biologically important questions by determining whether extinctions on one continent occurred at the same time (within a few months or years) as those on another or were asynchronous and separated by thousands of years.

Recognizing the limitations in documentation and correlation of the global record of change in the terrestrial vertebrate fauna during the Cretaceous–Tertiary transition, this study has been focused on a geographically limited area in the northern Western Interior of North America. Currently it provides the most complete record of terrestrial faunal change during the Cretaceous–Tertiary transition, but no claim can be made that patterns of faunal change in this area were typical of global patterns. They are simply the most detailed available for study. Without doubt, details of the following analysis will be modified by studies under way, but in view of the considerable research that already has been completed, it would be surprising if this assessment of the basic patterns will be falsified.

In the northern Western Interior, the last records of dinosaurs usually are employed to define the end of the Cretaceous, and the demise of these reptiles has been characterized by some as a "mass extinction" of the terrestrial fauna. The term *mass extinction* has yet to be defined operationally (Padian et al., 1984) and currently carries several different connotations. In short-term analyses, it can imply the abrupt demise of a relatively large proportion of the biota or some of its dominant members. In longer-term analyses, mass extinctions can be defined in terms of changes in rates of origination and extinction. This could, in the simplest case, recognize an increase in the rate of extinction over a normal or background rate. Also, it might involve recognition of time of decrease in taxonomic diversity resulting from decrease in rate of origination with the rate of extinction remaining little changed.

A simple evaluation of the relative incidence of extinction of members of the

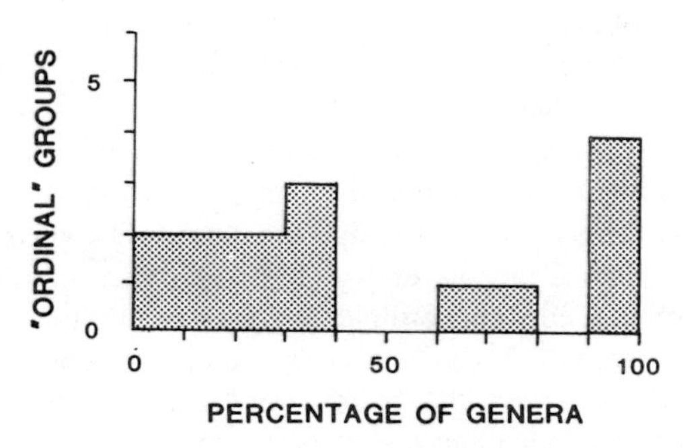

Figure 2. Percentage of extinction of genera within "ordinal"—including some superordinal—groups of terrestrial vertebrates in the Western Interior study area during the Cretaceous–Tertiary transition (see Appendix 2 for data base).

fauna of the northern Western Interior shows that it varied greatly among the ordinal or superordinal groups (Figure 2, Appendix 2). In most (9 of 15) of the groups, less than 40% of the latest Cretaceous (Lancian) genera lacked Paleocene descendants. In 10 of these 15 groups, 25% or less of the families died out. In contrast, all the saurischian, ornithischian, and pterosaurian lineages became extinct as did the eolacertilian lineage represented by *Litakis*. The marsupials suffered a high level of extinction with 75% of the genera and 66% of the families of marsupials lacking Paleocene descendants. Sharks and rays, the chondrichthyan fishes, lost three of the five Cretaceous genera. Given the high probability that these fishes were adapted to brackish and marine waters and the regression of the Western Interior Sea in the latest Cretaceous, the relatively high proportion of extinctions is not surprising. Considering the entire fauna, only 43% of the 117 known latest Cretaceous genera and 34% of the 65 recognized families lacked Paleocene descendants. To be sure, the number of taxa involved in these calculations is relatively small and the familial rankings of some groups are highly suspect. However, the relative proportion of extinctions does not appear extraordinary, and the concentration of the majority of extinctions of genera and families in a few groups indicates that their causal factors were highly selective.

The pattern of survival and extinction of terrestrial vertebrates in the northern Western Interior has intriguing parallels in the evolution of broadly contemporaneous marine invertebrates. All ordinal or superordinal groups of marine invertebrates were not equally decimated (e.g., Kauffman, 1979). Ammonites are an example of a group that underwent a striking decrease in diversity prior to its extinction. On land, ornithischian dinosaurs of the Western Interior study area appear to have suffered a similar decrease in diversity. In contrast, both the marine foraminifera and terrestrial marsupials, which were diversifying during the latest Cretaceous, were driven almost to the point of extinction at the end of the period.

Turning from short-term analyses of patterns of extinction and survival at the end of the Cretaceous, it would be desirable to determine if the rates of extinction in terrestrial vertebrate lineages near or at the end of the Cretaceous were unusually high or if rates of origination were depressed. Such an analysis requires similar measures of the evolution of related groups throughout their Phanerozoic ranges, but unlike the record of change in diversity of the marine fauna (Raup and Sepkoski, 1982), the record of change of the terrestrial fauna is much less complete and usually ordered on a coarser temporal scale. Padian and Clemens (1985), using data on terrestrial vertebrates drawn from Romer (1966) and other sources, present a preliminary analysis of the patterns of change in terrestrial vertebrates through the Phanerozoic. A time scale with a tripartite subdivision of the Cretaceous and epochal divisions of the Cenozoic and global summaries of orders of terrestrial vertebrates were employed. The ordinal analysis shows not a decrease but an increase in diversity from the Late Cretaceous to the Paleocene. At finer time scales and taxonomic levels, it is clear that a temporal lag existed between extinctions of Cretaceous lineages and appearances of new groups of mammals.

Saurischian and ornithischian reptiles are sometimes portrayed either as a stable or an increasingly diverse, dominant element of the terrestrial fauna, and their extinction came without "warning." What constituted a warning often is left undefined, but decrease in diversity (e.g., Alvarez, 1983) or restriction of geographic range might be taken as expressions of the waning of a group prior to its extinction. Total saurischian and ornithischian diversity in the Western Interior study area, assessed by comparison of the numbers of families recorded from the Judithian and the Lancian, was effectively stable. As already noted, the apparent decrease in generic diversity of saurischians might be negated by description of new Lancian taxa. However, the change in generic diversity of ornithischians—a decrease from 20 to 14—is profound. The biogeographic ranges of at least two genera of ornithischians were restricted southward prior to the end of the Cretaceous. Therefore, simply to assert that the dinosaurs, particularly the ornithischians, were a stable group that became extinct "without warning" is unwarranted.

Another pertinent line of evidence emerging from this preliminary analysis is the clear suggestion of a high rate of generic turnover among saurischians and ornithischians. Note that 50% of the genera of Judithian saurischians and more than 75% of the ornithischians are unknown in Lancian faunas of the Western Interior study area (Appendix 1). If these samples provide an accurate estimate of the tempo of evolution of these groups, they point to high rates of extinction and origination of genera. An obvious but currently unanswerable question is: Did the extinction of the saurischians and ornithischians result from an increase in the rate of extinction or a decrease in rate of origination (Bakker, 1977)?

Without doubt, extinction of the dinosaurs and pterosaurs significantly modified the composition of terrestrial vertebrate faunas, but analysis of faunal evolution in the northern Western Interior indicates that their extinction was not accompanied by extraordinary decimation of most contemporary groups of terrestrial vertebrates. Turning from studies of terminal Cretaceous obituaries, it would be highly desirable to be able to assess the pattern and tempo of terrestrial vertebrate evolution through analyses of changes in rates of origination and extinction through the Cretaceous. Unfortunately, the data necessary for detailed studies at the family or, preferably, generic level are not yet available. The possibility remains that extinction of these lineages may not prove to be extraordinary in rate or magnitude when the overall pattern and tempo of change in diversity through the Phanerozoic is adequately assessed.

The current limitations of studies of relative frequency or rates of origination or extinction of terrestrial vertebrates have fostered other lines of research. One has involved searches for common morphological or physiological traits characterizing groups that became extinct or survived. For example, in some hypotheses concerning the causes of extinction at the end of the Cretaceous, the fate of lineages is simply correlated with "large" or "small" body size (e.g., Pollack et al., 1982). In attempting to evaluate these hypotheses, the difference between large and small must be specified. In some discussions, the figure of 25-kg body weight is employed as an arbitrary division between a large and a small terrestrial vertebrate. But authors frequently do not specify whether reference is being made to average adult body size or the range growth stages beginning at hatching or birth.

Analysis of the fauna of the northern Western Interior focusing on adult or maximum body size reveals a familiar pattern. Among the fishes that survived into the Tertiary, at least some species of gars and amiids included individuals that probably reached weights of over 25 kg (i.e., large). Most, but possibly not all, of the other species of Cretaceous fish probably would be ranked as small. Among the amphibians, most would fall in the small category. However, during the Cretaceous, individuals of the salamander *Habrosaurus* are estimated to have grown to 1.5–2 m in length (Estes, 1981) and possibly exceeded 25 kg in weight. *Habrosaurus* survived into the Paleocene. In all the groups of terrestrial reptiles, there were Cretaceous species whose adult body weight appears to have exceeded 25 kg. In contrast to the large saurischians and ornithischians, large champsosaurs, crocodilians, and turtles survived into the Tertiary. As far as known, all Late Cretaceous mammals fall in the small category of adult body size, but the three major groups had very different patterns of survival and extinction. Thus, the history of latest Cretaceous terrestrial vertebrates reveals a bias favoring a greater incidence of extinction of species of large adult body size, but the correlation is far from absolute. This is not an uncommon evolutionary pattern.

If the proposed selective factor was *individual* body size, evaluation of hypotheses cannot be focused simply on average or maximum size of the adults. An estimate must be made of the range of individuals' size within a population. What was the average size of individuals when they were born or hatched? How rapidly did they grow? What was the population structure? For example, we can do little more than speculate about the size of Late Cretaceous reptiles, particularly the saurischian and ornithischian dinosaurs, at time of hatching. On the basis of the sizes of fossil eggs and hatchlings, Horner (1982, 1983) tentatively suggested that the Late Cretaceous, ornithischian dinosaur *Maiasaura* weighed less than a kilogram (about 1.5 lb) at hatching. The average structure of populations of *Maiasaura* or other extinct terrestrial vertebrates can only be modeled from our knowledge of distantly related modern forms (e.g., lizards and mammals). It is reasonable to suggest that at any point in time populations of Late Cretaceous vertebrates of large adult body size included small, young individuals. If extinction of latest Cretaceous vertebrates was strictly size related, with individuals of more than 25 kg body weight perishing, why did not the young of various species of large adult body size survive? Whether average adult or individual body size is considered, a particular range of body size is not a common feature of all groups that survived or became extinct at the end of the Cretaceous.

Suggestions that the survival or extinction of groups of terrestrial vertebrates was directly linked to their physiological mechanisms for regulation of body temperature require a brief evaluation. These hypotheses suggest that cold-blooded or ectothermic vertebrates were not able to cope with changes in temperature or other attributes of the climate that would be tolerated by their warm-blooded or endothermic contemporaries. Changes in climate involving either an increase or decrease in temperature have been invoked as causal factors of terminal Cretaceous extinctions. An evaluation of these hypotheses need not become involved in speculation about whether the saurischians or ornithischians were warm-blooded or cold-blooded. Frequencies of extinction of families or genera were low among osteichthyan fishes, amphibians, chelonians, and crocodilians, for example, and the members of all these groups probably were cold-blooded. The decimation of the marsupials contrasts with the high level of survival of eutherians; both groups of mammals probably were warm-blooded. Presumed physiological methods of control of body temperature and either extinction or survival at the end of the Cretaceous are not directly correlated.

This is not to say that climatic change might not have influenced the pattern of evolution of terrestrial vertebrates. Some aspects, if not the total pattern of faunal change, during the Cretaceous–Tertiary transition in the Western Interior study area can be interpreted as the result of decrease in mean annual temperature, effective temperature (based on temperature and duration of the summer), or equability (reflecting an increase in difference between maximum and minimum annual temperature), (Axelrod and Bailey, 1968). From the Judithian to the Lancian, the ornithischians, but possibly not the carnivorous saurischians, decreased in generic diversity within the Western Interior study area. Two Judithian genera of ornithischians are unknown in Lancian deposits of the study area but survived during the Lancian to the south in New Mexico. These changes suggest a trend in climatic change (cooling or loss of equability) prior to the end of the Cretaceous. Estes (1970) suggested a climatic change as a probable explanation for the decrease in diversity of lizards in the study area during the Cretaceous–Tertiary transition. Hutchison (1982) advanced the same hypothesis to explain changes in the diversity of champsosaurs, crocodilians, and the turtles during this interval. These paleontologic data suggest long-term ecologic changes and defy catastrophic explanations. Also, they support the thesis forcefully argued by Stanley (1984) that change in temperature has been a prominent agent of extinction.

To turn the question around, is there evidence—independent of the fossil record of terrestrial vertebrates—that the global climate was cooling or losing its equability during the transition from the Cretaceous to the Tertiary? During the closing millenia of the Cretaceous, there was a major regression of the shallow seas that had covered large portions of many continents (Matsumoto, 1980). Reduction in the areal extent of shallow epicontinental seas and coastlines has been identified by some workers (Hallam, 1984) as a major causal factor of terminal Cretaceous extinctions of marine organisms. Regression of these seas would have resulted in expansion of continental areas that could have provided additional habitat for dinosaurs and other terrestrial vertebrates. Schopf (1982) argued to the contrary, suggesting that the dinosaurs of the latest Cretaceous were limited to coastal areas around the Western Interior Sea. Van Valen (1984) has pointed out a number of serious flaws in this hypothesis. Expansion of continental areas during the latest Cretaceous and early Paleocene probably promoted the dispersion and intermingling of members of previously isolated faunas that could tolerate the environments of the newly emergent lands. Also, changes in the distribution of land and sea should be expected to have modified patterns of oceanic circulation and stratification and produced changes in the world's climate (Berner, et al., 1983). Studies of concentrations of oxygen and carbon isotopes preserved in the shells of marine organisms (e.g., Savin, 1977) indicate that oceanic temperatures decreased during this interval. Known changes in oceanic circulation and temperature could have contributed to the changes in continental temperature, rainfall, and/or equability that are documented by physiognomic changes in the leaves of terrestrial plants (Axelrod and Bailey, 1968).

Any analysis of the patterns of extinction and survival of terrestrial vertebrates during the Cretaceous–Tertiary transition must take into account the independent evidence that global climates were cooling and decreasing in equability during this period. Some groups decreased in abundance and diversity or suffered significant modifications of geographic range prior to their extinction, suggesting operation of long-term causal factors. These groups must be segregated from those that appear to abruptly disappear. As far as possible, the effects of climatic change during the Cretaceous–Tertiary transition must be isolated. If any groups remain whose patterns of extinction and survival do not appear to be related to climatic change and give evidence of exceptionally high rates of extinction at the end of the Cretaceous, there is reason to suspect the operation of an extraordinary causal factor. Even then, an understanding of the nature of the causal factors of extinctions during the Cretaceous–Tertiary transition will come from a summation of answers concerning the probable causes of extinction of individual lineages, not from the imposition of a hypothesis that dictates the cause of their extinction.

CONCLUSIONS

Our current perception of the pattern of extinction and survival of terrestrial vertebrates at the close of the Cretaceous is based on a geographically limited fossil record calibrated on a coarse temporal scale. The most detailed evolutionary history comes from analyses of fossils from sites in Montana, Wyoming, and Alberta, which constitute the core of the northern Western Interior study area.

The extinction of those terrestrial vertebrates used to mark the end of the Cretaceous did not involve a similar level of decimation of all groups. Extinctions were concentrated among the ornithischian, saurischian, and pterosaurian reptiles and the marsupial mammals. Extinction of lineages was not closely correlated with body size, either average adult size or range of size during the animal's life span. Also, the fate of Cretaceous lineages was not directly correlated with presumed endothermy or ectothermy.

Data necessary to accurately assess the rates of origin and extinction typical of groups of terrestrial vertebrates during the Cretaceous are not available; however, there is a suggestion that among the dinosaurs both were high. Therefore, it cannot be determined whether terminal Cretaceous extinctions were the product of increase in rates of extinction, decrease in rates of origination, or both.

In the northern Western Interior, ornithischian dinosaurs appear to have undergone a decrease in diversity and a southward shift in geographic range of some genera prior to their demise at the end of the Cretaceous. Also, during the Cretaceous–Tertiary transition, several lineages of amphibians and reptiles became extinct in this area but survived into the Tertiary in more southern regions. These evolutionary patterns are predictable consequences of cooling and loss of equability of the climate.

Evidence of cooling and loss of equability of the climate can be drawn from sources other than the evolutionary history of terrestrial vertebrates. Several lines of investigation suggest that changes in distribution of land and sea during the latest Cretaceous regression were the ultimate cause of the modification of the climate. Any attempt to develop a hypothesis identifying the causal factors of the extinctions utilized to mark the end of the Cretaceous must recognize their biological effects. The question remaining is: Were these climatic changes sufficient cause for all the observed changes in the biota?

An understanding of the causal factors of extinctions during the transition from the Cretaceous to the Tertiary will come from a summation of answers concerning the probable causes of extinction of individual lineages, not from the imposition of a hypothesis that prescribes the cause of their extinction.

APPENDIX 1: COMPARISION OF JUDITHIAN AND LANCIAN SAURISCHIANS AND ORNITHISCHIANS KNOWN FROM THE WESTERN INTERIOR STUDY AREA

Judithian	Lancian
Order Saurischia	
Family Ornithomimidae	
Dromiceiomimus	
Ornithomimus	Cf. *Ornithomimus*
Struthiomimus	
Family Dromaeosauridae	
Dromaeosaurus	Cf. *Dromaeosaurus*
Macrophalangia	
Saurornitholestes	
Family Saurornithoididae	
	Pectinodon
Stenonychosaurus	*Stenonychosaurus*
Family Caenognathidae	
Caenognathus	
Family Tyrannosauridae	
Albertosaurus	Cf. *Albertosaurus*
Dasypletosaurus	
	Tyrannosaurus
?Theropoda	
Cf. *Aublysodon*	*Aublysodon*
Paronychodon	*Paronychodon*
Order Ornithischia	
Family Hypsilophodontidae	
Genus indeterminable	*Thescelosaurus*
	Genus indeterminable

APPENDIX 1 (*cont.*)

Judithian	Lancian
Family Fabrosauridae	
Genus indeterminable	Genus indeterminable
Family Hadrosauridae	
	Anatosaurus
Brachylophosaurus	
Corythosaurus	
	Edmontosaurus
Hadrosaurus	
Lambeosaurus	
Parasaurolophus	
Prosaurolophus	
Family Pachycephalosauridae	
Gravitholus	
Ornatotholus	
Pachycephalosaurus	*Pachycephalosaurus*
Stegoceras	*Stegoceras*
	Stygimoloch
Family Nodosauridae	
Panoplosaurus	*Panoplosaurus*
Family Ankylosauridae	
	Ankylosaurus
Euoplocephalus	
Family Protoceratopsidae	
	Leptoceratops
Family Ceratopsidae	
Anchiceratops	
Centrosaurus	
Chasmosaurus	
	Diceratops
Eoceratops	
Monoclonius	
Styracosaurus	
	Triceratops
	Torosaurus

Of the taxa listed above, some are based on only isolated teeth (e.g., *Paronychodon*) or parts of feet (e.g., *Macrophalangia*). Others are founded on large samples of skeletal material (e.g., *Triceratops*), but analyses of morphologic variation to establish their limits and test their homogeneity are lacking. Also, until recently most of the field parties charged with collecting dinosaurian remains had the filling of exhibit halls rather than thorough faunal analyses as their primary goals. The reader must recognize that these faunal lists are preliminary compilations that will be modified by modern analyses of existing collections and future collecting.

The goal of this analysis is to assess changes in diversity from the Judithian to the Lancian within one paleobiogeographic region, the Late Cretaceous coastal lowlands of the northern Western Interior. Therefore, Judithian taxa from the Two Medicine Formation and Lancian taxa known only from sites south of Wyoming and the Denver Basin of Colorado are excluded. These sites appear to have been in different paleobiogeographic regions. Because of the smaller size of the samples and their geographically limited occurrences within the Western Interior study area, the dinosaurian faunas of Edmontonian age (Table 1) are not included in this study. Compilations of Edmontonian dinosaurian faunas are given by L. S. Russell (1983) and D. A. Russell (1984b).

The faunal lists presented here originally were based on records of Judithian and Lancian taxa found within the study area and cited in Glut's (1982) primarily bibliographic summary. This compilation has been revised in light of Russell's (1984b) checklist; however, unpublished records of new genera cited by Russell were not included. Without doubt, the current enthusiastic study of Lancian dinosaurs will result in the recognition of several new taxa. But, when undertaken, similar detailed systematic studies of samples of Judithian faunas also might reveal new dinosaurian taxa.

APPENDIX 2: FREQUENCY OF EXTINCTION AMONG TERRESTRIAL VERTEBRATES IN THE WESTERN INTERIOR STUDY AREA

	Total Genera	Extinct Genera	Total Families	Extinct Families	Extinct Genera (%)	Extinct Families (%)
"Fish"						
Chondrichthyans	5	3	4	1	60	25
Osteichthyans	13	5	10	1	38	10
Amphibia	12	4	8	1	33	13
Reptilia						
Chelonia	18	2	4	0	11	0
Eosuchia	1	0	1	0	0	0
Crocodilia	4	1?	1	0	25?	0
Eolacertilia	1	1	1	1	100	100
Lacertilia	15	4?	7	0	27?	0
Serpentes	2	0?	1	0	0?	0
Saurischia	8	8	5	5	100	100

APPENDIX 2 (*cont.*)

	Total Genera	Extinct Genera	Total Families	Extinct Families	Extinct Genera (%)	Extinct Families (%)
Ornithischia	14	14	8	8	100	100
Pterosauria	?	?	?	?	100	100
Aves	?	?	?	?	?	?
Mammalia						
Multituberculata	11	4	8	2	36	25
Marsupialia	4	3	3	2	75	66
Placentalia	9	1	4	1	11	25

ACKNOWLEDGMENTS

In compiling and analyzing the data presented in Appendix 1 I received considerable help from L. Bryant, R. Estes, J. H. Hutchison, R. A. Long, K. Padian, D. A. Russell, and S. P. Welles. However, the final responsibility to include or exclude taxa rests with the author. Their assistance and the comments on various drafts of this chapter made by J. D. Archibald, D. T. Clemens, L. Dingus, R. Estes, K. Padian, and other reviewers are gratefully acknowledged. Some of the research reported here was sponsored by grants from the National Science Foundation (DEB 77-24610 and 81-19217), the donors of the Petroleum Research Fund, American Chemical Society (PRF no. 12487-AC-2), and the Annie M. Alexander Endowment, Museum of Paleontology, University of California, Berkeley. These sources of support are acknowledged with thanks.

REFERENCES

Alvarez, L. W., 1983, Experimental evidence that an asteroid impact led to the extinction of many species 65 million years ago, *Proc. Natl. Acad. Sci.*, **80**, 627–642.

Alvarez, L. W., Alvarez, W., Asaro, F., and Michel, H. V., 1980, Extraterrestrial cause for the Cretaceous–Tertiary extinction, *Science*, **208**, 1095–1107.

Archibald, J. D., 1977, Fossil Mammalia and testudines of the Hell Creek Formation, and the geology of the Tullock and Hell Creek Formations, Garfield County, Montana, Ph.D. dissertation, University of California, Berkeley, 694 pp.

Archibald, J. D., 1982, A study of Mammalia and geology across the Cretaceous–Tertiary boundary in Garfield County, Montana, *Univ. Calif. Publ. Geol. Sci.*, **122**, 288 pp.

Archibald, J. D., 1984, Bug Creek Anthills (BCA), Montana: Faunal evidence for Cretaceous age and non-catastrophic extinctions, *Geol. Soc. Am. Abstr. Progr.*, **16**, 432.

Archibald, J. D., and Clemens, W. A., 1982, Late Cretaceous extinctions, *Am. Sci.*, **70**, 377–385.

Archibald, J. D., Butler, R. F., Lindsay, E. H., Clemens, W. A., and Dingus, L., 1982, Upper Cretaceous-Paleocene biostratigraphy and magnetostratigraphy, Hell Creek and Tullock Formations, northeastern Montana, *Geology*, **10**, 153–159.

Axelrod, D. I. and Bailey, H. P., 1968, Cretaceous dinosaur extinction, *Evolution*, **22**, 595–611.

Bakker, R. T., 1977, Tetrapod mass extinctions: A model of the regulation of speciation rates and immigration by cycles of topographic diversity, in Hallum, A., ed., *Patterns of Evolution, As Illustrated in the Fossil Record, Developments in Palaeontology and Stratigraphy*, No. 5, Amsterdam, Oxford, and New York, Elsevier, pp. 439–468.

Behrensmeyer, A. K., 1983, Time resolution in fluvial vertebrate assemblages, *Paleobiology*, **8**, 211–227.

Berner, R. A., Lasaga, A. C., and Garrels, R. M, 1983, The carbonate-silicate geochemical cycle and its effect on atmospheric carbon dioxide over the past 100 million years, *Am. J. Sci.*, **283**, 641–683.

Brodkorb, P., 1963, Birds from the Upper Cretaceous of Wyoming, *Proc. 13th Int. Ornithol. Cong.*, 55–70.

Butler, R. F., 1984, Paleomagnetic and rock-magnetism of continental deposits, San Juan Basin, New Mexico, *Geol. Soc. Am. Abstr. Progr.*, **16**, 217.

Butler, R. F., and Lindsay, E. H., 1983, Magnetic mineralogy of continental deposits, San Juan Basin, New Mexico, *EOS, Trans. Am. Geophys. U.*, **64**, 683.

Charig, A., 1979, *A New Look at the Dinosaurs*, New York, Mayflower Books, 160 pp.

Clemens, W. A., 1973, Fossil mammals of the type Lance Formation, Wyoming, *Univ. Calif. Publ. Geol. Sc.*, **94**, 102 pp.

Clemens, W. A., 1982, Patterns of extinction and survival of the terrestrial biota during the Cretaceous/Tertiary transition, in Silver, L. T., and Schultz, P. H., eds., *Geological Implications of Impacts of Large Asteroids and Comets on the Earth*, Geological Society of America Special Paper 190, pp. 401–406.

Clemens, W. A., 1984, Evolution of marsupials during the Cretaceous–Tertiary transition, in Reif, W.-E., and Westphal, F., eds., *Third Symposium on Mesozoic Terrestrial Ecosystems, Short Papers*, Tübingen, Attempto Verlag, pp. 47–52.

Clemens, W. A., Lillegraven, J. A., Lindsay, E. H., and Simpson, G. G., 1979, Where, when, and what: A survey of known Mesozoic mammal distribution, in Lillegraven, J. A., Kielan-Jaworowska, Z., and Clemens, W. A., eds., *Mesozoic Mammals: The First Two-Thirds of Mammalian History*, Berkeley, University of California Press, pp. 7–58.

Davis, M., Hut, P., and Muller, R. A., 1984, Extinction of species by periodic comet showers, *Nature*, **308**, 715–717.

Dingus, L., 1983a, A stratigraphic review and analysis of selected marine and terrestrial sections spanning the Cretaceous–Tertiary boundary, Ph.D. dissertation, University of California, Berkeley, 156 pp.

Dingus, L., 1983b, Expected temporal completeness of fluvial and pelagic sections spanning the K-T boundary, with implications for resolving catastrophic rates of extinction, *Geol. Soc. Am. Abstr. Progr.*, **15**, 558.

Dingus, L., and Sadler, P. M., 1982, The effects of stratigraphic completeness on estimates of evolutionary rates, *System. Zool.*, **31**, 400–412.

Estes, R., 1964, Fossil vertebrates from the Late Cretaceous Lance Formation, eastern Wyoming, *Univ. Calif. Publ. Geol. Sci.*, **49**, 187 pp.

Estes, R., 1970, Origin of the Recent North American lower vertebrate fauna: An inquiry into the fossil record, *Forma et Functio*, **3**, 139–163.

Estes, R., 1981, Gymnophiona, Caudata, in Wellnhofer, P., ed., *Teil 2, Handbuch der Palaeoherpetologie*, Stuttgart and New York, Gustav Fischer Verlag, 115 pp.

Estes, R., 1983a, The fossil record and early distribution of lizards, in Rhodin, A. G. J., and

Miyota, K., eds., *Advances in Herpetology and Evolutionary Biology: Essays in Honor of E. E. Williams*, Cambridge, Massachusetts, Museum of Comparative Zoology, Harvard, pp. 365–398.

Estes, R., 1983b, Sauria terrestria, Amphisbaenia, in Wellnhofer, P., ed., *Teil 10A, Handbuch der Palaeoherpetologie*, Stuttgart and New York, Gustav Fischer Verlag, 249 pp.

Estes, R., and Sanchiz, B., 1982, New discoglossid and palaeobatrachid frogs from the Late Cretaceous of Wyoming and Montana, and a review of other frogs from the Lance and Hell Creek Formations, *J. Vert. Paleontol.*, **2**, 9–20.

Fox, R. C., 1984, The dentition and relationships of the Paleocene primate *Micromomys* Szalay, with description of a new species, *Can. J. Earth Sci.*, **21**, 1262–1267.

Glut, D. F., 1982, *The new dinosaur dictionary*: Secaucus, New Jersey, Citadel Press, 288 pp.

Hallam, A., 1984, Pre-Quaternary sea-level changes: *Annual Review of Earth and Planetary Sciences*, **12**, 205–243.

Horner, J. A., 1982, Coming home to roost, *Montana Outdoors*, **13**, 2–5

Horner, J. A., 1983, Cranial osteology and morphology of the type specimen of *Maiasaura peeblesorum* (Ornithischia: Hadrosauridae), with discussion of its phylogenetic position, *J. Vert. Paleontol.*, **3**, 29–38.

Hsü, K. J., He, Q., et al., 1982, Mass mortality and its environmental and evolutionary consequences, *Science*, **216**, 249–256.

Hutchison, J. H., 1982, Turtle, crocodilian, and champsosaur diversity changes in the Cenozoic of the north-central region of western United States, *Palaeogeogr. Palaeoclimatol. Palaeoecol.* **37**, 149–164.

Johnston, P. A., 1980, First record of Mesozoic mammals from Saskatchewan, *Can. J. Earth Sci.*, **17**, 512–519.

Kauffman, E. G., 1979, The ecology and biogeography of the Cretaceous–Tertiary extinction event, in Christensen, W. K., and Birkelund, T., eds., *Cretaceous–Tertiary Boundary Events Symposium: II. Proceedings*, University of Copenhagen, pp. 29–37.

Lehman, T., 1981, The Alamo Wash local fauna: A new look at the old Ojo Alamo fauna, in Lucas, S., Rigby, K., Jr., and Kues, B., eds., *Advances in San Juan Basin Paleontology*, Albuquerque, University of New Mexico Press, pp. 189–221.

Lillegraven, J. A., 1969, Latest Cretaceous mammals of upper part of Edmonton Formation of Alberta, Canada, and review of marsupial-placental dichotomy of mammalian evolution, University of Kansas, Paleontological Contributions, Art. 50 (Vert. 12), 122 pp.

Lillegraven, J. A., Kielan-Jaworowska, Z., and Clemens, W. A., 1979, *Mesozoic Mammals: The First Two-Thirds of Mammalian History*, Berkeley, University of California Press, 311 pp.

Luck, J. M., and Turekian, K. K., 1983, Osmium-187/Osmium-186 in manganese nodules and the Cretaceous–Tertiary boundary, *Science* **222**, 613–615.

Matsumoto, T., 1980, Inter-regional correlation of transgressions and regressions in the Cretaceous Period, *Cretac. Res.*, **1**, 359–373.

Newell, N. D., 1982, *Mass Extinctions: Illusions or Realities*?, Geological Society of America Special Paper 190, pp. 257–263.

Officer, C. B., and Drake, C. L., 1983, The Cretaceous–Tertiary transition, *Science*, **219**, 1383–1390.

Ostrom, J. H., 1978, *Leptoceratops gracilis* from the "Lance" Formation of Wyoming, *J. Paleontol.*, **52**, 697–704.

Padian, K., and Clemens, W. A., 1985, Terrestrial vertebrate diversity: Episodes and insights, in Valentine, J. W., ed., *Phanerozoic Diversity Patterns: Profiles in Macroevolution*, Princeton, Princeton University Press (in press).

Padian K., Alvarez, W., Birkelund, T., Fütterer, D. K., Hsü, K. J., Lipps, J. H., McLaren,

D. J., Shoemaker, E. M., Smit, J., Toon, O. B., and Wetzel, A., 1984, The possible influences of sudden events on biological radiation and extinctions, in Holland, H. D., and Trendall, A. F., eds., *Patterns of Change in Earth Evolution*, Berlin, Springer Verlag, pp. 77–102.

Pollack, J. B., Toon, O. B., Ackerman, R. P., Turco, R. P., and McKay, C. P., 1982, Environmental effects of an impact generated dust cloud: Implications for the Cretaceous–Tertiary extinctions, *Science*, **219**, 287–289.

Rage, J.-C., 1984, Serpentes, in Wellnhofer, P., ed., *Teil 11, Handbuch der Palaeoherpetologie*, Stuttgart and New York, Gustav Fisher Verlag, 80 pp.

Raup, D. M., and Sepkoski, J. J., Jr., 1982, Mass extinctions in the marine fossil record, *Science*, **215**, 1051–1053.

Romer, A. S., 1966, *Vertebrate Paleontology*, 3rd ed., Chicago, University of Chicago Press, 468 pp.

Russell, D. A., 1984a, The gradual decline of the dinosaurs: Fact or fallacy?, *Nature*, **307**, 360–361.

Russell, D. A., 1984b, A check list of the families and genera of North American dinosaurs, *Syllogeus*, no. 53, 35 pp.

Russell, L. S., 1983, Evidence for an unconformity at the Scollard-Battle contact, Upper Cretaceous strata, Alberta, *Can. J. Earth Sci.*, **20**, 1219–1231.

Savage, D. E., and Russell, D. E., 1983, *Mammalian paleofaunas of the world*, London, Addison-Wesley Publishing Company, 432 pp.

Savin, S. M., 1977, The history of the earth's surface temperature during the past 100 million years, *Ann. Rev. Earth Planet. Sci.*, **5**, 319–355.

Schopf, T. J. M., 1982, Extinction of dinosaurs: A 1982 understanding, in Silver, L. T., and Schultz, P. H., eds., *Geological Implications of Impacts of Large Asteroids and Comets on the Earth*, Geological Society of America Special Paper 190, pp. 415–422.

Smit, J., and van der Kaars, S., 1984, Terminal Cretaceous extinctions in the Hell Creek area, Montana: Compatible with catastrophic extinction, *Science*, **223**, 1177–1179.

Stanley, S. M., 1984, Marine mass extinction: A dominant role for temperature, in Nitecki, M. H., ed., *Extinctions*, Chicago and London, University of Chicago Press, pp. 69–117.

Steel, R., 1975, *Die Fossilen Krokodile*, Die Neue Brehm-Buecherei, Lutherstadt, A. Ziemsen Verlag, 76 pp.

Van Valen, L. M., 1984, Catastrophes, expectations, and the evidence, *Paleobiology*, **10**, 121–137.

Zoller, W. H., Parrington, J. R., and Kotra, J. M. P., 1983, Iridium enrichment in airborne particles from Kilauea volcano: January 1983, *Science*, **222**, 1118–1120.

5

EXTINCTION DYNAMICS IN PELAGIC ECOSYSTEMS

JERE H. LIPPS

Department of Geology
University of California
Davis, California

INTRODUCTION

"Mass extinctions" have attracted much attention in the past few years because spectacular organisms, such as dinosaurs and ammonites, disappeared at particular times in the geologic record. The extinctions are reportedly cyclic (Raup and Sepkoski, 1984) at an interval of 26 m.y. The cyclicity remains to be tested with more robust data and statistical methods but has nevertheless attracted much attention. The extinctions are thus popular for several reasons.

These large, spectacular organisms that became extinct and their ecosystems, however, are generally poorly represented at any particular geologic horizon for two major reasons: reduction of sample size for the larger fossils as boundaries are approached and, more importantly here, artificial range truncations, or the probable failure of preservation and discovery of individuals representing the last biotic occurrence (Signor and Lipps, 1982). These two sampling effects make determination of precise extinction horizons nearly impossible, except where the sediments themselves are made of the fossils; this occurrence is especially common in pelagic sedimentary sequences. For these reasons, the record of extinction in pelagic microfossils is particularly useful. Furthermore, the extinctions in the pelagic ecosystems were more inclusive than perhaps in other ecosystems (Thierstein, 1982), and they are impressive for that reason as well.

During the last 150 m.y., there have been three major extinction episodes in the pelagic marine environment. These episodes occurred at the Cenomanian–Turonian boundary (91 m.y. ago), the Cretaceous–Tertiary (K/T) boundary (65 m.y. ago) and near the Eocene–Oligocene (E/O) boundary (36.6 m.y.ago). Of these episodes, the K/T and E/O have been known for a long time and hence are the best described. Pelagic biotic changes at the K/T boundary are especially well documented (e.g., Bramlette, 1965; Smit, 1982; Perch-Nielsen, McKenzie, and He, 1982; Thierstein, 1982; among many others). Although extinctions also occurred in terrestrial and some shallow marine ecosystems in near synchroneity with the pelagic events, their documentation is less precise and will not be considered in detail. The simultaneous extinctions in such diverse ecosystems, however, indicate that the causes of pelagic extinctions were worldwide and linked to almost the entire biosphere.

Prior to 150 m.y. ago, other major extinctions took place (Raup and Sepkoski, 1982; Sepkoski, 1982), and these surely had impact on the pelagic ecosystems as well. Sepkoski (1982) records 12 mass extinctions of varying magnitude in the Phanerozoic other than the 3 considered here. The detailed record of the pelagic microbiota at the 11 older times is not as clear, simply because the skeletal microplankton did not begin to dominate the fossil record until later and because the sedimentary record of pelagic environments is poorly understood. Other groups (ammonites, for example) show significant extinctions at some of these times, but they represent a small portion of the pelagic biota that must have been present. In the later extinctions discussed here, groups representing most pelagic trophic levels, from primary producers and microcarnivores to large terminal herbivores and carnivores, are preserved as fossils, commonly together. Thus, a more complete basis for interpretation of the dynamics of extinction is available. Furthermore, most of the fossil groups have living representatives, and while few species are identical, there is nevertheless biologic information available about similar living organisms that also provides bases for inference.

Many causes of the pelagic extinctions have been proposed, but most have been ruled out long ago (e.g., Lipps, 1970). New hypotheses have taken the place of those eliminated, and these have invigorated the field recently. The hypothesis that an extraterrestrial impact by asteroid or comet occurred at the K/T has received much determined work (L. Alvarez et al., 1980; W. Alvarez et al., 1984b; see papers in Silver and Schultz, 1982). The hypothesis that extraterrestrial impacts have caused extinctions has been extended with supporting geochemical evidence to other times as well; for example, the Upper Devonian event at the Frasnian–Famennian stage boundary (McLaren, 1983; Playford et al., 1984) and the Eocene–Oligocene boundary (Ganapathy, 1982; Alvarez et al., 1982; Asaro et al., 1982). This hypothesis has met with general and specific criticism (Officer and Drake, 1983; Corliss et al., 1984). The controversy is not resolved in general, although the case for an impact-associated extinction at the K/T seems convincing at present (Alvarez et al., 1984a).

Herein I examine the paleobiologic characteristics surrounding these events in order to understand the dynamics of pelagic extinctions and to provide bases for constraining (and in some cases eliminating) hypotheses. Even if one hypothesis, such as the extraterrestrial impact hypothesis, is judged acceptable, a paleobiologic approach to the extinction will aid in the understanding of the dynamics involved in the extinctions.

EXTINCTION DEFINED

Various phrases or words have been used to designate the extinction events observed in geologic history. Most of these designations carry with them implications about the nature or origin of the event.

Mass extinction is perhaps the most commonly used term to describe the extinctions. To many people, these words suggest that many or most groups suffered the extinction. While this may be generally true in some cases, each extinction is selective with regard to systematic groups, habitats, trophic levels, biogeography, and functional types. Thus, mass extinction, although not too objectionable, nevertheless does not acccurately describe the extinctions and indeed may give an erroneous impression of the fossil record. Certainly the K/T was not a mass extinction in many environments, for example, the deep sea (Douglas and Woodruff, 1981; Lipps and Hickman, 1982; Corliss et al., 1984) or the temperate coast of western North America (Saul, 1983), or within certain feeding types, for example, epifaunal suspension feeding mollusks declined significantly while others did not (Jablonski and Bottjer, 1983).

"Catastrophic extinction" is more objectionable because the phrase indicates that some sort of catastrophe befell the biota suddenly. While this too may be true of some extinctions, evidence of catastrophes does not exist for the majority of the extinctions. It automatically implies causality, which is unnecessary and undesirable. Even when evidence for a catastrophe coincides with an extinction, as at the K/T boundary, the catastrophic event may be coincidental.

Other terms have occasionally been used as well that have implied meanings. I choose to use a descriptive term, such as extinction episode or event, with the intent of simply indicating that a large proportion of the biota became extinct during a particular time interval without inferring cause or intensity of the extinction.

EXTINCTIONS IN PELAGIC ECOSYSTEMS

The Cretaceous–Tertiary extinction event and the Eocene–Oligocene event have certain biologic similarities. These include similarities in morphologic and functional groups of plankton affected, their biogeographic patterns, species diversity declines (although not degree), large pelagic carnivore or herbivore declines, and the decline of siliceous phytoplankton. These events also show different extinction patterns and durations for the K/T and E/O.

Prior to each of these extinction episodes, the pelagic biota was characterized by high species diversities, provincial distributions, both simple and complex morphologies within each group of fossil plankton, abundant siliceous phytoplankton and deposition of biogenic silica in parts of the world's oceans, and numerous large, vertebrate consumers, for example, mosasaurs or whales. After the extinction episodes, plankton species diversity was low, plankton distributions were widespread, morphologies were simple, siliceous phytoplankton decreased in abundance and siliceous biogenic sedimentation decreased, and large vertebrate consumers were absent.

Although there were these biologic similarities between extinction episodes, the nonbiologic aspects were significantly different. For example, the K/T event happened relatively suddenly, and the postextinction paleobiologic characteristics of the pelagic biota endured for about 3 m.y. (Smit, 1982; Boersma and Premoli Silva, 1983), whereas the E/O episode was steplike, with the extinction characteristics of the biota developing and remaining for 8–12 m.y. (Keller, 1983; Keller, D'Hondt, and Vallier, 1983). Similar steplike events are recorded in the deep-sea ostracode faunas (Benson, Chapman, and Deck, 1984).

Pelagic Species Diversity

Pelagic plants and animals of the open ocean were diverse in the Late Cretaceous and in the later Eocene before the extinction episodes. These plants and animals included both siliceous and calcareous microplankton, cephalopds, and large vertebrates. At both times, major diversity declines occurred in all well-documented pelagic groups.

At the end of the Cretaceous, the planktonic biota "became extinct while in full development" (Smit, 1982). In the latest Cretaceous, planktonic algae were represented by at least 59 species of calcareous nannoplankton (Thierstein, 1982), 130 species of diatoms from the Maastrichtian of California alone (Hanna, 1927, 1934; Long, Fuge, and Smith, 1946) and 89 from the same age rocks in western Siberia (Strelnikova, 1975), 11 silicoflagellate species (Bukry, 1981), and 26 dinoflagellates (Hansen, 1979). Among the microzooplankton, there were 36 species of foraminifera (Pessagno, 1967) and 77 radiolaria (Foreman, 1968). Ammonites, although undergoing a decline in diversity throughout the Upper Cretaceous (Wiedman, 1973; Ward and Signor, 1983), still had 11 genera at the end of the Cretaceous (Hancock, 1967), 7 of which are found in the very latest Maastrichtian beds at Stevns Klint, Denmark (Alvarez, et al., 1984b). Several genera of belemnites persist into the Maastrichtian as well (Donovan and Hancock, 1967). Marine reptiles were diverse in the Upper Cretaceous too—33 genera of pleisiosaurs, 2 of ichthyosaurs, and 20 of mosasaurs (Romer, 1966)—although not all of these lived to the end of the Cretaceous.

In contrast, the earliest Tertiary sedimentary rocks (Danian Stage) contain far fewer taxa of most pelagic groups and some do not extend across the K/T boundary at all. Ammonites and marine reptiles became extinct as groups at that boundary, while the oceanic microplankton were reduced in standing diversity by 2/3ds (Thierstein, 1982). In the lowermost planktonic foraminiferal zone of the Danian, only one species of planktonic foraminifera, *Guembelitria cretacea*, is present (Smit, 1982), indicating that the diversity recorded for the lower Danian as a whole, although much reduced, is still higher than that immediately following the K/T boundary.

The Eocene–Oligocene extinction episode was not as marked as the K/T in terms of species diversity reduction. The microplanktonic biota was reduced in species diversity in a series of successive events (Corliss et al., 1984).

After each of these extinction episodes, species diversity gradually climbed, reaching high values within a few million years. In the Paleocene, the total planktonic foraminiferal species diversity increased from 11 species in the lowest zone (*Planorotalites eugubinus* Zone) to 32 species 6 or 7 m.y. later (*Morozovella uncinata* Zone) and finally to 64 species 10 m.y. later (*Morozovella velascoensis* Zone) (Boersma and Premoli Silva, 1983). Following the somewhat lower diversity in the Oligocene, Miocene planktonic foraminiferal species diversity had recovered within 10 m.y. as well.

Morphology

General morphologic characteristics of planktonic organisms, especially planktonic foraminifera, underwent marked changes across both the K/T and E/O extinction episodes. Prior to these extinctions, the planktonic foraminifera were morphologically complex with tests having a wide variety of structural features (Cifelli, 1969; Lipps, 1970; Smit, 1982; Corliss et al., 1984). In the foraminiferal biotas as a whole, these features included keels, apertural lips and modifications, secondary calcification, rugosities and striae on the test surfaces, elongated or radial chambers, flattened to spherical tests, and other such characters (Figure 1). Of course, the

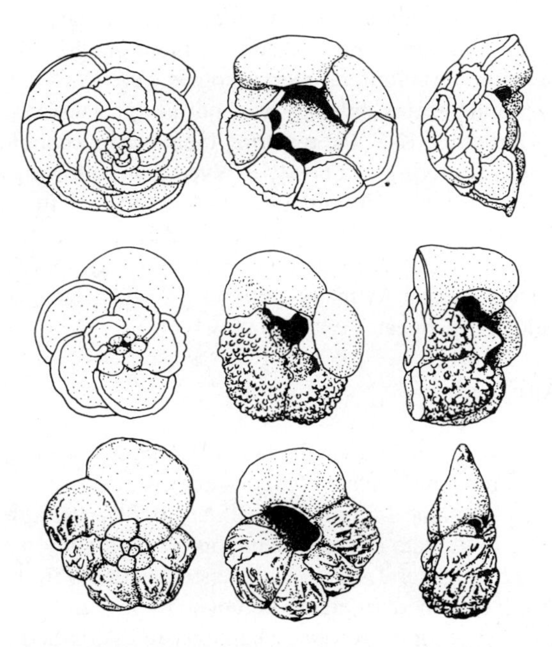

Figure 1. Complex morphologies in Late Cretaceous planktonic foraminifera. During much of the evolutionary history of planktonic foraminifera, complex but similar morphologies appeared. At these times, keels, apertural modifications, surficial ridges and striae, elongated chambers, and other complex morphologies coexisted. Investigations of living planktonic foraminifera show that these morphologies are related to vertical segregation in the water column.

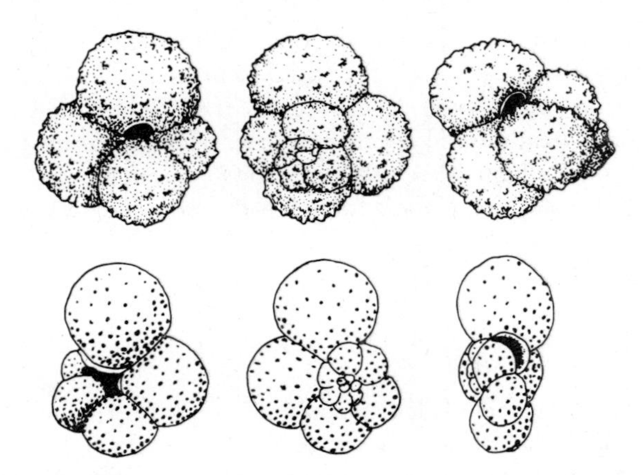

Figure 2. Simple morphologies in Early Paleocene planktonic foraminifera. At certain periods, marked by extinctions in planktonic foraminifera, morphologies consisted of trochospirals with generally spherical chambers. Such morphologies today dominate in high-latitude seas with homogeneous vertical water structure.

species in the latest Cretaceous and the late Eocene were different (so were higher taxa), but the kinds of morphologic complexities were similar.

After the sudden extinction at the K/T and the more gradual but stepped events near the E/O, planktonic foraminiferal morphologies were simple. Even though the planktonic foraminiferal species diversity was not as spectacularly reduced in the events near the E/O as at the K/T, the resulting fauna was morphologically distinct from the earlier ones. Surviving and newly evolved taxa following each extinction episode had trochospiral tests with smooth surfaces (except for spine bases and pustules), little or no secondary calcification, no keels or apertural modifications of any sort, and more or less spherical chambers (Figure 2). Only after reradiation of many new species several to many millions of years later did the complex morphologies gradually reappear.

Paleobiogeography

The plankton paleobiogeography during the later Cretaceous (Late Campanian through Late Maastrichtian) remained relatively constant for calcareous nannoplankton (Thierstein, 1981), planktonic foraminifera (Sliter, 1972), and apparently diatoms (Strelnikova, 1975). In the late Eocene, plankton biogeography resembled the late Cretaceous patterns in a general sense. The general patterns for foraminifera and nannofossils are similar to modern plankton biogeography. In the Late Cretaceous, the Eocene, and the Miocene to Recent, plankton are distributed generally in latitudinal bands corresponding, in the modern case (Figure 3), to water characteristics determined in large part by circulation patterns (McGowan, 1971). The Cretaceous and Eocene biogeographies were similarly related to water masses.

At the K/T boundary, the biogeography that had persisted for so long was

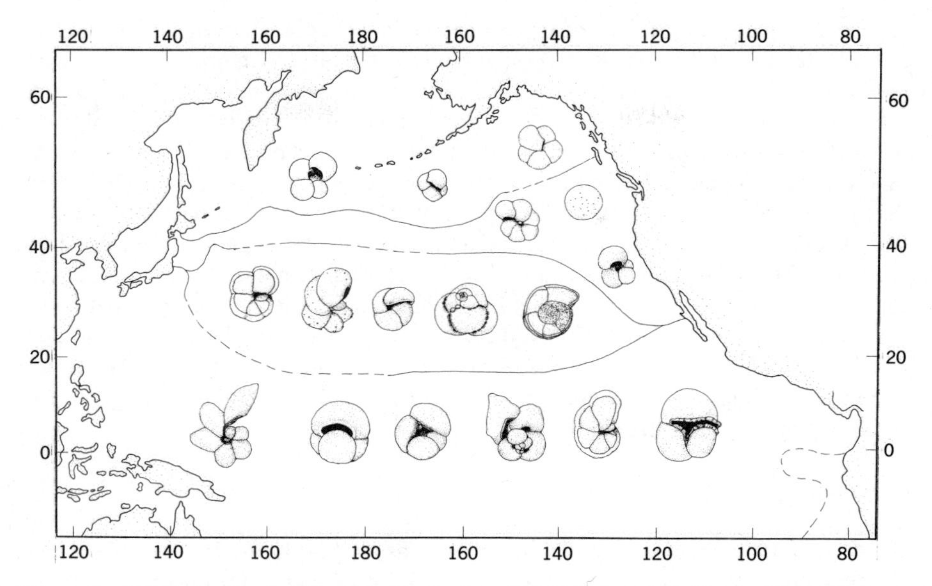

Figure 3. Biogeography of modern planktonic foraminifera in the North Pacific Ocean (based on Bradshaw, 1959). Characteristic species for each biogeographic province are shown. In the thermally stratified Central and Equatorial Provinces, species with complex morphologies are common; in higher-latitude seas, simple morphologies dominate. These provinces are maintained by the sea surface circulation and are similar for all oceans (Bé, 1977).

completely reorganized. In the planktonic foraminifera (Cifelli, 1969) and calcareous nannoplankton (Haq, Premoli-Silva, and Lohmann, 1977), species distributions were remarkably widespread in the initial part of the Paleocene. Identical planktonic foraminiferal assemblages of low diversity were found in all latitudes from Greenland to the tropical regions. Calcareous nannoplankton showed greater geographic diversity, but most of that diversity can be attributed to proximity of localities to ancient shorelines (Haq et al., 1977) rather than to alteration of the general cosmopolitan pattern.

Two generalized pelagic biogeographic patterns can be discerned for most of the last 100 m.y. of geologic time. On one hand, planktonic organisms were distributed in latitudinal fashion determined largely by surface circulation patterns. This pattern dominates the record, as might be expected from the factors governing the earth's atmospheric and oceanic circulation. Contrarily, during the early Paleocene and much of the Oligocene, distributions were broader and fewer biogeographic groupings can be identified (Cifelli, 1969; Boersma and Premoli Silva, 1983; Haq, 1981; Thunell and Belyae, 1982; among others).

Large Carnivores

During much of the time since the evolution of a complex skeletonized pelagic biota, large carnivores were prominent members of these ecosystems. At the end of the Cretaceous, a variety of carnivorous marine reptiles became extinct as did ammonites that fed on smaller zooplankton. Later in the Cenozoic, the cetacean suborder Archeoceti, a rather primitive group of whales, similarly became extinct at near the end of the Eocene, having first evolved in the early part of that epoch (Barnes and Mitchell, 1978). During the Oligocene, archeocetes declined severely and are virtually unknown in most of the epoch (Lipps and Mitchell, 1976). In the early Miocene, however, the modern cetaceans, the Mysticeti or baleen whales and the Odontoceti or toothed whales, including the smaller porpoises and dolphins, evolved. These were more advanced aquatic forms than the Archeoceti, having telescoped skulls and nostrils on top of the head rather than at the proximal end of the muzzle as in archeocetes. Thus, the extinction of whales near the E/O, although not as sudden or complete as the extinction of marine reptiles at the K/T, was nevertheless important because a radiation of new adaptive types evolved from a more aquatically primitive suborder.

Siliceous Phytoplankton

During the Late Cretaceous and Late Eocene, as well as during the entire Neogene, siliceous phytoplankton were important primary producers and sources of biogenic silica deposits. These phytoplankton were mostly diatoms but included silicoflagellates and ebridians as well. Diatoms were so abundant at these times that thick deposits of diatomites were formed in many places around the world, for example, in the Pacific Ocean at California, New Zealand, Japan, and the equatorial zone (Ingle, 1975; Orr, 1972; Bramlette, 1946; Winterer, 1973). However, diatomite formation ceased, as far as the known geologic record shows, in the early Paleocene and in the late Eocene and early Oligocene (Berger, 1973; Leinen, 1979; Winterer, 1973). In areas where such deposits might be expected, such as the North Pacific rim and the equatorial zone, biogenic siliceous deposits of these ages are virtually unknown.

PALEOBIOLOGY OF EXTINCTION EPISODES

The paleobiologic characteristics of the pelagic biota before and after extinction episodes can be interpreted by reference to the Recent pelagic ecosystems. This method of inference is a common one in paleontology, and it provides a powerful, but complex, tool. It assumes that the basic physiologic and ecologic requirements of pelagic organisms have not changed significantly over time; indeed, if they had, the fossil record would surely have been much different, and no similarities could be seen. Certain biologic attributes of plankton, for example, temperature tolerances, food or nutrient requirements, availability of sunlight, and many others, make these paleontologic systems powerful bases for inference. Objections have been raised that there are no species in common between the fossils and the modern species, and hence the method cannot be employed, but this argument is defeatist and makes paleontologic interpretations in general futile. Paleontologists have had enormous success using exactly this methodology with organisms as diverse as dinosaurs, ammonites, trilobites, brachiopods, and many others chiefly because basic biologic functions have remained similar throughout time. It is equally useful in its interpretation of fossil plankton and other pelagic organisms. Furthermore, its power constrains hypotheses framed without regard to the fundamental biologic characteristics. Some of the paleobiologic interpretations can now be confirmed by independent geochemical techniques as well.

Each of the paleontologic characteristics associated with extinction episodes provide different constraints on interpretations of the extinction and radiation dynamics. These will be interpreted separately.

Species Diversity

The pelagic environment differs significantly from benthic and terrestrial environments in its much reduced number of ecological opportunities. As in most ecosystems, habitat diversity is an important factor controlling species diversity. Far fewer habitats are available in the water column, and hence fewer species are present in most groups (McGowan, 1971). Indeed, in foraminifera, modern planktonic species number 35 or so, while benthic species are estimated at over 4000 (Tappan, 1971).

Relatively low species diversity has commonly been associated with lower temperatures because of the pole-to-equator species diversity gradient observed in most marine organisms. Temperature itself, however, should have no limiting effect on species diversity, and it is unclear that the species diversity gradient is, in fact, related to decreasing temperatures (Dunbar, 1968). Certainly the high-diversity deep-sea fauna (Sanders and Hessler, 1969) occupying regions of low temperature confirms this view. Nevertheless, paleontologists dealing with fossil plankton continue to refer to low-diversity assemblages as representing low temperatures. If plankton species diversity is not related to temperature, then what is it related to?

In the modern seas, plankton species diversity is highest in the central and equatorial regions of the oceans where water column heterogeneity is greatest. In higher-latitude seas, plankton diversity decreases in most cases to a tenth or so of that of the lower latitudes probably because the water column is rather uniform.

Plankton species diversity is interpreted to be related to water column heterogeneity. Low-diversity plankton assemblages live in a more homogeneous water column, where the thermocline or halocline is reduced. High-diversity assemblages live in heterogenous water columns where a well-developed thermo- or halocline exists. Low-diversity assemblages in the fossil record, like those following extinction episodes, may then represent periods of lowered water column heterogeneity.

Morphology

Morphologic complexity in planktonic foraminifera is related to certain biologic functions that are depth correlated. Species of these foraminifera in modern seas occupy preferred depth habitats as adults (Berger, 1969; Bé, 1977). The chief depth habitat of these species varies, some preferring to live their entire lives in the uppermost parts of the water column, while others prefer to live at deeper depths (Hemleben and Spindler, 1983). Still others live at one depth and reproduce at another more narrow depth. This habitat segregation by depth is related to ontogeny (Hemleben and Spindler, 1983), with young individuals starting their lives in the surface waters and finally migrating to specific depth intervals as adults (Figure 4). Depth segregation seems related to one or more of several functional responses, such as to algal symbiosis, to changing food supplies, or to reproduction (Bé, 1977; Hemleben and Spindler, 1983).

Much of the morphologic variation observed in planktonic foraminifera is depth associated. Complex shell structures—secondary calcification, aberrant terminal chambers, and others—are present on adults that inhabit or migrate to deeper depths. For example, in spinose species undergoing reproduction, the spines are shed in shallower water, and as secondary calcification over the final whorl occurs, the foraminifera descend to the depth of reproduction (Hemleben and Spindler, 1983). This calcification is given the special term *gametogenic calcification* because of its association with the formation of gametes at depth. This reproductive process results in a vertical segregation of species at reproduction (Figure 4), a strong selective mechanism for the evolution of new species.

In modern seas, planktonic foraminifera with simple morphologies consisting of trochospiral tests with more or less spherical chambers occupy a variety of pelagic environments (Figure 3). They are most abundant and dominate in the higher-latitude seas where water column heterogeneity is reduced (Bradshaw, 1959). These planktonic foraminifera may be found throughout the water column (Berger, 1969) and they apparently do not segregate according to depth.

Modern planktonic foraminifera show a relationship between morphology and

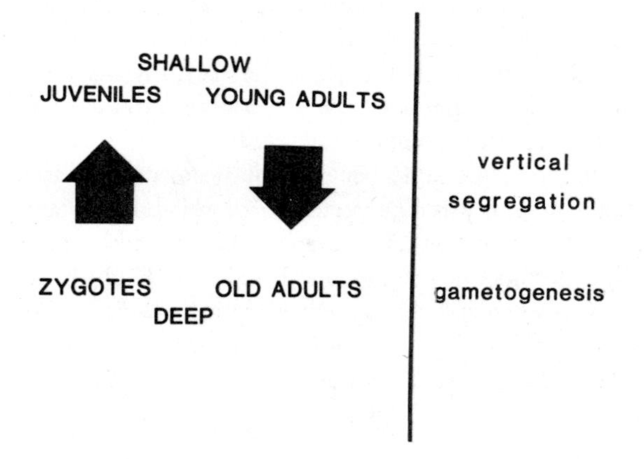

Figure 4. Generalized life history model for tropical and subtropical planktonic foraminifera. Vertical segregation of species in stratified seas takes place at times of reproduction. Young adults develop morphologic modifications and descend to various depths for development and release of gametes. Gametes fuse, presumably at depth, and the zygotes or young juvenile foraminifera ascend to the euphotic zone for feeding. Based on data from Bé (1977) and Hemleben and Spindler (1983).

water column complexity. In relatively homogeneous seas, morphologies are simple, whereas they are complex in heterogeneous seas (see Figure 3). The differences between these types occur because of the complexities of planktonic foraminiferal morphologic adaptations for habitat segregation according to depth (Bouvier-Soumagnac and Duplessy, 1985; Healy-Williams, Williams, and Ehrlich, 1985).

In the fossil record of planktonic foraminifera, complex morphologies suggest that the foraminifera were undergoing habitat segregation during ontogeny. Simple morphologies suggest that the foraminifera did not segregate by habitat. Of course, simple morphologic types do not necessarily indicate that vertical migration through the water column did not take place, only that the foraminifera did not occupy different habitats at any time during their ontogeny. Because pelagic habitats (i.e., temperature, salinity, density) are chiefly functions of depth, periods when simple morphologies dominated may have been times of decreased vertical heterogeneity, and times when complex morphologies were present may have had vertically heterogeneous oceans, at least in places. Certainly, the stable isotopic evidence supports this interpretation (Douglas and Savin, 1978).

Biogeography

Modern plankton are distributed in latitudinal belts corresponding in a general way to water mass distribution. Thus, there are nine biogeographic provinces in modern plankton throughout the world. In each hemisphere, there is a central, a temperate or transitional, a subpolar, and a polar province; these are repeated in mirror image in the opposite hemisphere. A single equatorial province is shared by both hemispheres. While these provinces may have different causes and may not correspond directly with the water masses recognized by physical oceanographers (Cifelli and Benier, 1976; McGowan, 1974), they are recognizable both in modern oceans and in the fossil record.

The pelagic provinces are related to oceanographic barriers. The provincial boundaries correspond mainly to oceanographic discontinuities of one kind or another. These discontinuities are commonly controlled by wind-driven currents, which are related to the earth's thermal gradient. Where currents are strong, the boundaries between provinces are sharp, and where the currents are weak or diffuse, the provincial boundaries are indistinct. Circulation intensity is then ultimately responsible for the biogeographic patterns in pelagic organisms (McGowan, 1971, 1974; Bradshaw, 1959).

Paleobiogeography can then be interpreted with reference to sea surface circulation. When provinces were numerous and distinct, circulation was well developed with strong currents and separate water masses. When provinces were fewer, circulation was correspondingly weaker.

The fossil record shows provincial distributions of pelagic microfossils prior to extinction episodes but widespread or cosmopolitan biotic distributions following these episodes. These biogeographic patterns indicate that extinction episodes were accompanied by a decrease in oceanic circulation intensity.

Large Carnivores

The large pelagic vertebrates, in particular cetaceans and marine reptiles but also larger fish, have specific trophic requirements. These animals have enormous food requirements that cannot be met in all parts of the oceans. Today cetaceans feed predominately in regions of upwelling where nutrients are brought within the photic zone, resulting in a short food chain with enormous quantities of phytoplankton and smaller herbivorous zooplankton and fish (Ryther, 1969). Active marine reptiles with large body size likewise would have required large quantities of food resources. The secondary producers, chiefly fish and larger zooplankton, are the main food of these large carnivores.

Upwelling sufficient for the production of large quantities of fish and zooplankton occurs in few places in the world's oceans, primarily at current divergences, along the equator, around Antarctica, and on the eastern sides of ocean basins. The intensity of upwelling, and hence the flow of nutrients into the photic zone, is largely a function of atmospheric and oceanographic circulation. When these are strong, upwelling is also strong and primary and secondary producers are abundant, as are the large carnivores that feed on the smaller organisms. As upwelling dies down seasonally, the carnivores migrate to other areas.

In the geologic past, the large carnivores underwent extinctions. A probable explanation for these extinctions is that the primary and secondary production on which large carnivores depend declined in abundance (Lipps and Mitchell, 1976). This trophic model for the extinction of the large carnivores indicates that nutrient supply became reduced within the photic zone, thus depressing primary production. As the chief regions of primary production are now, and probably always have been, related to the upwelling necessary to transport nutrients to the photic zone, Lipps and Mitchell (1976) inferred that upwelling intensity declined significantly in the world's oceans, resulting in the extinction of archeocete whales in the Oligocene. Similar reasoning can be applied to the extinction of marine reptiles near the end of the Cretaceous.

Siliceous Phytoplankton

Siliceous phytoplankton, chiefly diatoms but also silicoflagellates and ebriidians, have special requirements for the production of their skeletons. Silicon is depleted in surface waters because of its thorough uptake by diatoms in particular. Siliceous phytoplankton and the biogenic sediments that they produce are abundant only in regions of upwelling where the silica and other nutrients are carried to the photic zone. Thus, a decline in siliceous primary producers and the sediment they produce suggests a decline in upwelling to low levels.

EXTINCTION DYNAMICS

The evidence from and paleobiology of fossil pelagic organisms provide bases for reconstructing the dynamics of their extinction. Major extinctions in pelagic organisms occurred at three times in the last 150 m.y.—the Cenomanian–Turonian, the

Cretaceous–Tertiary, and the Eocene–Oligocene. Although each of these episodes of extinction varies in the time involved in the extinction and in the degree of extinction, certain paleobiologic characteristics are common to each of them. These common characteristics indicate similarity of ecologic conditions during and following the episodes. They may also provide constraints on mechanisms proposed to bring on such conditions.

The general paleobiologic patterns described above all indicate that the marine circulation became less intense during extinction episodes and for varying periods of time afterward. Species diversity declines indicate a reduction in habitat variability, while morphologic changes suggest that these habitat restrictions resulted from decreased heterogeneity with depth. The paleobiogeography of extinction events indicate similar conditions throughout the surficial waters of the world's oceans, and that circulation intensity was much reduced. Along with reduced circulation, upwelling declined in intensity as well, as evidenced from the extinction of large carnivores and a decrease in siliceous phytoplankton and their biogenic sedimentary deposits. In all paleobiologic respects, the oceans seem to have become less structured and more homogeneous both vertically and horizontally. The thermocline and/or halocline became diffuse, and surficial water masses degraded.

This general view of the world's oceans is corroborated by isotopic values measured on planktonic and benthonic foraminifera (Douglas and Savin, 1978; Boersma and Premoli Silva, 1983). In general, the oxygen isotopic evidence from surface-dwelling plankton and from benthonic foraminifera during extinction episodes indicates a reduced vertical thermal gradient (Douglas and Woodruff, 1981). The two lowest vertical thermal gradients recorded during the Tertiary are in the earliest Paleocene and at the end of the Eocene at the time of the two extinction episodes recorded for the Tertiary. For example, in the earliest Paleocene, after the K/T extinction, Boersma and Premoli Silva (1983) show thermal gradients varying from 10 to 3°C for the various parts of the Atlantic Ocean. These gradients widened in the later part of the Paleocene, just as species diversity and morphologic complexity increased. The gradients were controlled by sea surface temperatures, as the bottom water remained relatively constant. Similar controls on thermal gradients exist in modern seas as well. Likewise, near the end of the Eocene, the vertical thermal gradient declined for a short period of time, largely as a result of surface cooling.

Carbon isotopic values for the extinction episodes indicate reduced primary productivity. In the Paleocene, these values are lightest in the early part of the epoch just after the K/T extinction, indicating low levels of primary production (Williams et al., 1983), and then they increase in value (Boersma and Premoli Silva, 1983). This increasing carbon isotopic record indicates that productivity increased throughout this period as well, and this confirms the development of more vigorous circulation and upwelling in the world's oceans. In the late Eocene, the carbon isotopic values of planktonic and benthonic foraminifera began to decline and continued at lowered levels through the late Oligocene (Belanger and Matthews, 1984). If this decline also indicates a reduction in productivity, then these data confirm the paleobiologic interpretation (above) that the late Eocene and Oligocene were periods of low primary productivity.

Pelagic organisms thus underwent extinction for two primary reasons: habitat destruction and a reduction in nutrient and trophic resources. Smaller plankton became extinct principally as the vertical structure of the water column became homogeneous. This homogenization of the oceans through temperature restructuring eliminated the environmental variability required for the depth segregation of species, particularly during reproduction. As in many extinctions, species could not survive when habitats required by their life history strategies were destroyed. Larger pelagic organisms and phytoplankton with large nutrient requirements could not survive the decline in their nutrient and trophic supplies. When upwelling declined, nutrient supply and the enormous trophic structure built on that supply ceased or was much reduced. Organisms like the siliceous and other eutrophic phytoplankton and large carnivores then became extinct.

The extinction episodes in pelagic ecosystems are thus related to a changing oceanographic pattern throughout the world's oceans. These changes resulted in fewer depth-related habitats because of a reduction in the vertical heterogeneity, largely through temperature gradient changes, and in the degradation of provincial boundaries, again because of a reduction in latitudinal thermal gradients. As these thermal gradients were reduced, oceanographic circulation declined, resulting in decreased upwelling intensity and its associated primary productivity. The evolutionary history of pelagic ecosystems is closely linked with oceanographic opportunities enabling the development of alternative life history strategies. These strategies particularly relate to reproductive and trophic ecology.

These observations and interpretations eliminate extinction hypotheses that are based on nonselective mechanisms, such as poisoning of the oceans, or on mechanisms unrelated to oceanographic events. These events are linked to terrestrial ecosystems because of the close relationship between oceanographic and atmospheric circulation. Any hypothesis that fails to make this link is suspect.

The most discussed hypothesis today is that of an asteroid impact that set off a series of events resulting in the extinction of a large portion of the Earth's biota (Alvarez et al., 1980). Data from the pelagic environment indicates that the simplest explanation would be to make the ocean circulation more sluggish. An asteroid impact in the ocean should have just the opposite effect—the oceans would become more turbulent and more enriched in nutrients. Even after the period of total darkness that might ensue from such an impact, the oceans would remain eutrophic, and the surviving phytoplankton and zooplankton should fluorish. Just the opposite seems to have occurred. If an asteroid or comet impacted a continental area, a dust and aerosol cloud might be created that could cause an amelioration in atmospheric, and hence oceanographic, circulation. Modeling of such an event is, however, very complex and beyond the scope of this chapter. The paleobiologic events described herein and the oceanographic conditions that they demand, however, must be accounted for by any extinction hypothesis. To date, such an accounting has not been made.

REFERENCES

Alvarez, L. W., Alvarez, W., Asaro, F., and Michel, H. V., 1980, Extraterrestrial cause for the Cretaceous–Tertiary extinction, *Science*, **208**, 1095–1108.

Alvarez, W., Asaro, F., Michel, H. V., and Alvarez, L. W., 1982, Iridium anomaly approximately synchronous with terminal Eocene extinctions, *Science*, **216**, 886–888.

Alvarez, W., Alvarez, L. W., Asaro, F., and Michel, H. V., 1984a, The end of the Cretaceous: Sharp boundary or gradual transition?, *Science*, **223**, 1183–1186.

Alvarez, W., Kauffman, E. G., Surlyk, F., Alvarez, L. W., Asaro, F., and Michel, H. V., 1984b, Impact theory of mass extinctions and the invertebrate fossil record, *Science*, **223**, 1135–1141.

Asaro, F., Alvarez, L. W., Alvarez, W., and Michel, H. V., 1982, *Geochemical Anomalies Near the Eocene/Oligocene and Permian/Triassic Boundaries*, Geological Society of America Special Paper 190, pp. 517–528.

Barnes, L. G., and Mitchell, E., 1978, Cetacea, in Cook, H. B. S., and Maglio, V., eds., *African Fossil Mammals*, Cambridge, Harvard University Press, pp. 582–602.

Bé, A. W. H., 1977, An ecological, zoogeographic and taxonomic review of Recent planktonic foraminifera, in Ramsay, A. T. S., ed., *Oceanic Micropaleontology*, London, Academic Press, pp. 1–100.

Belanger, P. E., and Matthews, R. K., 1984, The foraminiferal isotopic record across the Eocene/Oligocene boundary at Deep-Sea Drilling Project Site 540, *Init. Rep. Deep-Sea Drilling Project*, **77**, 589–592.

Benson, R. H., Chapman, R. E., and Deck, L. T., 1984, Paleoceanographic events and deep-sea ostracodes, *Science*, **224**, 1334–1336.

Berger, W. H., 1969, Ecologic patterns of living planktonic foraminifera, *Deep-Sea Res.*, **16**, 1–24.

Berger, W. H., 1973, Cenozoic sedimentation in the eastern tropical Pacific, *Geol. Soc. Am. Bull.*, **84**, 1941–1954.

Boersma, A., and Premoli Silva, I., 1983, Paleocene planktonic foraminiferal biogeography and the paleooceanography of the Atlantic Ocean: *Micropaleontology*, **29**, 355–381.

Bouvier-Soumagnac, Y., and Duplessy, J.-C., 1985, Carbon and oxygen isotopic composition of planktonic foraminifera from laboratory culture, plankton tows and Recent sediment: Implications for the reconstruction of paleoclimatic conditions and of the global carbon cycle, *J. Foraminif. Res.*, **15**, 302–320.

Bradshaw, J. S., 1959, Ecology of living planktonic foraminifera in the north and equatorial Pacific, *Contrib. Cushman Found. Foraminif. Res.*, **10**, 25–64.

Bramlette, M. N., 1946, *The Monterey Formation of California and the Origin of its Siliceous Rocks*, U. S. Geological Survey Professional Paper 212, 57 pp.

Bramlette, M. N., 1965, Massive extinctions in biota at the end of Mesozoic time, *Science*, **148**, 1696–1699.

Bukry, D., 1981, *Synthesis of Silicoflagellate Stratigraphy for Maestrichtian to Quaternary Marine Sediment*, Society of Economic Paleontologists and Mineralogists Special Publication 32, pp. 433–444.

Cifelli, R., 1969, Radiation of Cenozoic planktonic foraminifera, *System. Zool.*, **18**, 154–168.

Cifelli, R., and Benier, C. S., 1976, Planktonic foraminifera from near the West African coast and a consideration of faunal parcelling in the North Atlantic, *J. Foraminif. Res.*, **6**, 258–273.

Corliss, B. H., Aubry, M.-P., Berggren, W. A., Fenner, J. M., Keigwin, L. D., Jr., and Keller, G., 1984, The Eocene/Oligocene boundary event in the deep sea, *Science*, **226**, 806–810.

Donovan, D. T., and Hancock, J. M., 1967, Mollusca: Cephalopoda (Coleoidea), in Harland, W. B., et al., eds., *The Fossil Record*, London, Geological Society of London, pp. 461–467.

Douglas, R. G., and Savin, S. M., 1978, Oxygen isotopic evidence for the depth stratification of Tertiary and Cretaceous planktic foraminifera, *Marine Micropaleontol.*, **3**, 175–196.

Douglas, R. G., and Woodruff, F., 1981, Deep-sea benthic foraminifera, in Emiliani, C., ed., *The Sea*, Vol. 7, *The oceanic lithosphere*, New York, Wiley-Interscience, pp. 1233–1327.

Dunbar, M. J., 1968, Ecological development in polar regions, Englewood Cliffs, N. J., Prentice-Hall, 119 pp.

Foreman, H. P., 1968, *Upper Maestrichtian Radiolaria of California, Palaeontological Association*, London, Special Papers in Palaeontology, **3**, 82 pp.

Ganapathy, R., 1982, Evidence for a major meteorite impact on Earth 34 million years ago: Implication for Eocene extinctions, *Science*, **216**, 885–886.

Hancock, J. M., 1967, Some Cretaceous–Tertiary marine faunal changes, in Harland, W. B., et al., eds., *The Fossil Record*, London, Geological Society of London, pp. 91–103.

Hanna, G. D., 1927, Cretaceous diatoms from California, *Occasional Papers of the Calif. Acad. Sci.*, **13**, 5–39.

Hanna, G. D., 1934, Additional notes on diatoms from the Cretaceous of California, *J. Paleontol.*, **8**, 352–355.

Hansen, J. M., 1979, Dinoflagellate zonation around the boundary, in Birkelund, T., and Bromley, R. G., eds., *Cretaceous–Tertiary boundary events symposium I*, Copenhagen, University of Copenhagen, pp. 136–151.

Haq, B. U., 1981, Paleogene paleoceanography: Early Cenozoic oceans revisited, *Oceanologica Acta*, 1981, 71–82.

Haq, B. U., Premoli-Silva, I., and Lohmann, G. P., 1977, Calcareous plankton paleobiogeographic evidence for major climatic fluctuations in the early Cenozoic Atlantic Ocean, *J. Geophys. Res.*, **82**, 3861–3876.

Healy-Williams, N., Williams, D. F., and Ehrlich, R., 1985, Morphometric and stable isotopic evidence for subpopulations of *Globorotalia truncatulinoides*, *J. Foraminif. Res.*, **15**, 242–253.

Hemleben, C., and Spindler, M., 1983, Recent advances in research on living planktonic foraminifera, *Utrecht Micropaleontol. Bull.*, **30**, 141–170.

Ingle, J. C., Jr., 1975, Summary of later Paleogene: Neogene insular stratigraphy, paleobathymetry, and correlations, Philippine Sea and Sea of Japan region, *Init. Rep. Deep-Sea Drilling Project*, **31**, 837–855.

Jablonski, D., and Bottjer, D. J., 1983, Soft-bottom epifaunal suspension-feeding assemblages in the Late Cretaceous: Implications for the evolution of benthic paleocommunities, in Tevesz, M. J. S., and McCall, P. L., eds., *Biotic Interactions in Recent and Fossil Benthic Communities*, New York, Plenum Press, pp. 747–812.

Keller, G., 1983, Biochronology and paleoclimatic implications of middle Eocene to Oligocene planktic foraminiferal faunas, *Marine Micropaleontol.*, **7**, 463–486.

Keller, G., D'Hondt, S., and Vallier, T. L., 1983, Multiple microtektite horizons in Upper Eocene marine sediments: No evidence for mass extinctions, *Science*, **221**, 150–152.

Leinen, M., 1979. Biogenic silica accumulation in the central equatorial Pacific and its implications for Cenozoic paleoceanography, *Geol. Soc. Am. Bull. Pt. II*, **90**, 1310–1376.

Lipps, J. H., 1970, Plankton evolution, *Evolution*, **24**, 1–22.

Lipps, J. H., and Hickman, C. S., 1982, Origin, age and evolution of Antarctic and deep-sea faunas, in Ernst, W. G., and Morin, J. G., eds., *The Environment of the Deep Sea*, Englewood Cliffs, N. J., Prentice-Hall, pp. 324–356.

Lipps, J. H., and Mitchell, E., 1976, Trophic model for the adaptive radiations and extinctions of pelagic marine mammals, *Paleobiology*, **2**, 147–155.

Long, J. A., Fuge, D. P., and Smith, J., 1946, Diatoms of the Moreno Shale, *J. Paleontol.*, **20**, 89–119.

McGowan, J. A., 1971, Oceanic biogeography of the Pacific, in Funnell, B. M., and Riedel, W. R., eds., *The Micropaleontology of the Oceans*, Cambridge, Cambridge University Press, pp. 3–74.

McGowan, J. A., 1974, The nature of oceanic ecosystems, in Miller, C. B., ed., *The Biology of the Oceanic Pacific*, Corvallis, Oregon State University Press, pp. 9–28.

McLaren, D. J., 1983, Bolides and biostratigraphy, *Geol. Soc. Am. Bull.*, **94**, 313–324.

Officer, C. B., and Drake, C. L., 1983, The Cretaceous–Tertiary transition, *Science*, **219**, 1383–1390.

Orr, W. N., 1972, Pacific northwest siliceous phytoplankton, *Palaeogeog., Palaeoclimatol., Palaeoecol.*, **12**, 95–114.

Perch-Nielsen, K., McKenzie, J., and He, Q., 1982, *Biostratigraphy and Isotope Stratigraphy and the "Catastrophic" Extinction of Calcareous Nannoplankton at the Cretaceous/Tertiary Boundary*, Geological Society of America Special Paper 190, pp. 353–371.

Pessagno, E. A., Jr., 1967, Upper Cretaceous planktonic foraminifera from the Western Gulf Coastal Plain, *Paleontographica Americana*, **5**(37), 245–445.

Playford, P. E., McLaren, D. J., Orth, C. J., Gilmore, J. S., and Goodfellow, W. D., 1984, Iridium anomaly in the Upper Devonian of the Canning Basin, western Australia, *Science*, **226**, 437–439.

Raup, D. M., and Sepkoski, J. J., Jr., 1982, Mass extinctions in the marine fossil record, *Science*, **215**, 1501–1503.

Raup, D. M., and Sepkoski, J. J., Jr., 1984, Periodicity of extinctions in the geologic past, *Proc. Natl. Acad. Sci., U.S.A.*, **81**, 801–805.

Romer, A. S., 1966, *Vertebrate Paleontology*, Chicago, University of Chicago Press, 468 pp.

Ryther, J. H., 1969, Photosynthesis and fish production in the sea, *Science*, **166**, 72–76.

Sanders, H., and Hessler, R. R., 1969, Ecology of the deep-sea benthos, *Science*, **163**, 1419–1424.

Saul, L. R., 1983, Turitella zonation across the Cretaceous–Tertiary boundary, California, *Univ. Calif. Publ. Geol. Sci.*, **125**, 1–165.

Sepkoski, J. J., Jr., 1982, *Mass Extinctions in the Phanerozoic Oceans: A Review*, Geological Society of America Special Paper 190, pp. 283–289.

Signor, P. W., III, and Lipps, J. H., 1982, *Sampling bias, Gradual Extinction Patterns and Catastrophes in the Fossil Record*, Geological Society of America Special Paper 190, pp. 291–296.

Silver, L. T., and Schultz, P. H., eds., 1982, *Geological Implications of Impacts of Large Asteroids and Comets on the Earth*, Geological Society of America Special Paper 190, 528 pp.

Sliter, W. V., 1972, Upper Cretaceous planktic foraminiferal zoogeography and ecology: Eastern Pacific margin, *Palaeogeog., Palaeoclimatol., Palaeoecol.*, **12**, 15–31.

Smit, J., 1982, *Extinction and Evolution of Planktonic Foraminifera After a Major Impact at the Cretaceous/Tertiary Boundary*, Geological Society of America Special Paper 190, pp. 329–352.

Strelnikova, N. I., 1975, Diatoms of the Cretaceous Period, *Nova Hedwigia*, **53**, 311–321.

Tappan, H., 1971, Foraminiferida, *McGraw-Hill Encyclopedia of Science and Technology*, 5, 467–475.

Thierstein, H. R., 1981, *Late Cretaceous Nannoplankton and the Change at the Cretaceous–Tertiary Boundary*, Society of Economic Paleontologists and Mineralogists Special Publication 32, pp. 355–394.

Thierstein, H. R., 1982, *Terminal Cretaceous Plankton Extinctions: A Critical Assessment*, Geological Society of America Special Paper 190, pp. 385–399.

Thunell, R., and Belyea, P., 1982, Neogene planktonic foraminiferal biogeography of the Atlantic Ocean, *Micropaleontology*, **28**, 381–398.

Ward, P. D., and Signor, P. W., III, 1983, Evolutionary tempo in Jurassic and Cretaceous ammonites, *Paleobiology*, **9**, 183–198.

Wiedmann, J., 1973, Evolution or revolution of ammonoids at Cretaceous System boundaries, *Biol. Rev.*, **48**, 159–194.

Williams, D. R., Healy-Williams, N., Thunell, R. C., and Leventer, A., 1983, Detailed stable isotope and carbonate records from the upper Maestrichtian-Lower Paleocene section of hole 516F (Leg 72) including the Cretaceous/Tertiary boundary, *Init. Rep. Deep-Sea Drilling Project*, **72**, 921–929.

Winterer, E. L., 1973, Sedimentary facies and plate tectonics of the equatorial Pacific, *Am. Assoc. Petrol. Geol. Bull.*, **57**, 265–282.

PART III

EXTINCTIONS IN THE GEOLOGICAL RECORD: QUATERNARY

6

REFUTING LATE PLEISTOCENE EXTINCTION MODELS

PAUL S. MARTIN

Department of Geosciences
University of Arizona
Tucson, Arizona

Around 11,000 yr B.P. (years before present), a mysterious catastrophe destroyed two-thirds of America's large mammals. Had they survived, mammoths, mastodonts, scimitar cats, ground sloths, glyptodonts, toxodonts, and similar big beasts would now be featured attractions in zoos and game parks. The ice age losses severely depleted the kinds of big game available to human hunters and the potential supply of emblems for the conservation movement. Any contemporary Greenpeace campaign to "save the vanishing mylodon," complete with bumper stickers of winsome ground sloths, would be 10,000 years too late. The losses had another consequence: they distorted our perception of modern ecosystems. Investigators of plant–animal coevolution who seek explanations for fruit or seed dispersal in extinction-ravaged ecosystems may, in the absence of mastodonts, giant ground sloths, and other members of the large herbivore guild, turn to free-ranging horses, cattle, and other livestock for their proxies (Janzen and Martin, 1982; Janzen, 1984).

LATE PLEISTOCENE EXTINCTION PATTERNS

The loss of the mammoths and other megamammals rivals dinosaur extinction in popular appeal, and one might expect paleontologists to study both events seeking traces of some common patterns or even a common cause. This is rarely done, perhaps because those who have tried (e.g., Newell, 1963, 1967) seem to have found little profit in it. The door is open for ad hoc explanations for both events.

Table 1. Late Pleistocene Extinct and Living Genera of Terrestrial Megafauna (>44 kg adult body weight)[a]

	Extinct	Living	Total	Percent Extinct
Africa	7	42	49	19.3
North America	33	12	45	73.7
South America	45	12	57	78.9
Australia	19	3	22	86.4

[a]For list of scientific names, see Martin (1984a).

Table 2. Late Pleistocene Extinct Genera of Large Mammals[a]

	North America	South America
EDENTATA		
Dasypodidae	*Holmesina*	*Pampatherium* *Propraopus*
Glyptodontidae	*Glyptotherium*	*Chlamydotherium* *Doedicurus* *Glyptodon* *Hoplophorus* *Lomaphorus* *Neothoracophorus* *Panochthus* *Plaxhaplous* *Sclerocalyptus*
Megalonychidae	*Megalonyx*	*Ocnopus* *Nothropus* *Nothrotherium* *Valgipes*
Megatheriidae	*Eremotherium* *Nothrotheriops*	*Eremotherium* *Megatherium*
Mylodontidae	*Glossotherium*	*Glossotherium* *Lestodon* *Mylodon* *Scelidodon* *Scelidotherium*
LITOPTERNA		
Macraucheniidae		*Macrauchenia* *Windhausenia*
NOTOUNGULATA		
Toxodontidae		*Mixotoxodon* *Toxodon*

Table 2. (*Continued*)

	North America	South America
RODENTIA		
Castoridae	*Castoroides*	
Hydrochoeridae	*Hydrochoerus*[b]	*Neochoerus*
	Neochoerus	
CARNIVORA		
Felidae	*Smilodon*	*Smilodon*
	Homotherium	
	Acionyx[b]	
Ursidae	*Tremarctos*[b]	*Arctodus*
	Arctodus	
ARTIODACTYLA		
Tayassuidae	*Mylohyus*	*Platygonus*
	Platygonus	*Catagonus*
Camelidae	*Camelops*	*Eulamaops*
	Hemiauchenia	*Hemiauchenia*
	Palaeolama	*Palaeolama*
Cervidae	*Navahoceros*	*Agalmaceros*
	Sangamona	*Charitoceros*
	Cervalces	*Morenelaphus*
		Paraceros
Antilocapridae	*Tetrameryx*	
Bovidae	*Saiga*[b]	
	Euceratherium	
	Symbos	
	Bootherium	
	Bos[b]	
PERISSODACTYLA		
Equidae	*Equus*[b]	*Equus*[b]
		Hippidion
		Onohippidium
Tapiridae	*Tapirus*[b]	
PROBOSCIDEA		
Elephantidae	*Mammuthus*	
Mammutidae	*Mammut*	
Gomphotheriidae	*Cuvieronius*	*Cuvieronius*
		Haplomastodon
		Notiomastodon
		Stegomastodon
Total	33	45

[a]From Martin (1984[b]) and Marshall et al. (1984).
[b]Genus surviving on another continent.

The late Pleistocene revolution removed over 100 genera of large mammals from the continents (Table 1), including 33 from North America and even more, 45, from South America (Table 2). Radiocarbon dates chronicle extinction of 44 species of North American large mammals during the last 20,000 years (Kurtén and Anderson, 1980, pp. 364–365). The mode for last appearance dates, selected according to the protocol of Mead and Meltzer (1984), falls around 11,000 yr B.P., and there is great interest in degree of scatter.

Late Pleistocene extinction involves a high percentage of large mammals on the smaller continents (America, Australia). The loss of small vertebrates was pronounced only on oceanic islands (Diamond, 1984; Martin, 1967, 1984b). Because most phyla did not suffer more than background extinctions, the late Pleistocene will not qualify as a mass extinction event. It was, nevertheless, catastrophic for the large herbivore and large carnivore guilds of American and Australian land vertebrates.

The late Pleistocene extinctions are out of phase with the 26–28-m.y. cycle of mass extinction in marine families outlined in Raup and Sepkoski (1982). The last significant marine extinctions involved tropical benthic invertebrates blighted by cold water incursions (Stanley, 1984). This took place regionally in the Pliocene, too early to apply in this case. Furthermore, the late Pleistocene mammalian losses happened not at a time of initial cooling, but rather during the last of a long sequence of glacial terminations, each of which was marked by melting ice, a rising sea, and warming global temperatures.

Neither the asteroid impact model of extinction at the end of the Cretaceous (Alvarez et al, 1980) nor its rival, the volcanic model (Officer and Drake, 1985), will help explain the late Pleistocene losses of the megamammals. Both require an iridium anomaly that is unknown in the late Pleistocene. Butler and Hoyle (1979) proposed a reflective blanket of cometary particles and a sudden change in global albedo as a cause of mammoth extinction. Since the late Pleistocene extinctions are time transgressive (Martin, 1967, 1984b), they are discordant with any model of globally contemporaneous extinction. If freezing rain extinguished subarctic mammoths, one is left with the problem of accounting for the crisis among equatorial ground sloths followed by the loss thousands of years later of giant lemurs in Madagascar and of moas in New Zealand (Figure 1). Even mammoth extinction itself was not synchronous; from south to north, Eurasian mammoths disappeared over thousands of years (Martin, 1984b; Vereshchagin and Baryshnikov, 1984).

My claim that certain guilds of mammals of large size were uniquely affected was rejected by Graham and Lundelius (1984) on the ground that of 78 species of late Pleistocene mammals that became extinct in North America, perhaps as many as one-third were small. However, their view is not supported by the evidence at hand, and the matter invites a closer look.

I have adopted 100 lb (44 kg) as a practical boundary between "large" and "small" mammals and, with the aid of Margaret Hardy, have tabulated with respect to extinction all terrestrial species and genera listed in Kurtén and Anderson's checklist of North American vertebrates of the last 3 m.y. The result shows that

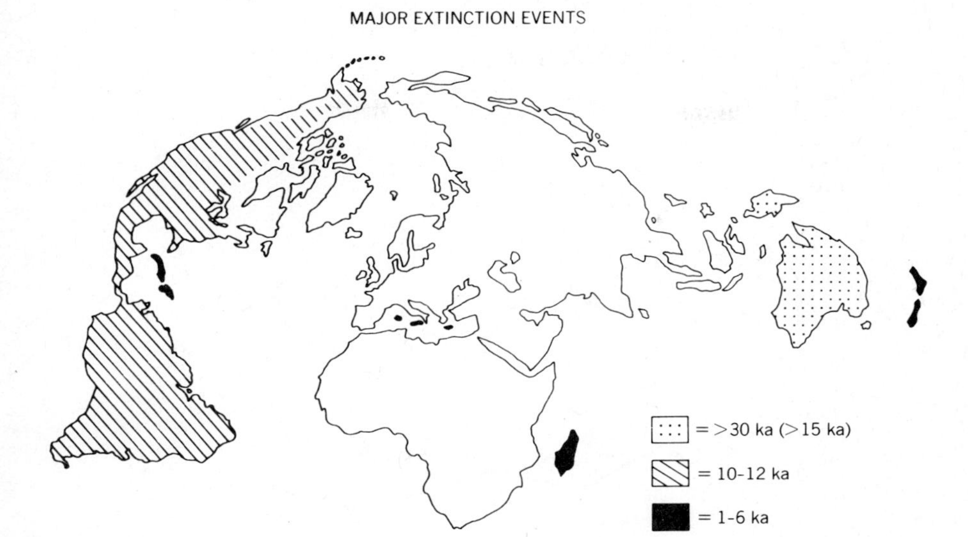

Figure 1. Timing of late Pleistocene extinctions in different parts of the globe. Australian extinctions preceded those in the Americas, which were followed by extinctions on oceanic islands. The pattern precludes any astronomical or other model requiring synchroneity (after Martin, 1984b).

2–3 m.y. ago (late Pliocene and early Pleistocene) small mammals suffered a severe loss of species relative to standing diversity (Figure 2). The late Pliestocene reveals no apparent impoverishment in numbers of *species or genera* of continental small mammals, and extinctions do not obviously exceed the background level (Figure 2). In the case of large mammals, there is a minor peak in extinctions around 2 m.y. ago, with a major peak only at the end of the Pleistocene (Figure 3). During late Wisconsin time, more species (and genera) were lost from North America than in the previous 3 m.y. combined (Figure 3). An intercontinental comparison of genera shows that South America and Australia resemble North America in that all three lost over 70% of their late Pleistocene megafauna, while Africa lost only 20% (Table 1) over a much longer time interval (Martin, 1984b). With occasional exceptions, such as the extinct skunk (*Brachypotoma*), the diminutive antilocaprid (*Capromeryx*), or the Australian muskrat kangaroo (*Propleopus*), the extinct continental genera were all large. Even the exceptions are not truly small (Norway rat-size or less).

Unlike the continents, many oceanic islands lost many small mammals, especially endemic rodents and flightless birds. As their fossil faunas become better known, extinctions of terrestrial fauna have even been extended to the land snails in the case of Hawaii (Kirch, 1983) and the Solomon Islands (Christensen and Kirch, 1981). Destruction of oceanic island biotas seems to have been more severe in the Holocene than it was historically (see chapters by Cassels, Diamond, and Olson and James in Martin and Klein, 1984).

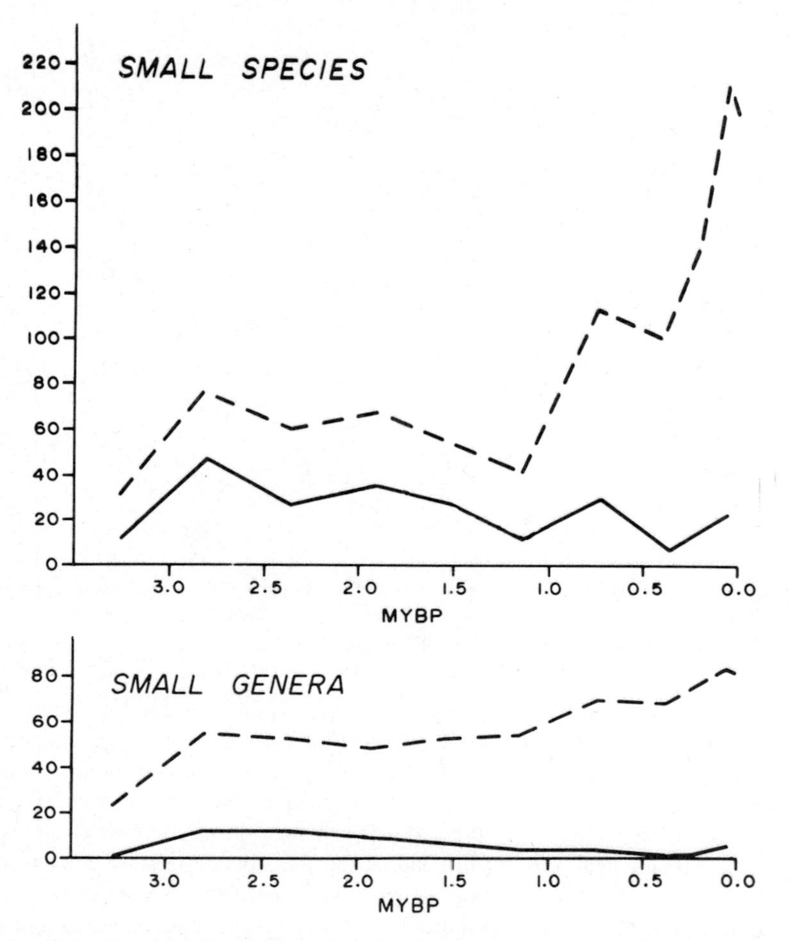

Figure 2. Extinction (solid line) and standing diversity or total taxa (broken line) curves for small mammals of the Plio-Pleistocene of North America; data points derived from Kurtén and Anderson (1980) (after Martin, 1984a).

REFUTATIONS OF OVERKILL

Over 45 yr ago (1937), Edwin Colbert, than a young paleontologist in the American Museum of Natural History, wrote: "Curiously enough, the extinction of the North American Pleistocene mammals would seem to coincide remarkably well with the establishment of man on this continent. Is it not possible therefore, that man was an instrumental force in causing the extinction of the horse, the camel, the mammoth, the ground sloth, and the Pleistocene bison? Is it not possible that the mammals of North America, not being in an ecological balance with man as were the Eurasiatic mammals, were to some extent unable to adapt themselves to new conditions

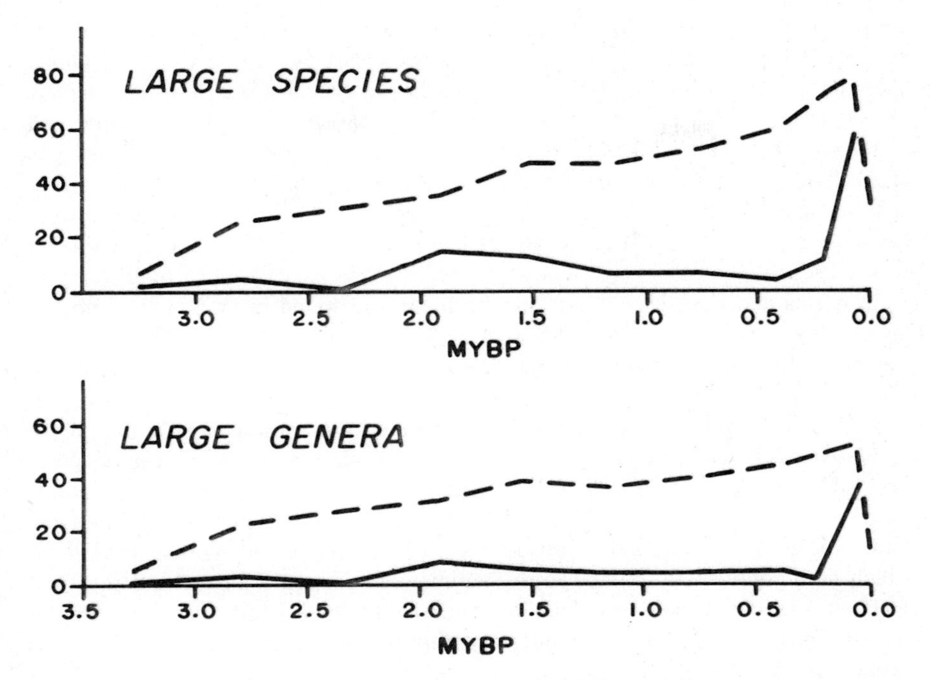

Figure 3. Extinction (solid line) and standing diversity or total taxa (broken line) curves for large mammals at the Plio-Pleistocene of North America; data points derived from Kurtén and Anderson (1980) (after Martin, 1984a).

imposed by the arrival of man on this continent, with the result that certain forms died out?" The analysis sounds fresh even now.

Colbert subsequently dismissed human effects in favor of the view that mammalian extinctions in the late Pleistocene were the result of a more complicated series of events. Such synthetic models, which have the gratifying property of being acceptable to most theorists, are unfortunately less vulnerable to testing and to demarcation. Given the scant possibility of proving anything in paleontology, it is useful to countermand the tradition and to probe simplified analytical positions, not abandoning them prematurely. The fact that Butler and Hoyle's (1979) astronomical model of mammoth extinction can be refuted (at least in my hands) does not detract from its potential value when applied to earlier extinctions or to its possible revision for the late Pleistocene. "Extremism in the defense of a simple model is no vice," to paraphrase Senator Barry Goldwater from Arizona. The vice lies in failing to be equally ardent in seeking falsifications. With this caveat, I review the large-mammal extinction chronology with an eye toward test refutations of extinction models.

On the continents, the differential loss of large mammals may be linked to prehistoric cultural differences. Continents of human origin (Africa, Asia) lost relatively few genera, and those relatively gradually (Martin, 1984b). For example, the giant deer, *Megaloceros*, disappeared from Asia before its ultimate extinction in western Europe, including Ireland (Vereshchagin and Baryshnikov, 1984). The last known proboscideans from the Levant predate the Mousterian, roughly 100,000 yr B.P. (Tchernov, 1984). Woolly mammoth were lost by 18,000 yr B.P. in China (Liu and Li, 1984), by 14,000 yr B.P. or earlier in Britain (Stuart, 1982), and by 13,000 yr B.P. in Sweden (Berglund et al., 1976). Mammoths finally disappeared 12,000–10,000 yr B.P. in Siberia (Vereshchagin and Baryshnikov, 1984). The ranges of the living elephants of Africa and Asia have been shrinking steadily since prehistoric time. Curiously, the last surviving species of extinct proboscideans were dwarfs occupying the Aegean Islands as late as the middle Holocene (Bachmayer et al., 1976).

In Africa and Eurasia, extinct faunal remains are found in Paleolithic archaeological sites over tens of thousands of years or more (Klein, 1980, 1984). In America, a much larger number of late Pleistocene megamammals were lost (Table 1). Whether or not all losses were truly simultaneous (Agenbroad, 1984a; Semken, 1984), native American proboscideans disappeared much more rapidly than those of Eurasia, with unmistakable archaeological associations evident over no more than 1000 yr. In the American Southwest, fossil Shasta ground sloth populations found at different elevations from caves 1500 km apart disappeared at about the same time as the mammoth (Thompson et al., 1980). Unlike Old World excavations, which commonly yield extinct large mammals associated with cultural remains, such associations in the Americas are remarkably few. Their ages are virtually coeval (Haynes, 1984). Thus, in intensity, timing, and cultural manifestations, the extinction pattern of Africa and Eurasia is distinctly different from that of America (Table 3).

American deposits rich in extinct late Pleistocene megafauna rarely yield evidence of early man. For example, Rancho la Brea, California, "is an ideal location for evidence of hunting and butchering practices from 15,000 to 40,000 years ago" (Meighan, 1983, p. 456). While Rancho la Brea has yielded some two million catalogued items, Meighan notes that early man seems to have been absent prior to 9000 yr B.P. The Hot Springs Mammoth Site in western South Dakota (Figure 4) has yielded a minimum of 30 individual Columbian mammoths from a time interval of slightly over 20,000 yr B.P., a time when Ukranian deposits of mammoth bone are typically found in an archaeological context (Klein, 1973). Boney Springs and

Table 3. Late Pleistocene Mammalian Extinction Patterns[a]

	Continents of Human Origin	Continents of Human Invasion	Endemics on Prehistorically Inhabited Oceanic Islands
Loss of large species	Moderate	Many	Total
Loss of small species	Few	Few	Many
Rate of disappearance	Gradual	Sudden	Mostly sudden
Extinction order	Oldest	Intermediate	Youngest
Associated archaeology	Common	Rare	Typically rare

[a]Africa and Asia are continents of human origin; America and Australia are continents of human invasion.

Figure 4. Excavation at the Mammoth Site, Hot Springs, South Dakota (see text). No evidence of human presence was found in a deposit of at least 30 individual mammoths (Agenbroad, 1984b). (Photograph by L. Agenbroad.)

other adjacent spring bog deposits in Missouri are rich in fossil mastodons, none of them culturally associated, contrary to the claims of Albert Koch, their nineteenth-century discoverer (Saunders, 1977). For the United States, Lundelius et al. (1983) list 178 late Pleistocene terrestrial local faunas, not all including megafauna; associated human remains are reported in only 25 cases.

Not until the Holocene, the last 10,000 yr, do we encounter numerous cultural deposits in America. Here the situation changes. When large mammal bones are

Figure 5. The Hudson-Meng Site, Crawford, Nebraska (see text). Bison butchering is inferred by the lack of skulls with horn cores, as well as the following lithics: Alberta projectile points, butchering tools, and scrapers (Agenbroad, 1978). No mammoth or other extinct genera were found. (Photograph by P. S. Martin.)

found, they are almost exclusively living genera such as bison, wapiti, or deer. For example, at the Hudson-Meng site in western Nebraska (Figure 5), approximately 600 bison were killed and butchered roughly 10,000 yr B.P. (Agenbroad, 1978); no extinct genus of large mammals was found. The Agate Basin site in Wyoming represents a possible exception where apparent Folsom-age finds of *Platygonus* (Walker, 1982) either represent a remnant surviving population or redeposition from older units. The same may be said for camel bones (*Camelops*) at the Casper Site, Wyoming (Frison, 1974). Despite their abundance in the fossil record right up to the time of Clovis hunters, no well-defined archaeological site indicates processing of camelids for food by big game hunters of any sort.

These observations can be marshalled to refute one version of overkill. Since there is no lengthy fossil record in the Americas revealing intense cultural activity associated with the extinct large mammals, any gradual depletion and ultimate extermination of large mammals at the hands of a population of hunters slowly increasing in numbers over thousands of years, as Budyko (1974) modeled for Eurasia, is highly unlikely. The major difference between the chronology and magnitude of Old World extinction (slow, moderate) compared with the New World (sudden, massive) is a basic feature of the fossil record.

When human predation is modeled for maximum prey destruction in minimum time, a sudden predatory burst, or "blitzkrieg," results (Mosimann and Martin, 1975). Chronologically, the blitzkrieg model requires North American megamammal extinction above the background level to occur within at most a millennium. The event is set to begin with the arrival of Clovis big game hunters at 11,500 yr B.P. Any earlier cryptic occupation by a small number of people not focused on hunting the megafauna, if such occurred, need not be fatal to the model of devastating destruction uniquely at the hands of the Clovis hunters. The value of the blitzkrieg model lies in the opportunity it affords for refutation by radiocarbon dating. When remains of the extinct megafauna are available for measurement, all huntable species are predicted to have disappeared at about the same time. The time of disappearance must coincide with that of the first well-defined big game hunters in the area. The chronological demand is so precise that few believe that it will hold.

Ideal material for radiocarbon analysis can be obtained directly from dung, keratin, anatomical soft parts, and bone collagen, all of which may be preserved in arid and arctic regions. In the western United States, dates on the Shasta ground sloth (*Nothrotheriops shastensis*) were run on 33 samples of dung balls from eight dry caves between 300 and 1800 m in elevation along a 1500-km transect between southern Nevada and west Texas. By sampling dung balls at or very near the surface of the deposits, the results establish the time of ground sloth extinction (summarized in Martin, 1984a; Martin et al., unpublished). Four dates rejected on methodological grounds are discussed in Long and Martin (1974). The mode falls between 11,000 and 11,500 yr B.P. (histogram in Figure 6), and the youngest dates fall at or within two sigma of 10,800 yr B.P. The age of the first well-defined big game hunters in the region (Clovis culture) is indicated by an arrow at the bottom of Figure 6 and a shaded panel centered on 11,000 yr B.P..

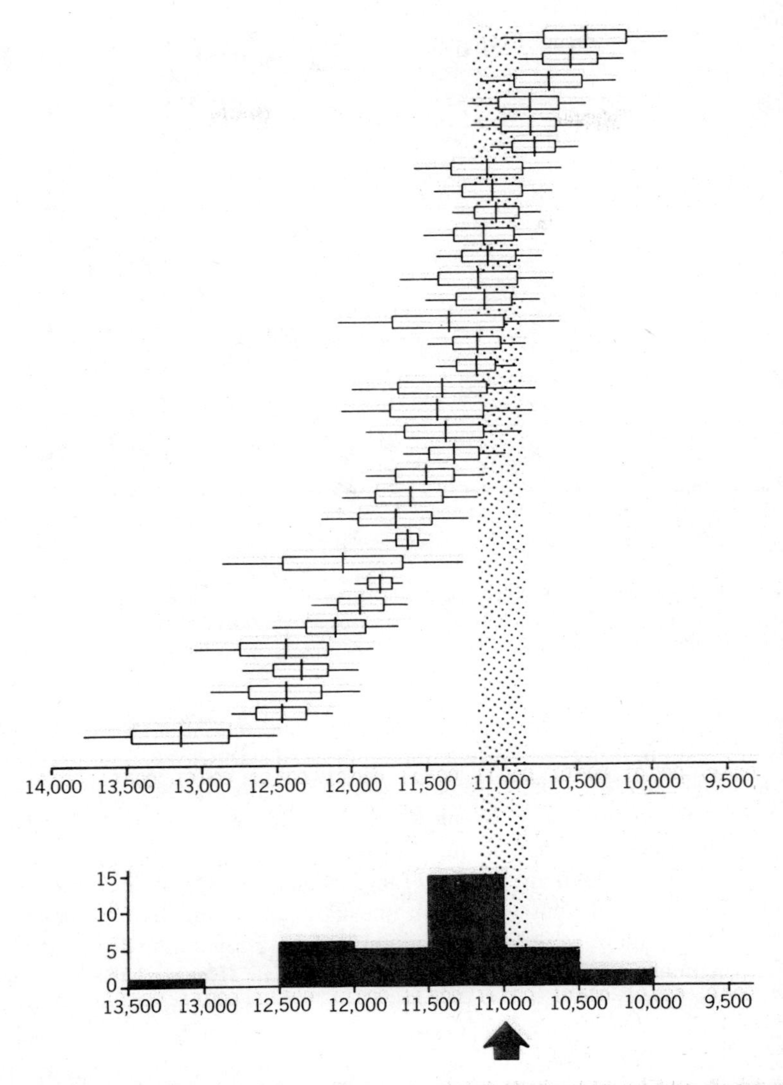

Figure 6. Youngest radiocarbon dates on Shasta ground sloth (*Nothrotheriops*) dung. Each horizontal line represents $\pm 2^{\sigma}$. Each block represents $\pm^{\sigma}$; the cross bar is the mean. The time of Clovis hunters is indicated by grid and arrow centered at 11,000 yr B.P..

Another independent chronology is available on amino acid residues extracted from the oil-impregnated, ideally preserved bone of Rancho la Brea (Marcus and Berger, 1984). The Rancho la Brea samples were collected at all levels (including the surface) in the upper 10 m of the deposit. The results (Figure 7) indicate survival of the extinct machairodonts (*Smilodon*), bison (*Bison*), and horse (*Equus*) into, but not beyond, the millennia coinciding with Clovis hunters in the region (arrow

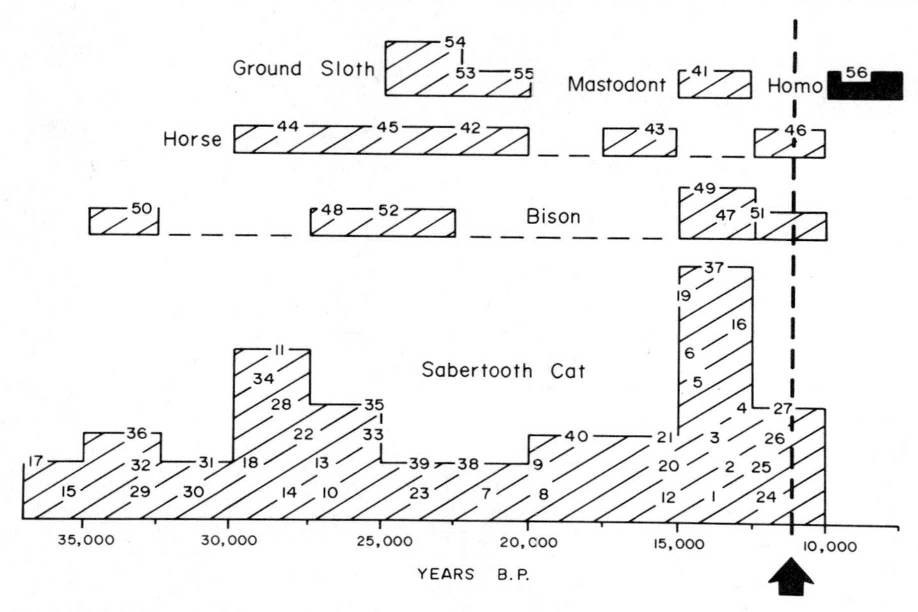

Figure 7. Radiocarbon dates on amino acid residues from bones of Rancho la Brea megafauna (from Marcus and Berger, 1984). Numbers refer to samples tabulated in Martin (1984a); the arrow marks the time of Clovis hunters. Reprinted from Nitecki, M., ed., *Extinction*, Chicago, University of Chicago Press, 1984.

in Figure 7). Three dates on ground sloth bones and one on a mastodont predate Clovis occupation; however, these animals survived into Clovis time elsewhere.

In a third study, perishable remains (dung and horn sheaths) of the extinct Harrington's mountain goat, *Oreamnos harringtoni*, found in caves in the Grand Canyon, have been subjected to the dating process. Again, the youngest dates terminate at around Clovis time, 10,800 B.P. (Mead, 1983; Robbins et al., 1984).

Cowboy Cave, Utah, has yielded an unusual example involving stratified perishable remains (Figure 8). A layer of mummified dung containing dry fecal remains of bison, mammoth, and other extinct large mammals (Hansen, 1980; Spaulding and Petersen, 1980) underlies 1.5 m of dry sediments containing perishable remains of a prehistoric gathering people (Jennings, 1980). A shallow eolian sand deposit overlies the dung layer (Figure 8). No contemporaneous cultural material was found in the dung layer, and no extinct megafaunal remains were recovered with the overlying archaeological material. The oldest date on the archaeological unit was 8700 yr B.P.; the youngest on the dung layer was 11,100 yr B.P. Cowboy Cave is a particularly good example of the separation between extinct fauna and cultural deposits in stratified cave deposits. While dung or bones of extinct genera may directly underlie or intrude into cultural horizons, implying use of the caves by the extinct mammals at or before human invasion, there is no evidence of actual coexistence. At the same time, the lack of evidence of human activity in Cowboy Cave prior to 8700 yr B.P. cannot be taken to mean that humans were missing until after megafaunal extinctions; their presence is established regionally in the Clovis

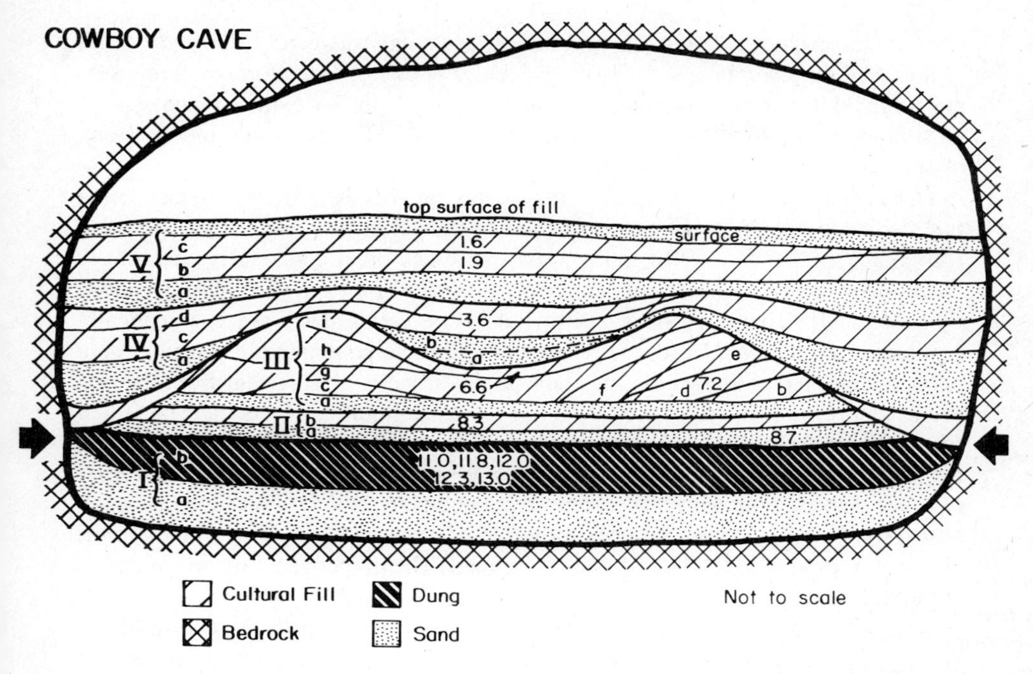

Figure 8. Diagrammatic sketch of Cowboy Cave, Utah (after Jennings, 1980). The dung unit (Ib) yielded excrement, bones, and hair of extinct large mammals without evidence of cultural material, which was richly represented in units II–V above. Radiocarbon dates are rounded off to the nearest hundred and expressed in thousands of years (Ka); extinction (arrow) may be inferred when the deposition of the dung blanket ended.

sites (Haynes, 1984). The unusual occurrence of a dung layer at Cowboy Cave has been followed by the discovery of another, even richer mammoth dung layer in Bechan Cave, southern Utah (Davis et al., 1984).

The stratigraphy of extinction at the Lubbock Lake site, Texas, yields a final case in point. Bones of large animals are preserved in alluvial and lacustrine sediments deposited at intervals throughout the late Quaternary (Stafford, 1981; Holliday et al., 1983). Before 10,800 yr B.P., the bones found in Yellow House Draw and its tributaries included mammoth, horse, camel, and bison. Afterward, they were exclusively of living species, particularly bison (Stafford, 1981), now associated with artifacts. The chronological relationship between large mammals and cultural activity illustrated at the Lubbock Lake site, Cowboy Cave, Bechan Cave, and Rancho la Brea is so commonplace in the archaeology of North America that it rarely elicits comment. Nevertheless, its implications are profound and should not be underestimated. In each site, there is a switch from the deposition of the extinct megafauna with scant evidence of artifacts to abundant evidence of prehistoric people and their food remains, exclusive of the extinct species.

In the examples cited above, the extinction of Shasta ground sloths, sabertooth cats, and Harrington's mountain goats converge on the time of Clovis hunters associated with mammoth in the same region. Given the small number of stratified

Clovis sites (Haynes, 1984), their very narrow age range between 11,500 and 10,500 yr B.P., and the scarcity and equivocal character of any older cultural material, one must conclude that any significant human contact with the extinct fauna was at most brief. From this evidence, the only plausible overkill extinction model is one of sudden and sweeping blitzkrieg.

Occasional Holocene age radiocarbon dates on members of the extinct fauna fuel doubts about the theory of extinction by blitzkrieg (see Semken, 1984, and Kurtén and Anderson, 1980). In my view, these and other claims of younger ages do not meet critical standards of sample acceptability, for example, those advanced by Mead and Meltzer (1984). Until or unless the young dates are systematically replicated, the putative Holocene records of extinct large mammals are suggestive only. A detached and objective position on this issue is no different from the one needed concerning claims of pre-12,000-yr human antiquity in America. The search for the latest records of extinct megafauna generates less passion than the search for the earliest human beings in the New World. Nevertheless, in both cases, patience, critical judgment, and careful replications by independent field teams are necessary. While no claim is necessarily free of observer bias, no one field report, however improbable, can be rejected out of hand. The proper approach to validation is through replications.

The blitzkrieg model will be disproved if judicious radiocarbon dating of the last occurrences of the more common extinct genera shows that their disappearance is significantly younger than the age of the Clovis culture. To be plausible, blitzkrieg must occur rapidly, within 10 yr locally and a few hundred years regionally. One would not expect to find many Clovis kill sites in such a short interval (Mosimann and Martin, 1975; Grayson, 1984b). Only terminal ages on extinct animals significantly younger than 11,000 yr B.P. or extinct fauna being hunted extensively thousands of years before or after 11,000 yr B.P. would falsify the blitzkrieg hypothesis.

REFUTATIONS OF CLIMATE

A model attributing extinctions to climatic change is the primary alternative to blitzkrieg; climatic change is highly effective in explaining many aspects of the fossil record, especially glacial-age events. Models of climatic change are favored by those scientists prone to doubt the involvement of humans or other explanations. Paleontologists appear to be as reluctant to attribute late Pleistocene mammal extinctions to human intervention as they are to attribute the extinction of dinosaurs to an asteroid.

None doubt that Late Pleistocene climates were changeable and changing; late Pleistocene faunas may reflect these changes in major range displacements and in the so-called "disharmony" or anomalous mixture of certain fossil faunas (Graham and Lundelius, 1984). Species that no longer coexist are believed on stratigraphic grounds to have once lived together. Examples are boreal forest shrews and voles with grassland mice and ground squirrels in the late Pleistocene of the Ozark

Mountains and northern shrews with southern cottonrats in the late Pleistocene of the High Plains (Lundelius et al., 1983, pp. 341, 345). Biogeographers devoting themselves to studying mammal range changes within the Pleistocene and across the Pleistocene–Holocene boundary view the ice age and its climatic consequences as the controlling variable. It is certainly logical to conclude that the glacial retreat at the end of the last ice age, accompanied by a rise in sea level, the dessication of pluvial lakes, and many changes in the ranges of plants and animals, *including the extinction of large mammals*, was a collective response to climatic change. The timing of the arrival of people themselves in both America and Australia was strongly influenced, if not completely determined, by climate. The main difficulty with the climatic model has been philosophical. Being popular, it was rarely subjected to tests. Such may no longer be the case (Grayson, 1984a); I will review two plant fossil records from the same region in which extinctions were critically measured (above) to see if major climate-induced vegetation changes may have driven the extinction.

In the last two decades, a vast enrichment of the late Pleistocene record by fossils recovered from ancient packrat middens has reshaped the biogeography of western North America. Features revealed by the fossil midden record (Spaulding et al., 1983; Thompson, 1984; Wells, 1983; Van Devender, in press) will help disclose what happened to plant communities before, during, and after the critical centuries around 11,000 yr B.P. when the megafauna lost so many species.

Deposits of plant macrofossils and some animal remains, including bones, dung, and even lizard epidermal scales (Cole and Van Devender, 1976) accumulated by packrats (*Neotoma*) and indurated by their urine, are found in caves, rock shelters, crevices, and fissures in western mountains. The fossil middens incorporate vegetation growing on or adjacent to bedrock and are not restricted to sedimentary rocks. Plants growing naturally in undisturbed upland habitats are presumed to be primarily under climatic control. Packrats have food preferences; however, broad selectivity is indicated by the impressive variety of forage plants found in any single fossil or modern midden. Modern middens closely reflect the modern vegetation. Even grasses, poorly known from most Pleistocene fossil floras, are preserved for the midden analyst (e.g., Rhodes Canyon, New Mexico, Van Devender and Toolin, 1983).

In the 20 yr since the method was developed, a total of 920 middens of all ages have been dated, the majority of these by the University of Arizona radiocarbon laboratory; over 640 middens were older than 8000 yr B.P. (Webb, 1985). With the recent development of tandem accelerator mass spectrometry (TAMS) technology, geochemists are now able to provide age estimates from samples as small as a few milligrams, and vexing questions about apparent contemporaneity of plant species not expected to associate can be investigated independent of stratigraphic assumptions.

From the beginning of midden analysis (Wells and Jorgensen, 1964), it was evident that the shifts in vegetation over the last 20,000 yr did not occur strictly at or before 11,000 yr B.P., the time of the well-dated regional megafaunal extinctions. Until about 8500 yr B.P., juniper woodland persisted in its present range both in the Mojave Desert (Wells and Berger, 1967) and in parts of the Sonoran Desert (Van Devender, 1977). Middens of the western Grand Canyon yielded an instructive example of the relationship between the last major shift in vegetation and the extinction of one member of the North American megafauna. Along with other extralocal species, single-leaf ash (*Fraxinus anomala*) and juniper (*Juniperus*) continued to grow outside Rampart Cave several thousand years after deposition of ground sloth dung in the cave had ceased and the sloths had disappeared (Phillips, 1984). Junipers and single-leaf ash were not eaten by the browsing Shasta ground sloths (Hansen, 1978) and in themselves these trees would contribute little to the forage available to native large mammals. Nevertheless, they are local indicators of cooler environment and of the productivity typical of the Pleistocene; their ultimate withdrawal in the early Holocene occurred 2000 yr after the extinction of the Shasta ground sloths and Harrington's mountain goats at a time when no megafaunal extinctions can be demonstrated.

A series of fossil middens collected with care from within a single environmental type may yield a lengthy chronosequence of the woody plants gathered by late Pleistocene packrats. Time gaps inevitably occur in the sampling process. Nevertheless, with ample radiocarbon control, a suite of midden samples may disclose the sort of biotic change revealed in a continuous core of lake sediments yielding fossil pollen.

Such a record for the 1300-m contour in the Hueco Mountains, west Texas, is now available (Van Devender, in press). From 33 individual middens, Van Devender recorded parts of six time intervals: middle Wisconsin, full glacial, late glacial, and early, middle, and late Holocene. Various types of woodland, with mountain mahogany (*Cercocarpus*), algerita (*Berberis trifoliolata*), sumac (*Rhus*), pinyon (*Pinus*), juniper (*Juniperus*), four-wing saltbush (*Atriplex canescens*), and oak (*Quercus*), prevailed prior to 8400 yr B.P. (Figure 9). Four middle Wisconsin samples contained sagebrush (*Artemisia*) and winterfat (*Ceratoides*), genera that disappeared by the full glacial. Three samples of the full glacial, along with the middle Wisconsin middens, contain rock spirea (*Petrophytum*) and sandpaper bush (*Mortonia*). These disappeared in the late glacial, while pinyon declined.

Four late glacial samples from 12,800 to 11,360 yr B.P. represent the environment approaching or during the time of megafaunal extinction. Sagebrush and snow berry (*Symphoricarpos*) are last encountered in this interval in small amounts; banana yucca (*Yucca baccata*) returned from a full glacial absence.

The early Holocene (five samples, 10,750–9520 yr B.P.) is readily distinguished from the middle Holocene (seven samples, 8420–6360 yr B.P.) by the abundance of spiny greasewood (*Forsellesia*), sumac, winterfat, scrub oak (*Quercus pungens*), yucca, and Mormon tea (*Ehpedra*). The late Holocene featured the arrival or abundant representation of desert shrubs as lechuguilla (*Agave lechuguilla*), ocotillo (*Fouquieria splendens*), and creosote bush (*Larrea*), and the site now would be mapped as Chihuahuan Desert.

Plant community changes reflected in the Hueco Mountains midden record occurred around 11,000 yr B.P. and are certainly noteworthy (Figure 9); at the same time, they are preceded and followed by other important changes and do not by themselves

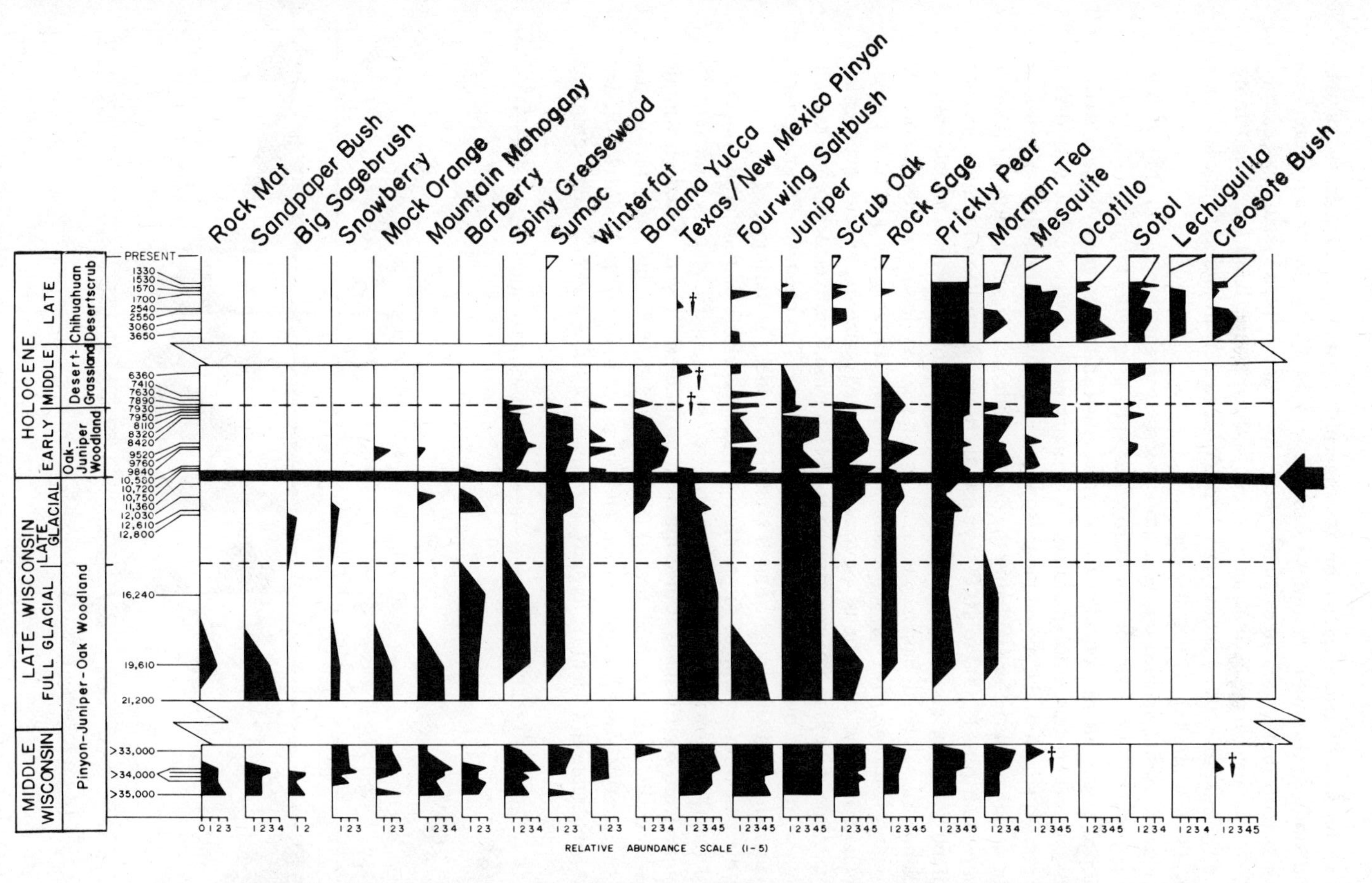

Figure 9. Packrat midden chronology from the Hueco Mountains, west Texas (after Van Devender, in press). The arrival, departure, and relative abundance of various species is shown. Five episodes of change in fossil assemblages can be recognized (see text), only one of which closely coincides with the time of local extinction of large mammals (arrow). Daggers indicate suspected contaminants.

seem to be a sufficient explanation for the contemporaneous extinctions of the ground sloths, mammoths, and machairodonts. If the Hueco Mountain chronosequence mirrors climatic changes, and if the changes determined megafaunal extinctions, the pattern suggests that large animals should have been lost much more gradually, and these losses should have been concentrated around five time intervals during the last 30,000 yr, not just one.

Conceivably the species changes among woody plants around 11,000 yr B.P. were particularly unfavorable for herbivores. For Beringia, Guthrie (1984) has developed a brilliant theoretical analysis proposing a shift from grassy steppe tundra to muskeg and ericads, implying a severe primary productivity decline with disastrous consequences to the megafauna. However, the Hueco Mountains sequence is rich in woody plants before and after the critical millenia. An inspection of Figure 9 indicates that from 12,000 to 10,000 yr B.P. there was a shift from pinyon–juniper woodland with rosaceous shrubs, sagebrush, and barberry (*Berberis*) to oak-juniper woodland with salt bush, banana yucca, and mesquite. While supporting the concept of climatic warming, the oak woodland assemblage still implies adequate forage production after the extinctions. Several of the newly arriving species (winterfat, salt bush, mesquite) are favored browse species for modern livestock or are known to have been eaten by ground sloths (Hansen, 1978). The changes do not portend a productivity crisis that might match Guthrie's model for Beringia.

In the preceding example, the relative abundance of midden material was estimated visually following disaggregation, wet screening, and drying. A more refined technique from the Eleana 2 Range middens in southern Nevada based on weight of macrofossils is shown in Figure 10 (Spaulding, 1983). In this case, 11 samples from one stratified midden from a rock shelter ranged in age from 17,100 to 10,400 yr B.P. Interstation variability is thus eliminated; all the packrats were sampling

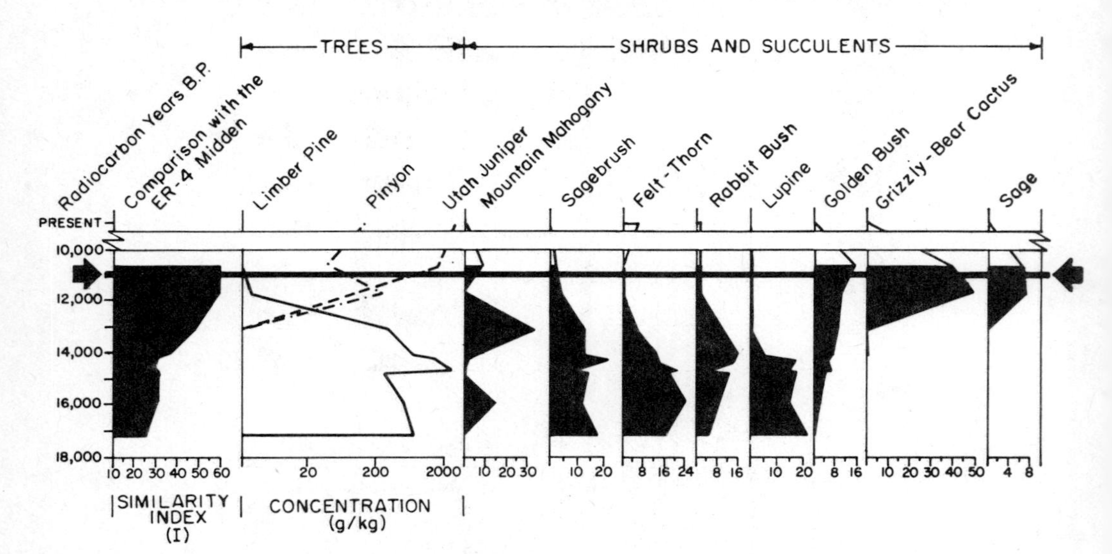

Figure 10. Packrat midden chronology from the Eleana 2 Range, Nevada (after Spaulding, 1983). The arrow indicates the time of local extinction of large mammals.

vegetation at the mouth of the same cave. A Pleistocene equid hoof fragment was found in Unit I near the top of the midden and was TAMS dated at 11,170 + 460 yr B.P. (AA-364). This is close to the time of last-known occurrence of native equids and other extinct herbivores in the region. From 17,000 to 11,700 yr B.P., the Eleana 2 Range chronosequence yields indices of similarity (ratio of plant species presently found in a modern midden at the site to species in the fossil midden) that rise from 40 to 80% as the late glacial modernization of the flora took place. The concentration of limber pine (*Pinus flexilis*) macrofossils (apparently the dominant species of the full glacial) dropped from 1000 to less than 10 g/kg, while the representation of pinyon and juniper rose correspondingly (Figure 10). The vegetation in the vicinity of the Eleana 2 Range site is currently pinyon–juniper woodland.

In Searles Lake, an adjacent basin, lake levels fell around 14,000 yr B.P., rose at 11,000, and fell again by 10,000 yr B.P. Evidently, a change in species similarity, a switch in abundance from limber pine to pinyon–juniper, and the initial drop in the level of Searles Lake all occurred several thousands of years *before* regional megafaunal extinction of native equids, ground sloths, Harrington's mountain goat, and machairodonts, as discussed earlier. Major vegetation changes and fluctuations in playa lake levels were taking place before any faunal extinction became apparent (Spaulding et al., in press).

Taking 11,000 yr B.P. as the time for all southwestern extinctions, it can be shown that regional changes in late Pleistocene vegetation were underway before, during, and after the critical millenium. The change in vegetation recognized at or near 11,000 yr B.P. is not an unusual event, certainly not one that might explain an apparent mass extinction of the large herbivore guild in this region at this time.

Plant changes shown from midden analyses will not disprove all conceivable models of extinction by climatic change, and one detailed record (Cole, 1985) supports them. While the midden record yields samples of natural vegetation before and after the 11,000 yr B.P. time of extinction, as accepted here, it lacks the sensitivity needed to disclose just what the climate was during those critical years of loss. Conceivably, there could have been a few fatally cold years throughout North America just as Butler and Hoyle (1979) proposed to explain extinction of the mammoths. Nevertheless, to match the catastrophic regional extinction chronology for ground sloths, Harrington's mountain goats, sabertooth cats, and mammoths, some highly unusual vegetation changes of the same age might be expected. Such are not apparent in the midden records of the Hueco Mountains of west Texas (Figure 9) and the Eleana 2 Range of southern Nevada (Figure 10).

Fossil packrat midden analysis appears to be an especially effective method of determining vegetation change in arid regions. Further applications of the new technique to the problem of megafaunal extinction can be expected.

SUMMARY

Ecologists and paleontologists only peripherally interested in Neogene extinctions will note that the late Pleistocene offers unusual analytical advantages. In radiocarbon years, the timing of extinction is measurable in millennia or centuries. The character of the local environment and the taphonomy of the faunas soon to become extinct can be studied in unusual detail. Even the dung deposits of certain large herbivores are available just prior to extinction. For most paleontologists the loss of late Pleistocene megafauna has not been and does not promise to become a key to explaining mass extinctions earlier in the Phanerozoic. The reason is simple; late Pleistocene extinctions were not driven by any of the natural causes operating earlier.

Recent developments in radiocarbon dating and paleoecological analysis can help refute extinction models proposed for the late Pleistocene. Over 70% of the large North American mammals were lost. Proportional to their representation in the total continental fauna, small mammals were only slightly affected. In western North America, extinction chronologies based on radiocarbon dates on amino acid residues of machairodont bone, dung, and other soft tissue reveal the widespread occurrence of now extinct populations of large animals before 11,500 yr B.P. with few, if any, survivors after 10,500 yr B.P. A gradual destruction of machairodonts, ground sloths, and Harrington's mountain goats over tens of thousands of years that would match the gradual decline of large extinct mammals in Eurasia is not supported. There was no gradual overkill in the New World. An inferred model of sudden overkill (blitzkrieg) requires all major extinction of huntable large animals to be coincidental with Clovis time. Such a scenario should be easy to refute by radiocarbon dating. Refutations of climatic models of extinction are less satisfactory. I attempt one using two fossil packrat midden sequences in arid America. The middens show important environmental changes occurring between 12,000 and 10,000 yr B.P., coinciding with the time of the extinctions. However, it is not certain that any of the environmental changes seen in the fossil plant records would have been sufficiently lethal by themselves to trigger the extinction of any animal species, or even to reduce local carrying capacity. The middens show that changes of comparable magnitude occurred both earlier and later. If these were accompanied by large animal extinctions, such losses have yet to be adequately demonstrated.

Late Pleistocene extinction was insignificant in the oceans; severe losses of endemic small mammals and birds occurred only on oceanic islands. The global sequence of megafaunal extinction was time transgressive with different continents being affected at different times. A much greater intensity of megafaunal extinctions occurred on continents of human invasion than on those of human origin. When quality radiocarbon dates are available, waves of extinction of continental large mammals and insular small animals are seen to coincide with human colonizations (Figure 1). All these features point toward prehistoric human activity as a controlling variable in the late Pleistocene extinction process, one that is potentially refutable on a case-by-case basis as a more refined radiocarbon chronology is attained.

ACKNOWLEDGMENTS

Julio Betancourt, Mark Boyce, Owen K. Davis, Karl Flessa, Russ Graham, David Jablonski, Holmes Semken, Geof Spaulding, and Thomas R. Van Devender reviewed the manuscript; Charles Sternberg aided with the illustrations; Larry Agenbroad provided Figure 4 and Thomas R. Van Devender and W. Geoffrey Spaulding supplied the midden diagrams used in Figures 9 and 10. Deborah Gaines and Betty Fink helped type and rescue the manuscript editorially. This research was supported in part by NSF Grant BSR 82-14939.

REFERENCES

Agenbroad, L. D., 1978, *The Hudson-Meng Site: An Alberta Bison Kill in the Nebraska High Plains*, University Press of America, 230 pp.

Agenbroad, L. D., 1984a. New World mammoth distribution, in Martin, P. S., and Klein, R. G., eds., *Quaternary Extinctions: A Prehistoric Revolution*, Tucson, University of Arizona Press, pp. 90–108.

Agenbroad, L. D., 1984b, Hot Springs, South Dakota: Entrapment and taphonomy of Columbian mammoth, in Martin, P. S., and Klein, R. G., eds., *Quaternary Extinctions: A Prehistoric Revolution*, Tucson, University of Arizona Press, pp. 113–127.

Alvarez, L. W., Alvarez, W., Asaro, F., and Michel, H. V., 1980, Extraterrestrial cause for the Cretaceous–Tertiary extinction, *Science*, **208**, 1095–1108.

Bachmayer, F., Symeonidis, N., Seeman, R., and Zapfe, H., 1976, Die Ausgrabungen in der Zwergele fantenhole "Charkadio" auf den insel Tilos, *Ann. Naturhistor. Mus. Wien*, **80**, 113–144.

Berglund, B., Hakansson, S., and Lagerlund, E., 1976, Radiocarbon-dated mammoth (*Mammuthus primigenius*) Blumenbach finds in South Sweden, *Boreas*, **5**, 177–191.

Budyko, M. I., 1974, *Climate and Life*, New York, Academic Press.

Butler, E. J., and Hoyle, F., 1979, On the effects of a sudden change in the albedo of the earth, *Astrophys. Space Sci.*, **60**, 505–511.

Christensen, C. C., and Kirch, P. V., 1981, Non-marine molluscs from archaeological sites on Tikopia, southeastern Solomon Islands, *Pacific Sci.*, **35**, 75–88.

Colbert, E. H., 1937, The Pleistocene mammals of North America and their relations to Eurasian forms, in MacCurdy, G. G., ed., *Early Man*, Philadelphia, New York, and London, J. B. Lippincott Co., pp. 173–184.

Cole, C. J., and Van Devender, T. R., 1976, Surface structure of fossil and Recent epidermal scales from North American lizards of the genus *Sceloporus* (Reptilia, Iguanidae), *Bull. Am. Mus. Nat. Hist.* **150**, 455–513.

Cole, K., 1985, Past rates of change, species richness, and a model of vegetational inertia in the Grand Canyon, Arizona, *Am. Natural.*, **125**, 289–303.

Davis, O. K., Agenbroad, L., Martin, P. S., and Mead, J. I., 1984, The Pleistocene dung blanket of Bechan Cave, Utah, in Genoways, H., and Dawson, M., eds., Pittsburgh, Pa., John Guilday Memorial Volume, Carnegie Museum of Natural History Special Publication 8, pp. 267–282.

Diamond, J., 1984, Historic extinctions: A Rosetta Stone for understanding prehistoric extinctions, in Martin, P. S., and Klein, R. G., eds., *Quaternary Extinctions: A Prehistoric Revolution*, Tucson, University of Arizona Press, pp. 824–861.

Frison, G. C., ed., 1974, *The Casper Site: A Hell Gap Bison Kill on the High Plains*, New York, Academic Press.

Graham, R. W., and Lundelius, E., 1984, Coevolutionary disequilibrium and Pleistocene extinctions, in Martin, P. S., and Klein, R. G., eds., *Quaternary Extinctions: A Prehistoric Revolution*, Tucson, University of Arizona Press, pp. 227–249.

Grayson, D. K., 1984a, Explaining Pleistocene extinctions: Thoughts on structure of a debate, in Martin, P. S., and Klein, R. G., eds., *Quaternary Extinctions: A Prehistoric Revolution*, Tucson, University of Arizona Press, pp. 807–823.

Grayson, D. K., 1984b, Archaeological associations with extinct Pleistocene mammals in North America, *J. Archaeol. Sci.*, **11**, 213–221.

Guthrie, D., 1984, Mosaics, allelochemics, and nutrients: On ecological theory of late Pleistocene megafaunal extinctions, in Martin, P. S., and Klein, R. G., eds., *Quaternary Extinctions: A Prehistoric Revolution*, Tucson, University of Arizona Press, pp. 259–298.

Hansen, R. M., 1978, Shasta ground sloth food habits, Rampart Cave, Arizona, *Paleobiology*, **4**, 302–319.

Hansen, R. M., 1980. Late Pleistocene plant fragments in the dungs of herbivores at Cowboy Cave, in Jennings, J. D., ed., *Cowboy Cave*, Anthropological Papers No. 104, Salt Lake City, University of Utah Press, pp. 179–189.

Haynes, C. V., 1984. Stratigraphy and late Pleistocene extinction in the United States, in Martin, P. S., and Klein, R. G., eds., *Quaternary Extinctions: A Prehistoric Revolution*, Tucson, University of Arizona Press, pp. 345–353.

Holliday, V. T., Johnson, E., Haas, H., and Stuckenrath, R., 1983, Radiocarbon ages from the Lubbock Lake site, 1950–1980: Framework for cultural and ecological change on the southern high plains, *Plains Anthropol.*, **28**, 165–182.

Janzen, D. H., 1984, Dispersal of small seeds by big herbivores: Foliage is the fruit, *Am. Natural.* **123**, 338–353.

Janzen, D. H., and Martin, P. S., 1982, Neotropical anachronisms: The fruits the Gomphotheres ate, *Science*, **215**, 19–27.

Jennings, J. D., 1980, *Cowboy Cave*, Anthropological Papers No. 104, Salt Lake City, University of Utah Press, 224 pp.

Kirch, P. V., 1983, Man's role in modifying tropical and subtropical Polynesian ecosystems, *Archaeol. Oceania*, **18**, 26–31.

Klein, R. G., 1974, *Ice Age Hunters of the Ukraine*, Chicago, Ill., University of Chicago Press, 140 pp.

Klein, R. G., 1980, Late Pleistocene hunters, in Sherrall, A., ed., *The Cambridge Encyclopedia of Archaeology*, Cambridge, England, Cambridge University Press, pp. 87–95.

Klein, R. G., 1984, Mammalian extinctions and Stone Age people in Africa, in Martin, P. S., and Klein, R. G., eds., *Quaternary Extinctions: A Prehistoric Revolution*, Tucson, University of Arizona Press, pp. 553–573.

Kurtén, B., and Anderson, E., 1980, *Pleistocene Mammals of North America*, New York, Columbia University Press, 442 pp.

Liu, T., and Li, X., 1984, Mammoths in China, in Martin, P. S., and Klein, R. G., eds., *Quaternary Extinctions: A Prehistoric Revolution*, Tucson, University of Arizona Press, pp. 517–527.

Long, A., and Martin, P. S., 1974, Death of American ground sloths, *Science*, **186**, 638–640.

Lundelius, E. L., Jr., Graham, R. W., Anderson, E., Guilday, J., Holman, J. A., Steadman, D. W., and Webb, S. D., 1983, in Wright, H. E., Jr., ed., *Late-Quaternary Environments of the United States*, Vol. 1, *The Late Pleistocene* (S. C. Porter, ed.), pp. 311–353.

Marcus, L. F., Berger, R., 1984, The significance of radiocarbon dates for Rancho la Brea, in Martin, P. S., and Klein, R. G., eds., *Quaternary Extinctions: A Prehistoric Revolution*, Tucson, University of Arizona Press, pp. 159–183.

Marshall, L. G., Berta, A., Hoffstetter, R., Pascual, R., Reig, O. A., Bombin, M., and Mones, A., 1984, Mammals and stratigraphy: Geochronology of the continental mammal-bearing Quaternary of South America, *Palaeovertebrata, Mem. Extr.*, pp. 1–80.

Martin, P. S., 1967, Prehistoric overkill, in Martin, P. S., and Wright, H. E., Jr., eds., *Pleistocene Extinctions*, New Haven, Yale University Press, pp. 75–120.

Martin, P. S., 1984a, Catastrophic extinctions and late Pleistocene blitzkrieg: Two radiocarbon tests, in Nitecki, M. H., ed., *Extinctions*, Chicago, Ill., University of Chicago Press.

Martin, P. S., 1984b, Prehistoric overkill: The global model, in Martin, P. S., and Klein, R. G., eds., *Quaternary Extinctions: A Prehisotirc Revolution*, Tucson, University of Arizona Press, pp. 354–403.

Martin, P. S., and Klein, R. G., eds., 1984, *Quaternary Extinctions: A Prehistoric Revolution*, Tucson, University of Arizona Press, 892 pp.

Martin, P. S., Thompson, R. S., and Long, A., unpublished, Shasta ground sloth extinction: A test of the blitzkrieg model, Extinction Symposium, Center for the Study of Early Man, University of Maine, Orono.

Mead, J. I., 1983, Harrington's extinct mountain goat (*Oreamnos harringtoni*) and its environment in the Grand Canyon, Arizona, Ph.D. dissertation, University of Arizona, Tucson, 215 pp.

Mead, J. I., and Meltzer, D. J., 1984, North American late Quaternary extinctions and the radiocarbon record, in Martin, P. S., and Klein, R. G., eds., *Quaternary Extinctions: A Prehistoric Revolution*, Tucson, University of Arizona Press, pp. 440–450.

Meighan, C. W., 1983, Early man in the New World, in Masters, P. M., and Flemming, N. C., eds., *Quaternary Coastlines and Marine Archaeology*, New York, Academic Press.

Mosimann, J. E., and Martin, P. S., 1975, Simulating overkill by Paleoindians, *Am. Sci.* **63**, 304–313.

Newell, N. D., 1963, Crises in the History of Life, *Sci. Am.* **218,** 2–16.

Newell, N. D., 1967, *Revolutions in the History of Life*, Geological Society of America Special Paper 89, pp. 63–91.

Officer, C. B., and Drake, C. L., 1985, Terminal Cretaceous environmental events, *Science*, **277**, 1161–1167.

Phillips, A. M., 1984, Shasta ground sloth extinction: Fossil packrat midden evidence from the western Grand Canyon, in Martin, P. S., and Klein, R. G., eds., *Quaternary Extinctions: A Prehistoric Revolution*, Tucson, University of Arizona Press, pp. 148–158.

Raup, D. M., and Sepkoski, J. J., Jr., 1982, Mass extinctions in the marine fossil record, *Science*, **215**, 1501–1503.

Robbins, E. I., Martin, P. S., and Long, A., 1984, Paleoecology of Stanton's Cave, In Euler, R. C., ed., *The Archaeology, Geology and Paleobiology of Stanton's Cave, Grand Canyon National Park, Arizona*, Grand Canyon, Arizona, Grand Canyon Natural History Association.

Saunders, J. J., 1977, *Late Pleistocene Vertebrates of the Western Ozark Highlands, Missouri*, Illinois State Museum Reports of Investigators 33, 118 pp.

Semken, H. A., Jr., 1984, Holocene mammalian biogeography and climatic change in the eastern and central United States, in Wright, H. E., Jr., ed., *Late-Quaternary Environments of the United States*, Vol. 2, *The Holocene*, pp. 182–207.

Spaulding, W. G., 1983, *Vegetation and Climates of the Last 45,000 Years in the Vicinity of the Nevada Test site, South-Central Nevada*, U.S. Geological Survey Open-file Report 83-535, 205 pp.

Spaulding, W. G., Leopold, E. B., and Van Devender, T. R., 1983, Late Wisconsin paleoecology of the American Southwest, in Wright, H. E., Jr., ed., *Late-Quaternary Environments of the United States*, Vol. 1, *The Late Pleistocene* (S. C. Porter, ed.), pp. 259–293.

Spaulding, W. G., and Petersen, K. L., 1980, The late Pleistocene-early Holocene paleoecology of Cowboy Cave, in Jennings, J. D., ed., *Cowboy Cave*, University of Utah Anthropological Paper 104, pp. 163–177.

Spaulding, W. G., Robinson, S. W., and Paillet, F. L., in press, *A Preliminary Assessment of Late Glacial Vegetation and Climate Change in the Southern Great Basin*, U.S. Geological Survey Water Resources Research Report.

Stafford, T., 1981, Alluvial geology and archaeological potential of the Texas southern high plains, *Am. Antiqu.*, **46**, 548–565.

Stanley, S. M., 1984, Marine mass extinctions: A dominant role for temperatures, in Nitecki, M. H., ed., *Extinctions*, Chicago, Ill., University of Chicago Press, pp. 69–117.

Stuart, A. J., 1982, *Pleistocene Vertebrates in the British Isles*, London, Longmans, 212 pp.

Tchernov, E., 1984, Faunal turnover and extinction rate in the Levant, in Martin, P. S., and Klein, R. G., eds., *Quaternary Extinctions: A Prehistoric Revolution*, Tucson, University of Arizona Press, pp. 528–552.

Thompson, R. S., 1984, Late Pleistocene and Holocene Environments in the Great Basin, Ph.D. dissertation, University of Arizona, Tucson, 225 pp.

Thompson, R. S., Van Devender, T. R., Martin, P. S., Foppe, T., and Long, A., 1980, Shasta ground sloth (*Nothrotheriops shastense* Hoffstetter) at Shelter Cave, New Mexico: Environment, diet and extinction, *Quat. Res.*, **14**, 360–376.

Van Devender, T. R., 1977, Holocene woodlands in the southwestern deserts, *Science*, 198, 189–192.

Van Devender, T. R., in press, Pleistocene climates and endemism in the Chihuahuan Desert flora, in Barlow, J. C., Timmerman, B. N., and Powell, A. M., eds., *Second Annual Chihuahuan Desert Symposium*, Alpine, Texas, 1983.

Van Devender, T. R., and Toolin, L. J., 1983, Late Quaternary vegetation of the San Andres Mountains, Sierra County, New Mexico, in Eidenbach, P. J., ed., *The Prehistory of Rhodes Canyon: Survey and Mitigation*, Tularosa, New Mexico, Human Systems Research, pp. 33–54.

Vereshchagin, N. K., and Baryshnikov, G. F., 1984, Quaternary mammalian extinctions in northern Eurasia, in Martin, P. S., and Klein, R. G., eds., *Quaternary Extinctions: A Prehistoric Revolution*, Tucson, University of Arizona Press, pp. 483–516.

Walker, D. N., 1982, Early Holocene vertebrate fauna, in Frison, G. C., and Stanford, D. J., eds., *The Agate Basin Site: A Record of the Paleoindian Occupation of the Northwestern High Plains*, New York, Academic Press, pp. 274–308.

Webb, R. H., 1985, Spatial and temporal distribution of radiocarbon ages on fossil middens from the Southwestern United States, *Radiocarbon*, **25**, 1–8.

Wells, P. V., 1983, Paleobiogeography of montane islands in the Great Basin since the last glaciopluvial, *Ecol. Monogr.*, **53**, 341–382.

Wells, P. V., and Berger, R., 1967, Late Pleistocene history of coniferous woodland in Mohave Desert, *Science*, 155, 1640–1647.

Wells, P. V., and Jorgensen, C. D., 1964, Pleistocene woodrat middens and climatic change in Mohave Desert: A record of juniper woodlands, *Science*, **143**, 1171–1174.

7

PLANT–ANIMAL INTERACTIONS AND PLEISTOCENE EXTINCTIONS

RUSSELL W. GRAHAM
Quaternary Studies Center
Illinois State Museum
Springfield, Illinois

Crossing the high grass savanna woodland, we scarcely saw any game at all; . . . the scarcity of animals could not be blamed on poachers . . . the boundaries had been emptied of people years ago . . . . No, it was all this unburned grass that had driven the animals away. These tall tussock grasses produced new tissue only during the rainy season; once they had matured and flowered, they turned dry and stalky and lost all nutritive value . . . . the miombo *was a recent habitat type, no older than the post pluvial period, perhaps 12,000 years ago.*
(From Sand Rivers, Peter Matthiessen, 1981, pp. 83–84)

INTRODUCTION

The end of the Pleistocene, approximately 10,000 yr B.P., was a dynamic time of earth history. Changes in global climatic patterns caused massive continental glaciers to retreat in the northern hemisphere (Denton and Hughes, 1980; Shackleton and Opdyke, 1973); ablading glaciers choked drainage systems with vast quantities of cool meltwater ladened with glacial-fluvial sediments (Baker, 1983). Continental shelves were inundated by a eustatic rise in sea level (Bloom, 1983a, b; Cronin,

1983) and many pluvial lakes were depleted (Smith and Street-Perrott, 1983). New landscapes emerged in both glaciated and unglaciated terrains. The reorganization of terrestrial biotic communities resulted in the evolution of new ecosystems (Lundelius et al., 1983), and Paleolithic cultures first immigrated into the New World (Haynes, 1980). A major extinction event may have been the culmination of all these environmental changes.

The late Pleistocene extinction was a global phenomenon, but this extinction most adversely affected the terrestrial faunas of South America, North America, and Australia (Martin and Klein, 1984). The terminal Pleistocene extinction was not restricted to a particular adaptive zone, trophic class, size category, or taxonomic group (Graham and Lundelius, 1984), although the most severely impacted guilds were medium- to large-sized mammalian carnivores and herbivores. The elimination of some small mammalian, reptilian, and avian species also occurred, and their demise cannot be ignored or considered insignificant (Graham, 1979; Graham and Lundelius, 1984; Grayson, 1977).

There are two predominant schools of thought with respect to the cause of this extinction event. The overkill hypothesis proposes that the instantaneous and massive elimination of late Pleistocene species was the result of the zealous predatory habits of Upper Paleolithic hunters (Martin, 1967, 1973, 1984; Mosimann and Martin, 1975). However, many facets of the extinction are not adequately explained by overkill (Graham, 1979; Graham and Lundelius, 1984; Grayson, 1977; Martin and Neuner, 1978). For instance, the overkill hypothesis would predict that the late Pleistocene extinction in specific geographic areas should be directly correlated with the first appearance of efficient human hunters. In Australia, man appears more than 15,000 yr before the extinction event (Horton, 1984), and in Ireland humans first appear during the Mesolithic, more than 3000 yr after the extinction of the Irish elk, *Megaceros* (Stuart, 1982).

Environmental change encompasses most of the other divergent hypotheses proposed to explain the late Pleistocene extinction. Because the extinction of a single taxon in a local environment can be an extremely complex process (Lidicker, 1965), multiple-species extinctions over broad geographic areas may involve the interaction of numerous variables in both the biological and physical environments. The purpose of this chapter is to examine the role of plant–animal interactions in the late Pleistocene extinction. Habitat destruction is probably one of the dominant factors disrupting these interactive systems. Other factors like the perturbation of browsing/grazing strategies and a general reduction in the availability of nutritious vegetation also had profound effects on the late Pleistocene fauna.

CLIMATIC CHANGE AND BIOTIC REORGANIZATION

If environmental models are proposed as the cause for the late Pleistocene extinction, then it may be asked why extinction events are not associated with previous glacial/interglacial cycles? There have been at least 22 alternating stages of high and low ice volumes in the Northern Hemisphere during the last 870,000 yr (Shackleton and Opdyke, 1973). Within the same temporal interval, there have only been two major extinction events, at the end of the Irvingtonian and at the end of the Rancholabrean. For proponents of climate as a cause, the obvious answer must be that the Pleistocene/Holocene change was significantly different in some way from many of the other environmental fluctuations.

Analysis of oxygen isotopes ($^{18}O/^{16}O$) in the shells of fossil foraminifera from deep-sea cores provides proxy measures of paleotemperatures and ice volumes for

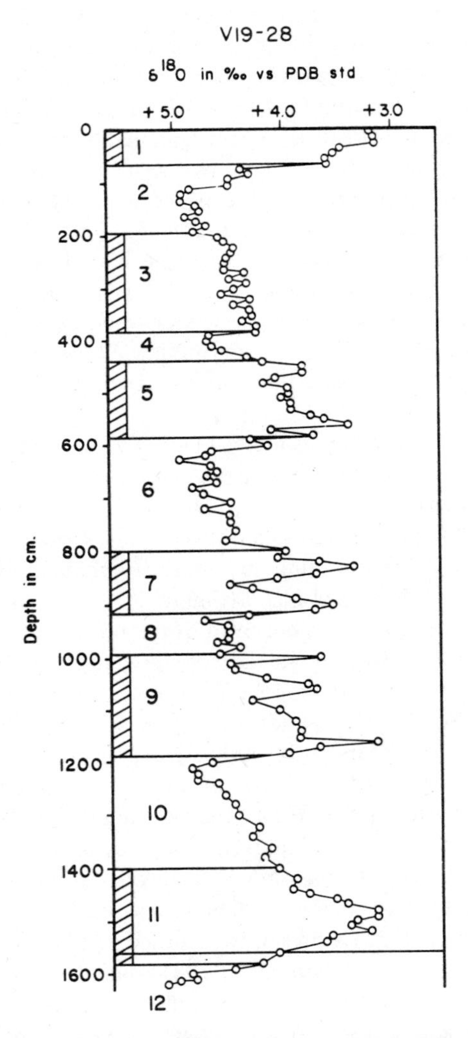

FIGURE 1. Late Quaternary variations in $\delta^{18}O$ measured in benthonic foraminifer *Uvigerina* in core (V19–28) for the eastern equatorial Pacific. Shaded isotope stages represent "warm" intervals with low volumes of continental ice, whereas unshaded stages represent "cold" intervals with high volumes of continental ice (adapted from Ruddiman and McIntyre, 1982).

the Pleistocene (Shackleton and Opdyke, 1973; Ruddiman and McIntyre, 1982). Interpretations of the $\delta\ ^{18}O$ signal is complex and foraminiferal curves may lag true ice volume by 1000–3000 yr and may misrepresent the amplitude of total ice volume change by as much as 30% (Mix and Ruddiman, 1984). In general, however, major alterations in the $\delta\ ^{18}O$ measurements define various glacial—interglacial (cold–warm) cycles. Inspection of these cycles (Figure 1) indicates that the oscillations have varying amplitudes and rates. The terminations (glacial–interglacial changes) in stages 8 and 7, for instance, are significantly smaller than the terminations of substage 5e and stage 1. In North America, substage 5e and stage 1 are equivalent to the Illinoian–Sangamonian and Wisconsinan–Holocene transitions, respectively.

The temperature changes between substage 5e and stage 1 appear to be identical, rapid, unidirectional warming trends (Shackleton and Opdyke, 1973). The rapid warming of the Holocene appears to be substantiated by proxy temperatures derived from palynological studies in the upper midwestern United States. Webb and Bryson (1972, p. 105, Figure 26) illustrate a steep rise in temperature (~6°F) during the terminal Pleistocene. Fairbridge (1982, p. 218, Figure 1) also illustrates this rapid rise in temperature at the Pleistocene–Holocene boundary in a generalized comparison of eustatic sea level and temperature. However, local climatic fluctuations must have been superimposed on this overall trend because numerous isolated advances in continental glacial ice during the late Pleistocene and early Holocene appear to be the result of climatic fluctuations rather than glacial surging (Wright, 1983).

Even though the morphology of the $\delta^{18}O$ curves are similar for substage 5e and stage 1, terrestrial faunas and floras suggest that the climates of the last interglacial were significantly warmer, perhaps 2°C, than those of the Holocene (Adam et al., 1981; CLIMAP, 1984, pp. 212–214; Gascoyne et al., 1981). During the Ipswichian (75,000–125,000 yr *B.P.*), the hippopotamus, *Hippopotamus amphibius*, existed as far north as North Riding in England (Figure 2), and in fact, *H. amphibius* may have extended even further north; but most of the northern sites, except for caves, have probably been destroyed by glacial erosion (Stuart, 1976, 1982). In historic times, the distribution of the hippopotamus was restricted to the Nile Valley and subsahara Africa (Nowak and Paradisio, 1983), so its northern range extension during the Ipswichian appears to be indicative of warmer climates. Geographic range shifts for the pond tortoise, *Emys obicularis*, in Europe also reflect that the last interglacial was warmer than the Holocene (Stuart, 1979).

In North America, climates of the last interglacial also appear to have been warmer than the Holocene. Sangamon age faunas from Florida (Reddick IA, Haile XIB, and Arrendondo IIA) contain vertebrate species that currently inhabit areas with warmer winters than those presently recorded in Florida (Webb, 1974). Vegetational shifts in California (Adam et al., 1981) during the Sangamon also appear to represent warmer climates than the Holocene. In essence, the Holocene cannot be used as a direct analog for terrestrial environments of the last interglacial. In fact, the Holocene may be significantly different from the other previous interglacial stages.

The presence of the "giant" tortoise (*Geochelone* spp.) in North America may

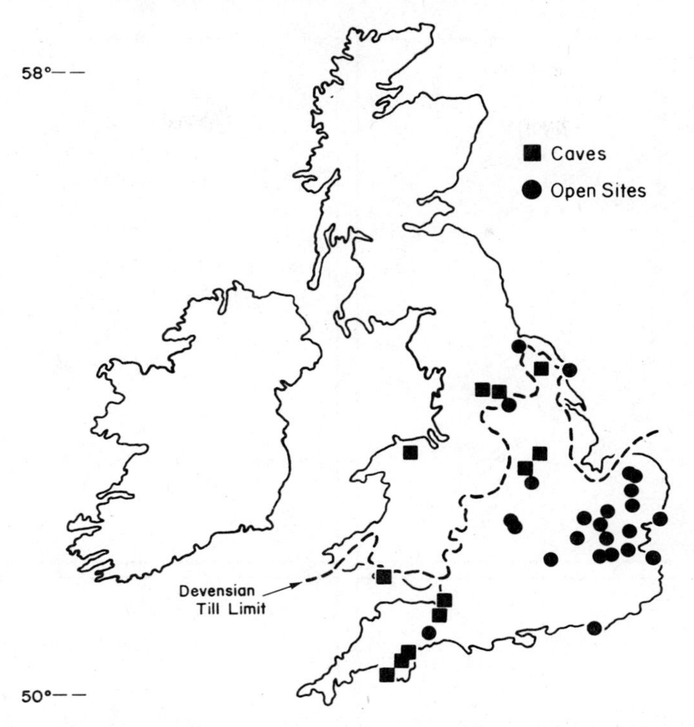

FIGURE 2. Distribution of *Hippopotamus amphibius* in the British Isles during the Ipswichian Interglacial. Note that fossil records north of the Devensian till limit are preserved in caves (modified after Stuart, 1982, p. 130; Figure 7.12).

provide evidence for mild winters and equable climates, especially during the Tertiary and pre-Wisconsinan parts of the Pleistocene (Hibbard, 1960). This inference is based on the observations that all modern species of *Geochelone* reside in equatorial areas and that in 1928 captive Galapagos tortoises (*Geochelone porteri*) expired when they were not provided with shelter from the "rigorous" winter temperatures in southern California, Arizona, southern Texas, Louisiana, and Florida (Hibbard, 1960). Apparently, the lower nocturnal temperatures sufficiently slowed digestion of forage gathered during the day and the tortoises died from enteritis (Hibbard, 1960).

Hibbard (1960) assumed that the extinct large species, like the living "giant" tortoises, could not burrow. Also, there was a lack of natural shelters (e.g., caves and crevices) for them on the Great Plains and Gulf Coastal Plain during the Tertiary and Quaternary. He reasoned that if winter temperatures were below freezing for any extended periods of time, then the unprotected large species of *Geochelone* would succumb to physiological failures created by the frigid temperatures.

There are a variety of complex mechanisms that turtles use for thermoregulation (Hutchison, 1979); the large tortoises of the Galapagos can insolate themselves from diurnal temperature fluctuations by inhabiting small pools of water during the

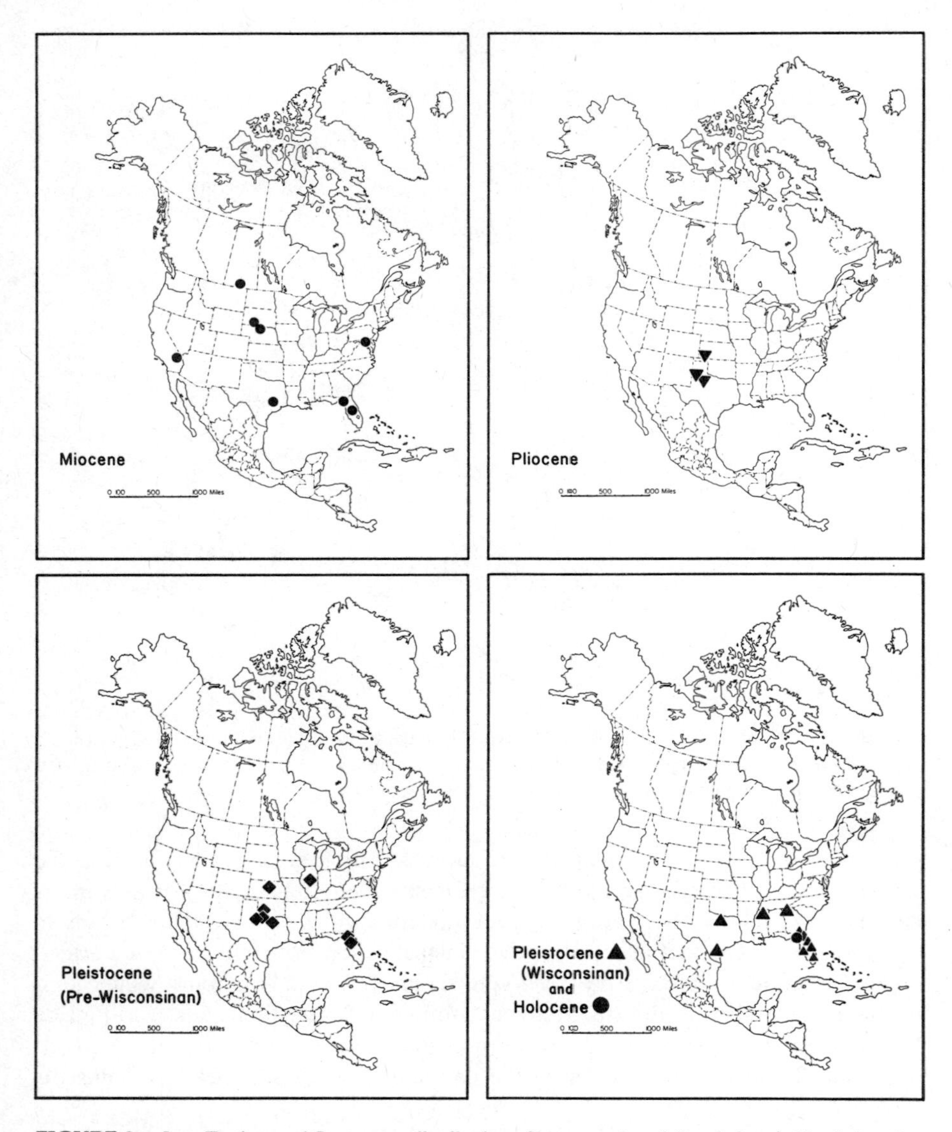

FIGURE 3. Late Tertiary and Quaternary distribution of large species of *Geochelone* in North America (excluding Mexico). Auffenberg (1963) has demonstrated that large size has evolved independently in both *Caudochelys* and *Hesperotestudo* clades (subgenera). Information on localities (age, species, references, etc.) is given in Appendix I.

nights. With extinct species, it is difficult to reconstruct all of the possible thermoregulatory mechanisms, especially behavioral ones. Therefore, extinct species of *Geochelone* may not necessarily be thermometers for frost-free winters as originally suggested by Hibbard (1960). However, even if the tolerance limits and methods of thermoregulation of the extinct species are not as narrow as the living species of *Geochelone*, it is quite unlikely that these large tortoises could survive, unprotected, the extended periods of frigid temperatures characteristic of the winters of the midcontinent today.

The late Cenozoic distribution of *Geochelone* in North America demonstrates that the Holocene climates are merely part of a long-term global trend of climatic deterioration and that the Holocene may be distinctly different from other interglacials. Throughout the late Tertiary and entire Quaternary, the distribution of the large species of *Geochelone* have gradually been restricted to the southeastern United States (Figure 3) with the latest records occurring in Florida. *Geochelone* was probably extirpated from the continent by the latest Pleistocene–earliest Holocene. This pattern has been interpreted as evidence for increasing continentality throughout this interval of time (Auffenberg and Milstead, 1965; Hibbard, 1960; Holman, 1971). The exclusion of *Geochelone* from northern latitudes during the Tertiary and Quaternary has occurred in other parts of the world as well (Crumly, 1982).

Analysis of geographic shifts has been restricted to the large *Geochelone* species because many of the smaller species like *G. wilsoni* and *G. incisa* might have been able to find natural shelters for overwintering. However, the contraction of the distributions of smaller extant tortoises like *Gopherus agassizi* and *Emys obicularis* and the extinction of *Geochelone wilsoni*, a small species, may be the result of increasing continentality at the end of the Pleistocene (Moodie and Van Devender, 1979; Stuart, 1982). Finally, Holman (1979) believes that closely spaced incremental growth lines on the specimens from Devil's Den, one of the latest known occurrences of *Geochelone*, might reflect increasing seasonality during the latest Pleistocene or earliest Holocene.

Geochelone is not the only climatic signal that suggests that the seasonal extremes of the Pleistocene were not as severe as those of the Holocene. The Tertiary and Quaternary distribution of the alligator (*Alligator mississippiensis*) essentially mirrors the range shifts of *Geochelone* in North America (Hibbard, 1960). However, unlike the extinct species of *Geochelone, A. mississippiensis* is still alive and its biology is fairly well known. Again, the contraction toward the Gulf Coast is apparently in response to colder winters. Also, at the end of the Pleistocene, tropical and subtropical species like *Neofiber* (round-tailed muskrat), *Desmodus* (vampire bat), *Hydrochaeris* (capybara), and so on, withdrew southward to warmer and/or moister climates (Figure 4). The congruence in the shifts of all of these species is probably the consequence of a primary forcing factor, like climate, rather than the result of an aggregate of multiple independent factors.

The differences between the Holocene climates and earlier climatic fluctuations are probably responsible for the environmental changes that precipitated the late Pleistocene extinction. The Hopwood local biota (King and Saunders, 1984) from Montgomery County, Illinois, contains stratified biotic components that document

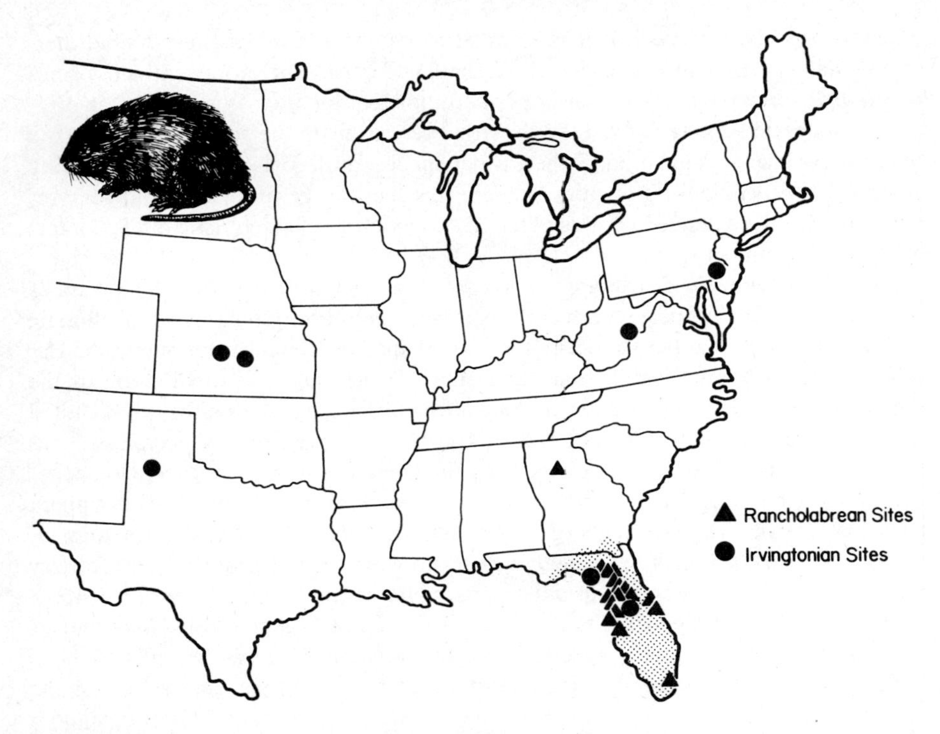

FIGURE 4. Fossil (dots and triangles) and modern (shaded area) distribution of the round-tailed muskrat (*Neofiber*).

the Illinoian–Sangamonian transition. King and Saunders (1984) feel that the Illinoian pollen assemblages from the midcontinent are quite different from the full and late glacial Wisconsinan pollen assemblages. Illinoian pollen assemblages resemble mid-Wisconsinan interstadial pollen records rather than the Wisconsinan full glacial (King and Saunders, 1984). This further underscores the differences in the environments and biotas of the late Wisconsinan and early Holocene from those of previous glacial–interglacial stages. In fact, King and Saunders (1984) believe that the Illinoian did not undergo as rapid a period of cooling and ice advance as did the onset of the Wisconsinan full glacial about 23,000 yr *B.P.*

The Sangamon fauna and flora from Hopwood are also distinctly different from the Holocene biota of this region. Specifically, the 68–83% nonarboreal pollen (NAP) in the Sangamon spectrum from Hopwood suggest an environment that is drier than anything that is inferred from comparable Holocene pollen diagrams (King and Saunders, 1984). Although, like the Holocene (King, 1981), the most xeric climates of central Illinois during the Sangamon occurred early in the interglacial stage (King and Saunders, 1984). The remains of the giant tortoise (*Geochelone*) in central Illinois (King and Saunders, 1984) also suggest that the Sangamon winter temperatures were warmer than present.

LATE PLEISTOCENE BIOTAS AND HABITAT DESTRUCTION

Late Pleistocene biotas throughout most of the world are characterized by the coexistence of species that today are allopatric and presumably ecologically incompatible (Graham and Lundelius, 1984; Lundelius, 1984). The terminal Pleistocene fauna (~11,000 yr *B.P.*) from New Paris No. 4, Pennsylvania (Guilday et al., 1964), contains species that today inhabit environments further to the north and west as well as species that still reside in the area (Figure 5). These fossil biotas, at least most of them, do not appear to be the result of bioturbation, redeposition, deflation, or other mixing processes that produce artificial assemblages (Graham and Lundelius, 1984). Instead, they represent viable communities that do not have modern analogs. Furthermore, these intermingled associations are not restricted to the mammalian fauna, but they have also been documented for late Pleistocene flora, terrestrial invertebrates, and other terrestrial vertebrates (Graham and Lundelius, 1984).

Many late Pleistocene communities had higher species densities than their modern counterparts (Foley, 1984; Graham, 1976, Guilday et al., 1978). The higher species densities are created, in part, by the existence in the late Pleistocene of a faunal component that is today extinct. However, the late Pleistocene faunas were also enriched by the integration of arctic and boreal biotas with those from the temperate region. The displacement of arctic and boreal biotas by glacial climates did not cause the relocation of more southern communities; rather all of these biotas were amalgamated to form new communities in the temperate zone.

These intermingled biotas with higher species densities were supported in some areas by late Pleistocene climates that were more equable than modern ones. Reduced seasonal extremes in temperature and effective moisture allowed species with disparate ecologies to coexist (Graham, 1976, 1979; Graham and Semken, 1976; Holman, 1976; Lundelius et al., 1983). Intermingled late Pleistocene biotas were also probably the result of greater heterogeneity in habitat types than exists today (Graham, 1985; Guthrie, 1982, 1984; Rhodes, 1984).

The biota from the mastodont bone bed at Christensen Bog (Graham et al., 1983) in central Indiana provides an excellent example. The vertebrate fauna and associated flora are from a time period (12,000–13,220 yr *B.P.*) just before the terminal Pleistocene extinction. Floral remains represent a diversified forest unlike modern ones (Whitehead et al., 1982). The vertebrate fauna contains, in part, mastodont, giant beaver, caribou, turtles, ducks, and turkey. Today, caribou inhabit boreal forest or tundra environments, whereas wild turkey and many of the turtles reside in environments further south. Their coexistence at Christensen Bog during the late Pleistocene was facilitated by the variety of habitats created in a more fine-grain vegetational mosaic.

Diverse habitat associations extend well back into the Pleistocene and occur in many different geographic areas and physiographic settings. Rhodes (1984) documents a more patchy environment in southwestern Iowa during both the Farmdalian (~23,000 yr B.P.) and Woodfordian (~15,000 yr B.P.). Again, these faunas

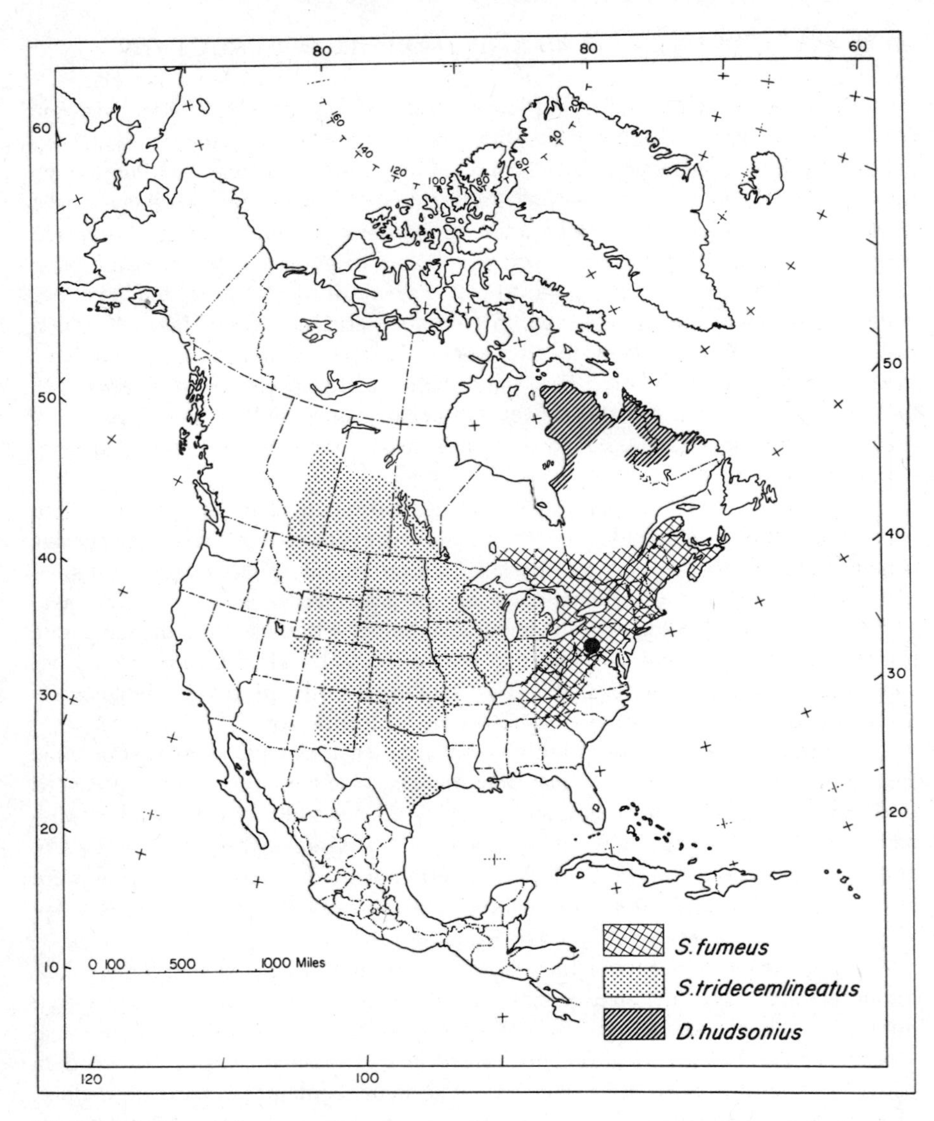

FIGURE 5. Modern distribution of species (*Sorex fumeus*, *Spermophilus tridecemlineatus*, and *Dicrostonyx hudsonius*), which occur together in 11,000-year-old cave deposits at New Paris No. 4 (star), Bedford County, Pennsylvania.

represent habitat types that no longer occur together, and in fact, the integrated environmental mosaic no longer exists. Walker (1982) and Guthrie (1982) illustrate greater habitat diversity during the late Pleistocene for the mountainous regions of Wyoming and the diverse environments of Beringia, respectively. It appears that this was a global phenomenon (Graham and Lundelius, 1984). Faunal studies in the Great Plains (Hibbard, 1970) and Florida (Webb, 1974) extend these complex environmental mosaics back to previous glacial and interglacial times in North America.

The terrestrial warming process of the late Pleistocene–early Holocene accentuated the latitudinal thermal gradient (Fairbridge, 1982). This may have created more continental climates in certain areas during the Holocene (Graham, 1976; Hibbard, 1960; Leopold, 1967). However, in other areas, like Beringia, dichotomies such as "continental" and "oceanic" may be inappropriate for the late Quaternary environments (Guthrie, 1982). Furthermore, the terminal Pleistocene environmental changes in parts of the Southern Hemisphere are in the opposite direction of those in the Northern Hemisphere. As grasslands evolved and expanded in North America during the Holocene, forests encroached on open environments in South America; both continents experienced significant mammalian extinctions. Therefore, the direction of environmental change may be negligible, but global alterations in seasonal patterns may be highly significant, especially as it affects the organization of biotic communities.

The evolution of modern biotic community patterns from those of the late Pleistocene is a consequence of species responding individually to environmental changes. Individualistic adjustments to environmental change have significant paleoecological and biological implications. If each species migrated independently, then the integrity of communities would not have been stable over long periods of time. Modern communities have probably evolved in the last 10,000 yr and they cannot be used as direct analogs in the reconstruction of Pleistocene environments. The individualistic dissolution of Pleistocene communities also reduced the predictability of the structure and composition of the resulting Holocene communities.

The individualistic pulling apart of the late Pleistocene communities reduced both vegetational and faunal diversity. New species associations and habitat types emerged as old ones disappeared. Destruction of the Pleistocene habitat mosaic undoubtedly had profound effects on the terrestrial vertebrate fauna, especially the large mammalian herbivores. Habitat destruction is probably the primary ecological cause of the Pleistocene extinction.

AVAILABILITY OF NUTRITIOUS VEGETATION

The availability of nutritious vegetation establishes the foundation upon which the herbivore guilds are built. Any significant perturbations in the nutritional quality of vegetation will reverberate throughout the food chain, especially at the primary consumer level. The environmental changes at the end of the Pleistocene initiated a reorganization of the vegetation mosaic, but these changes also altered the nutritional quality of forage.

The nutritional value of vegetation is dependent on many variables, for example, nutrient availability and recycling in the soil, length of growing season, and seasonal distribution of precipitation. Guthrie (1982) gives a comprehensive discussion of the interaction of these factors; he feels the critical variable is the duration of the peak in forage quality.

The different photosynthetic pathways used by various species of plants may also limit the nutritional quality of vegetation. Plants (C-3) that employ a three-carbon cycle of photosynthesis have a higher nutritive value than plants (C-4) that use a four-carbon cycle (Caswell et al., 1973). The distribution of plants with these photosynthetic pathways is strongly influenced by climatic parameters. The C-3 grasses generally inhabit cool, moist environments of the northern latitudes whereas warm, dry environments favor C-4 grasses, although there are exceptions in each category. It has been postulated that with the global warming at the end of the Pleistocene, the less nutritious C-4 grasses expanded their distributions at the expense of the more nutritive C-3 grasses (J. W. Wilson, in Kurtén and Anderson, 1980). This hypothesis can be tested by analyzing carbon isotope ratios in fossil bone (Land et al., 1980).

Also, modern tundra grasses, whose dominance emerged in the Holocene, contain a wide variety of alkaloids (Guthrie, 1982). The toxic nature of these secondary compounds would also significantly limit the availability of nutritious forage. The extensive reduction in the nutritive value of the graminoid resources in North America during the late Pleistocene–Holocene may have been responsible for the extinction of a large segment of the grazing herbivore guild (Guthrie, 1982; Kurtén and Anderson, 1980; Rhodes, 1984; Vereshchagin and Baryshnikov, 1982; Wilson, 1973). Many of the herbivores that survived the extinction have evolved nutritional adaptations to cope with problems of low forage quality (Batzli et al., 1981; Graham and Lundelius, 1984).

Browsing herbivores may have experienced a similar reduction in the availability of nutritious browse at the end of the Pleistocene. Today, subarctic browsing vertebrates prefer the relatively undefended mature-growth-form vegetative structures of competitive plant species (Table 1) that can regenerate parts destroyed by browsing (Bryant and Kuropat, 1980). In contrast, stress-tolerant plant species (Table 1) allocate relatively uniform quantities of carbon, in the form of resins, to defense throughout life and therefore always have low palatability (Bryant and Kuropat, 1980). Presently, stress-tolerant species in the northern latitudes are limited by soil nutrient deficiencies, whereas in the late Pleistocene moisture stress may have been more prevalent (Guthrie, 1982).

Late Pleistocene environmental changes may have favored stress-tolerant plants, which reduced the overall nutritive value of Holocene vegetation. The decreasing nutritive value in a competitive species/stress-tolerant species continuum is reflected in the forage preferences of snowshoe hares and mountain hares [willow > aspen > larch > dwarf birch > pine (jackpine = lodgepole pine = scots pine > white pine > red pine) > fir > spruce (white spruce > black spruce) > alder] (Bryant and Kuropat, 1980). In general, "one pattern of plant distribution in time that hints of a possible role of decreasing food quality in the late Wisconsin is the transition from willow to birch to alder to spruce at the end of the Wisconsin at high latitudes. If one were to predict a pattern of decreasing food quality based on what we know this would be the pattern predicted" (Bryant, written communication).

Indeed, this is the pattern that emerges from pollen profiles from various areas in Alaska. From 25,000 yr *B.P.* to slightly less than 15,000 yr *B.P.*, worm-wood

Table 1. Adaptive and Defensive Strategies of Subarctic Plants Used by Vertebrate Browsers

	Stress-Tolerant Plants	Competitive Plants
Site of growth	Extreme environmental stress (nutrient poor)	Low environmental stress (nutrient rich)
Growth rate	Slow	Rapid
Leaf and twig turnover rate	Low	High
Browsing tolerance	Low	High
Allelochemic defense	Throughout life	Juvenile stages
Palatability	Low throughout life	Low in juvenile stages; high in mature-growth form
Examples	White Pine (*Pinus albicaulis*) White Spruce (*Picea glauca*) Black Spruce (*Picea mariana*) Alder (*Alnus*)	Willow (*Salix*) Aspen (*Populus tremuloides*) Larch (*Larix*) Birch (*Betula*)

or *Artemesia*, grass–sedge–willow was the dominant vegetational type (Heusser, 1983). This was replaced by birch–willow–herbs (Heusser, 1983), which was eventually replaced by birch–willow–spruce at Birch Lake (Ager, 1975) about 10,000 yr *B.P.* Radiocarbon-dated spruce wood samples are no older than 13,600 yr *B.P.* in the unglaciated interior of Alaska and the adjacent Yukon Territory (Péwé, 1975).

Browsing may have also created a negative feedback. As the abundance of the highly nutritious species was reduced, the intensity of browsing on these limited resources would have increased. When mammals severely overbrowse their food supply, the toxic juvenile shoots that plants produce in response to browsing may play an important part in causing the decline and in regulating the rate of recovery of the browsing populations (Bryant et al., 1983). Consequently, the extensive browsing on diminishing resources during the late Pleistocene would have adversely affected megaherbivore populations and contributed to their demise.

Spruce forests covered most of the eastern half of the United States during the late Pleistocene (Watts, 1983). The widespread distribution of a stress-tolerant species (spruce) appears to contradict the notion of high-quality forage during the late Wisconsin. However, in many midwestern areas these "forests" seem to have been more open than contemporary spruce forests (King, 1981; Rhodes, 1984). In the upper midwest, Rhodes (1984) has referred to them as boreal grasslands. Open habitats would encourage the growth of competitive species as well as graminoids, herbs, and forbs, which would enrich the nutritive value of the vegetational mosaic. Also, browsing may have facilitated the maintenance of these open habitats.

The species composition of late Pleistocene spruce forests was quite different from modern boreal forests (Cushing, 1965, 1967; Davis, 1976). They contained a wider variety of competitive species such as ash, other deciduous species, and herbs (Amundson and Wright, 1979; Watts, 1983; Wright, 1981). Therefore, the late Pleistocene spruce forests of North America may have had a much higher nutritive value than their modern counterparts.

The rapid northern migration of spruce during the individualistic biotic reorganization (Davis, 1976) at the end of the Pleistocene formed a relatively homogeneous vegetation of stress-tolerant species with low palatability and nutritive value. Also, the late Pleistocene spruce "forests" of the eastern United States were replaced with dominant stands of stress-tolerant species like oak, hickory, and southern pines. These environments have low carrying capacities for large browsing herbivores as indicated in the Ozarks (Murphy and Crawford, 1970). Again, the duration of the peak in forage quality may be the critical factor since deer are primarily limited by the availability of forage during the winter (Murphy and Crawford, 1970). Also, the increasing dominance in Holocene forests of the eastern United States of species with well-developed allochemic defenses may have caused a shift from browsing of vegetative structures to dependence on mast production. This is an extremely variable resource base that would probably not support a diverse large browsing herbivore guild.

Vegetational changes described by Spaulding et al. (1983) would also indicate a significant reduction in nutritive vegetation at the end of the Pleistocene for the southwestern United States. The vegetational gradients of the southwest may have been determined by the seasonal distribution of precipitation, especially during the winter (Spaulding et al., 1983). Therefore, the reduction in the nutritive value of vegetation in the Southwest appears to be the result of shifts in seasonal patterns.

A fairly consistent pattern of reduction in the availability of nutritious forage is apparent for most of North America during the late Pleistocene. Vereshchagin and Baryshnikov (1982) have argued similarly for the extinction of elements of the mammoth fauna of the Eurasian arctic. Likewise, nutritive values were probably lowered when vegetational mosaics in parts of South America (Bradberry et al., 1981) and Central America (Leyden, 1984) changed from savannas to closed forests.

Finally, spatial heterogeneity of species distribution may be one of the most important features of ecological systems contributing to their integrity and continuity (McNaughton, 1983). Increased complexity of heterogeneous environments can enhance biomass stability (King and Pimm, 1983). The shift from the heterogeneous Pleistocene environments to the more homogeneous ones of the Holocene would have lowered biomass stability, which in turn might decrease carrying capacity.

DISRUPTION OF FEEDING STRATEGIES

The diversity of Pleistocene megaherbivores in North America as well as other areas poses some interesting ecological questions about how the herbivores partitioned the vegetational resource to avoid interspecific competition. The role of competition in modern communities is hotly contested (review in Schoener, 1982); however, a recent evaluation (Schoener, 1983) of field-experimental studies found a high percentage (90%) that documented competition in a variety of modern ecological systems. In any case, most modern communities are not directly analogous to Pleistocene communities, and therefore, competition may have played a significantly different role in community organization during the Pleistocene.

Four mechanisms that may have partitioned the environment for the Pleistocene herbivore guilds have been outlined by Guthrie (1982, p. 311):

1. specialization for plant form or species,
2. specialization for plant part,
3. specialization for different growth stages of plant succession, and
4. separation by habitat.

Graham and Lundelius (1984) and Guthrie (1984) have postulated similar criteria for the maintenance of herbivore diversity during the Pleistocene. The destruction of habitats at the end of the Pleistocene, as discussed earlier, would have dissolved any separation of herbivores by habitat. Any alterations in the other three mechanisms would have involved the disruption of feeding strategies.

Although the grazing succession exemplified by the modern diverse herbivore megafauna of the African savannas is not an exact analog for the late Pleistocene fauna, it can serve as a powerful tool in modeling Pleistocene ecosystems (Graham and Lundelius, 1984; Guthrie, 1982; Martin, 1982; Schweger and Martin, 1976). Today, diverse herbivore species migrate throughout the African savannas, each species grazing on a specific plant species, plant part, growth stage, or select group of plants. This successional grazing strategy stimulates the growth and development of other plant species or plant parts that will be harvested by subsequent waves of migrating herbivores. The food resource for one herbivore species may consequently be dependent on the grazing strategy of the preceding species (Bell, 1971; Gwynne and Bell, 1968; McNaughton, 1976; Vesey-Fitzgerald, 1960). Productivity patterns of the vegetation and the migrational behavior of ungulates is directly related to the annual climatic cycle (McNaughton, 1976; Vesey-Fitzgerald, 1960). A similar ecological separation has also been documented for browsing ungulates (Leuthold, 1978).

Maintenance of these diverse herbivore systems may require stability in the sequential structure of the plant–animal interactions. Changes in the migratory patterns of the herbivore species, adjustments in productivity schedules or distributions of plant species, or alterations in the climatic cycle will disrupt this succession. The individualistic reorganization of Pleistocene communities in response to climatic change would have diminished the predictability of the composition and structure of the new Holocene communities. Instead of merely tracking old environments, the vegetational deck of cards would have been randomly reshuffled, and each species would have been dealt a new hand of forage types. This may have lessened niche differentiation, heightened competition, and in some cases created detoxification problems for some plant predators (Graham and Lundelius, 1984).

In addition to climate, geology and topography, as they relate to edaphic conditions, are also primary factors that define and constrain the evolution of grazing communities (McNaughton, 1983). The emergence of new landscapes in response to deglaciation and all of its attendant parameters (e.g., development of extensive outwash plains and agrading/degrading cycles in alluvial valleys) would have altered the preexisting geology and topography of most of North America. Therefore, all of the important composite factors affecting grazing communities (McNaughton, 1983) would have been changed at the end of the Pleistocene.

CONCLUSIONS

Many believe environmental changes were responsible for the late Pleistocene extinction. Habitat destruction is probably the ultimate ecological cause of the extinction, but the reduction in the availability of nutritious vegetation and the disruption of feeding strategies also played important roles. Through a complex network of interactive systems, all of these factors can be related to climatic change at the end of the Pleistocene.

Climates did not change uniformly, and therefore, in the broad sense, it may not be appropriate to argue simply for a shift in one climatic regime (i.e., equability) to another (i.e., continentality) as the driving mechanism of extinction. However, changes in the seasonal patterns of different climatic parameters (i.e., precipitation, temperature, wind, insolation, cloudiness, effective moisture, etc.) have far-reaching effects on biotic systems. Variations in seasonal patterns determine the widths of the bottlenecks that limit biological populations.

Apparently, in the Holocene, the seasonal patterns were significantly different from those of previous glacial–interglacial stages. Further investigations into the unique properties of these differences, especially for the winter season, will undoubtedly elucidate our understanding of extinctions in general. Also, fluctuations in seasonal patterns may be independent of the overall direction of climatic change. Linking extinction with seasonal fluctuations introduces a strong probability factor into the extinction process. The application of quantitative models to the late Pleistocene extinction event has not been fully exploited, but these models should provide new insights and lines of inquiry.

ACKNOWLEDGMENTS

I wish to thank J. E. King and J. J. Saunders for access to unpublished data on the Hopwood Farm locality. Discussions with J. E. King, F. B. King, J. J. Saunders, J. R. Purdue, J. A. Holman, and W. F. Ruddiman were also constructive. I am indebted to J. P. Bryant for sharing his thoughts on the nutritive value of vegetation. P. S. Martin provided a valuable critique of an earlier draft of this manuscript. My sincere thanks to M. A. Graham for her continued support and assistance with manuscript preparation. I thank Julianne Snider for drafting the illustrations. This is contribution No. 77 of the Quaternary Studies Program of the Illinois State Museum.

APPENDIX I: LOCALITIES OF *GEOCHELENE* USED IN FIGURE 3.

Locality	Species of *Geochelone*	Epoch	Land Mammal Age	Glacial–Interglacial Stage	Reference
Garvin Gully, Grimes Co., TX	*G. (Caudochelys) williamsi*	Miocene	Arikareean		Auffenberg, 1974
Thomas Farm, Gilchrist Co., FL	*G. (Caudochelys) tedwhitei*	Miocene	Hemingfordian		Auffenberg, 1963
Rockglen, Saskatchewan, Canada	*G. (Caudochelys)* sp.	Miocene	Barstovian		Holman, 1971
Calvert Cliffs, Calvert Co., MD	*G. (Caudochelys) ducatelli*	Miocene	Barstovian		Auffenberg, 1974
Barstow Syncline, San Bernardino Co., CA	*G. (Caudochelys) milleri*	Miocene	Barstovian		Auffenberg, 1974
Big Spring Canyon, Bennet Co., SD	*G. (Hesperotestudo) orthopygia*	Miocene	Clarendonian		Hibbard, 1960
Valentine Formation, Brown Co., NE	*G. (Hesperotestudo) orthopygia*	Miocene	Clarendonian		Hibbard, 1960
Bone Valley, Polk Co., FL	*G. (Caudochelys) hayi*	Miocene	Hemphillian		Auffenberg, 1963
Rexroad, Meade Co., KS	*G. (Hesperotestudo) rexroadensis*	Pliocene	Blancan		Hibbard, 1960
Borchers, Meade Co., KS	*G. (Caudochelys)* sp.	Pliocene	Blancan		Hibbard, 1960
Deer Park, Meade Co., KS	*G. (Caudochelys)* sp.	Pliocene	Blancan		Hibbard, 1956

APPENDIX I: (*Continued*)

Locality	Species of *Geochelone*	Epoch	Land Mammal Age	Glacial–Interglacial Stage	Reference
Beck Ranch, Scurry Co., TX	(*G. Hesperotestudo*) *rexroadensis*	Pliocene	Blancan		Rogers, 1976
Blanco, Crosby Co., TX	*G.* (*Hesperotestudo*) *campester*	Pliocene	Blancan		Johnston and Savage, 1955
Gilliland, Knox Co., TX	*G.* (*Caudochelys*) sp.	Pleistocene	Irvingtonian		Hibbard and Dalquest, 1966
Holloman, Tillman Co., OK	*G.* (*Caudochelys*) sp.	Pleistocene	Irvingtonian		Hibbard and Dalquest, 1966
Kanopolis, Ellsworth Co., KS	*G.* sp.	Pleistocene	Irvingtonian		Holman, 1972
Reddick I, Marion Co., FL	*G.* (*Caudochelys*) *crassiscutata*	Pleistocene	Rancholabrean	Sangaman	Auffenberg, 1963
Reddick II, Martion Co., FL	*G.* (*Caudochelys*) *crassiscutata*	Pleistocene	Rancholabrean	Sangamon	Auffenberg, 1963
Rock Springs, Orange Co., FL	*G.* (*Caudochelys*) *crassiscutata*	Pleistocene	Rancholabrean	Sangamon	Affenberg, 1963
Hopwood Farm, Montgomery C., IL	*G.* (*Caudochelys*) *crassiscutata*	Pleistocene	Rancholabrean	Sangamon	King and Saunders, 1984
Clear Creek, Denton Co., TX	*G.* sp.	Pleistocene	Rancholabrean	Sangamon	Holman, 1963
Easley Ranch, Foard Co., TX	*G.* (*Caudochelys*) *crassiscutata*	Pleistocene	Rancholabrean	Sangamon?	Dalquest, 1962

Catalpa Creek, Clay/ Lowndes Cos., MS	*G. (Caudochelys) crassiscutata*	Pleistocene	Rancholabrean	Wisconsinan	Holman, 1976
Vero, Indian River Co., FL	*G. (Caudochelys) crassiscutata*	Pleistocene	Rancholabrean	Wisconsinan	Auffenberg, 1963
Ichetucknee River, Columbia Co., FL	*G. (Caudochelys) crassiscutata*	Pleistocene	Rancholabrean	Wisconsinan	Auffenberg, 1963
Melbourne, Brevard Co., FL	*G. (Caudochelys) crassiscutata*	Pleistocene	Rancholabrean	Wisconsinan	Auffenberg, 1963
Jupiter Inlet, Palm Beach Co., FL	*G. (Caudochelys) crassiscutata*	Pleistocene	Rancholabrean	Wisconsinan	Auffenberg, 1963
Little Salt Spring, Sarasota Co., FL	*G. (Caudochelys) crassiscutata*	Pleistocene	Rancholabrean	Wisconsinan	Holman and Clausen, 1984
Haile XIVA, Alachua Co., FL		Pleistocene	Rancholabrean	Wisconsinan	Martin, 1974
Ingleside, San Patricio Co., TX	*G. (Caudochelys) crassiscutata*	Pleistocene	Rancholabrean	Wisconsinan	Ludelius, 1972
Ladds, Bartow Co., Ga	*G. (Caudochelys) crassiscutata*	Pleistocene	Rancholabrean	Wisconsinan	Holman, personal communication
Medford Cave II, Marion Co., FL	*G. (Caudochelys) crassiscutata*	Pleistocene	Rancholabrean	Wisconsinan?	Auffenberg, 1957
Moore Pit, Dallas Co., TX	*G. (Caudochelys) crassiscutata*	Pleistocene	Rancholabrean	Wisconsinan?	Slaughter, 1966
Devil's Den, Levy Co., FL	*G. (Caudochelys) crassiscutata*	Holocene?			Holman, 1978

REFERENCES

Adam, D. P., Sims, J. D., and Throckmorton, C. K., 1981, 130,000-year continuous pollen record from Clear Lake, Lake County, California, *Geology*, **9**, 373–377.

Ager, T. A., 1975, *Late Quaternary Environmental History of the Tanana Valley, Alaska*, Ohio State University Institute of Polar Studies Report 54, pp. 1–117.

Amundson, D. C., and Wright, H. E., Jr., 1979, Forest changes in Minnesota at the end of the Pleistocene, *Ecol. Monogr.*, **49**, 1–16.

Auffenberg, W., 1957, A note on an unusually complete specimen of *Dasypus bellus* (Simpson) from Florida, *Quart. J. Florida Acad. Sci.*, **20**, 233–237.

Auffenberg, W., 1963, Fossil Testidunine turtles of Florida genera *Geochelone* and *Floridemys*, *Bull. Florida State Mus. Biol. Sci.*, 7(2), 53–98.

Auffenberg, W., 1974, Checklist of fossil land tortoises (Testudinae), *Bull. Florida State Mus. Biol. Sci.*, **18**, 121–251.

Auffenberg, W., and Milstead, W. W., 1965, Reptiles in the Quaternary of North America, in Wright, H. E., Jr., and Frey, D. G., eds., *The Quaternary of the United States*, Princeton, Princeton University Press, pp. 557–568.

Baker, V. R., 1983, Late-Pleistocene fluvial systems, in Porter, S. C., ed., *Late-Quaternary Environments of the United States*, Vol. 1, *The Late Pleistocene*, Minneapolis, University of Minnesota Press, pp. 115–129.

Batzli, G. O., White, R. G., and Bunnell, F. L., 1981, Herbivory: A strategy of tundra consumers, in Bliss, L. C., Heal. O. W., and Moore, J. J., eds., *Tundra Ecosystems: A Comparative Analysis*, Cambridge, Massachusetts, Cambridge University Press, pp. 359–375.

Bell, H. V., 1971, A grazing ecosystem in the Serengeti, *Sci. Am.*, **225**, 86–93.

Bloom, A. L., 1983a, Sea level and coastal changes, in Wright, H. E., Jr., ed., *Late-Quaternary Environments of the United States*, Vol. 2, *The Holocene*, Minneapolis, University of Minnesota Press, pp. 42–51.

Bloom, A. L., 1983b, Sea level and coastal morphology of the United States through the late Wisconsin glacial maximum, in Porter, S. C., ed., *Late-Quaternary Environments of the United States*, Vol. 1, *The Late Pleistocene*, Minneapolis, University of Minnesota Press, pp. 215–229.

Bradberry, J. P., Leyden, B., Salgado-Nobaurian, M., Lewis, W. M., Jr., Schubert, C., Binford, M. W., Frey, D. G., Whitehead, D. K., and Heibejabrn, F. H., 1981, Late Quaternary environmental history at Lake Valencia, Venezuela, *Science*, **214**, 1299–1305.

Bryant, J. P., and Kuropat, P. J., 1980, Selection of winter forage by subarctic browsing vertebrates: The role of plant chemistry, *Ann. Rev. Ecol. System.*, **11**, 261–285.

Bryant, J. P., Chapin, F. S., III, and Klein, D. R., 1983, Carbon/nutrient balance of boreal plants in relation to vertebrate herbivory, *Oikos*, **40**, 357–368.

Caswell, H., Reed, F., Stephenson, S. N., and Werner, P. A., 1973, Photosynthetic pathways and selective herbivory: A hypothesis, *Am. Natur.*, **107**, 465–480.

CLIMAP, 1984, The last interglacial ocean, *Quat. Res.*, **21**, 123–224.

Cronin, T. M., 1983, Rapid sea level and climate change: Evidence from continental and island margins, *Quat. Sci. Rev.*, **1**, 177–214.

Crumly, C. R., 1982, A cladistic analysis of *Geochelone* using cranial osteology, *J. Herpetol.*, **16**, 215–234.

Cushing, E. J., 1965, Problems in Quaternary phytogeography of the Great Lakes Region, in Wright, H. E., Jr., and Frey, D. G., eds., *The Quaternary of the United States*, Princeton, Princeton University Press, pp. 403–416.

Cushing, E. J., 1967, Late Wisconsinan pollen stratigraphy and the glacial sequence in Minnesota, in Cushing, E. J., and Wright, H. E., Jr., eds., *Quaternary Paleoecology*, New Haven, Yale University Press, pp. 59–88.

Dalquest, W. W., 1962, The Good Creek Formation, Pleistocene of Texas, and its fauna, *J. Paleontol.*, **36**, 568–582.

Davis, M. B., 1976, Pleistocene biogeography of temperate deciduous forests, *Geosci. Man*, **13**, 13–26.

Denton, G. H., and Hughes, T. J., 1980, *The Last Great Ice Sheets*, New York, Wiley.

Fairbridge, R. W., 1982, The Pleistocene-Holocene boundary, *Quat. Sci. Rev.*, **1**, 215–244.

Foley, R. L., 1984, *Late Pleistocene (Woodfordian) Vertebrates from the Driftless Area of Southwestern Wisconsin, The Moscow Fissure Local Fauna*, Illinois State Museum Reports of Investigations, No. 39, pp. 1–50.

Gascoyne, M., Currant, A. P., and Lord, T. C., 1981, Ipswichian fauna of Victoria Cave and the marine paleoclimatic record, *Nature*, **294**, 652–654.

Graham, R. W., 1976, Late Wisconsin mammal faunas and environmental gradients of the eastern United States, *Paleobiology*, **2**, 343–350.

Graham, R. W., 1979, Paleoclimates and late Pleistocene faunal provinces in North America, in Humphrey, R. L., and Stanford, D., eds., *Pre-Llano Cultures of the Americas: Possibilities and Paradoxes*, Washington, D.C., Washington Anthropological Society, pp. 49–69.

Graham, R. W., 1985, Diversity and community structure of the late Pleistocene mammal fauna of North America, *Acta Zool. Fenn*, **170**, 181–182.

Graham, R. W., and Lundelius, E. L., Jr., 1984, Coevolutionary disequilibrium and Pleistocene extinctions, in Martin, P. S., and Klein, R. G., eds., *Quaternary Extinctions: A Prehistoric Revolution*, Tucson, University of Arizona Press, pp. 223–249.

Graham, R. W., and Semken, H. A., 1976, Paleoecological significance of the short-tailed shrew (*Blarina*) with a systematic discussion of *Blarina ozarkensis*, *J. Mammal.*, **57**, 433–449.

Graham, R. W., Holman, J. A., and Parmalee, P. W., 1983, *Taphonomy and Paleoecology of the Christensen Bog Mastodon Bone Bed, Hancock County, Indiana*, Illinois State Museum Reports of Investigations, No. 38, pp. 1–29.

Grayson, D. K., 1977, Pleistocene avifaunas and the overkill hypothesis, *Science*, **195**, 691–693.

Guilday, J. E., Martin, P. S., and McGrady, A. D., 1964, New Paris No. 4, Pleistocene cave deposit in Bedford County, Pennsylvania, *Bull. Natl. Speleol. Soc.*, **26**, 121–194.

Guilday, J. E., Hamilton, H. W., Anderson, E., and Parmalee, P. W., 1978, The Baker Bluff cave deposit, Tennessee, and the late Pleistocene faunal gradient, *Bull. Carnegie Mus. Natur. Hist.*, **11**, 1–67.

Guthrie, R. D., 1982, Mammals of the mammoth steppe as paleoenvironmental indicators, in Hopkins, D. M., Matthews, J. V., Jr., Schweger, C. E., and Young, S. B., eds., *Paleoecology of Beringia*, New York, Academic Press, pp. 307–326,

Guthrie, R. D., 1984, Mosaics, allochemics, and nutrients: An ecological theory of late Pleistocene megafaunal extinctions, in Martin, P. S., and Klein, R. G., eds., *Quaternary Extinctions: A Prehistoric Revolution*, Tucson, The University of Arizona Press, pp. 259–298.

Gwynne, M. D., and Bell, R. H. V., 1968, Selection of vegetation components by grazing ungulates in the Serengeti National Park, *Nature*, **220**, 390–393.

Haynes, C. V., 1980, The Clovis culture, *Can. J. Anthropol.*, **1**, 115–121.

Heusser, C. J., 1983, Vegetational history of the northwestern United States including Alaska, in Porter, S. C., ed., *Late-Quaternary Environments of the United States*, Vol. 1, *The Late Pleistocene*, Minneapolis, University of Minnesota Press, pp. 239–258.

Hibbard, C. W., 1956, Vertebrate fossils from the Meade Formation of southwestern Kansas, *Pap. Mich. Acad. Sci., Arts, Lett.*, **41**, 145–203.

Hibbard, C. W., 1960, An interpretation of Pliocene and Pleistocene climates in North America, *Ann. Rep. Mich. Acad. Sci., Arts, Lett.*, **62**, 5–30.

Hibbard, C. W., 1970, Pleistocene mammalian local faunas from the Great Plains and Central Lowland provinces of the United States, in Dort, W., Jr., and Jones, J. K., Jr., eds., *Pleistocene and Recent Environments of the Central Great Plains*, Lawrence, University Press of Kansas, pp. 394–433.

Hibbard, C. W., and Dalquest, W. W., 1966, Fossils from the Seymour Formation of Knox and Baylor counties, Texas, and their bearing on the late Kansan climate of that region, *Contrib. Mus. Paleontol. Univ. Mich.*, **21**, 1–66.

Holman, J. A., 1963, Late Pleistocene amphibians and reptiles of the Clear Creek and Ben Franklin local faunas of Texas, *J. Grad. Res. Ctr.*, **31**, 152–167.

Holman, J. A., 1971, Climatic significance of giant tortoises from the Wood Mountain Formation (Upper Miocene of Saskatchewan), *Can. J. Earth Sci.*, **8**, 1148–1151.

Holman, J. A., 1972, Herpetofauna of the Kanapolis local fauna (Pleistocene: Yarmouth) of Kansas, *Mich. Acad.*, **5**, 87–98.

Holman, J. A., 1976, Paleoclimatic implications of "ecologically incompatible" herpetological species (late Pleistocene: southeastern United States), *Herpetologica*, **32**, 290–294.

Holman, J. A., 1978, The late Pleistocene herpetofauna of Devil's Den Sinkhole, Levy County, Florida, *Herpetologica*, **34**, 228–237.

Holman, J. A., and Clausen, C. J., 1984, Fossil vertebrates associated with Paleo-Indian artifact at Little Salt Spring, Florida, *J. Verteb. Paleontol.*, **4**, 146–154.

Horton, D. R., 1984, Red kangaroos: Last of the Australian megafauna, in Martin, P. S., and Klein, R. G., eds., *Quaternary Extinctions: A Prehistoric Revolution*, Tucson, University of Arizona Press, pp. 639–680.

Hutchison, V. H., 1979, Thermoregulation, in Horless, M., and Morlock, H., eds., *Turtles: Perspectives and Research*, New York, Wiley, pp. 207–228.

Johnston, C. S., and Savage, D. E., 1955, A survey of various late Cenozoic vertebrate faunas of the panhandle of Texas, Part I, *Univ. Calif. Publ. Geol. Sci.*, **31**, 27–49.

King, A. W., and Pimm, S. L., 1983, Complexity, diversity, and stability: A reconciliation of theoretical and empirical results, *Am. Natur.*, **122**, 229–239.

King, J. E., 1981, Late Quaternary vegetation history of Illinois, *Ecol. Monogr.*, **51**, 43–62.

King, J. E., and Saunders, J. J., Jr., 1984, *Hopwood Farm Paleoecology: A Geochelone-containing Illinoian-Sangamonian Biota from the Type Region, Central Illinois*, AMQUA Program and Abstracts, Boulder, University of Colorado, p. 68.

Kurtén, B., and Anderson, E., 1980, *Pleistocene Mammals of North America*, New York, Columbia University Press.

Land, L. S., Lundelius, E. L., Jr., and Volastro, S., 1980, Isotopic ecology of deer bones, *Paleogeog., Paleoclimatol., Paleoecol.*, **32**, 143–151.

Leopold, E. B., 1967, Late Cenezoic patterns of plant extinction, in Martin, P. S., and Wright, H. E., Jr., eds., *Pleistocene Extinctions: The Search for a Cause*, New Haven, Yale University Press, pp. 203–246.

Leuthold, W., 1978, Ecological separation among browsing ungulates in Tsavo East National Park, Kenya, *Oecologica*, **35**, 241–252.

Leyden, B. W., 1984, Guatemalan forest synthesis after Pleistocene aridity, *Proc. Natl. Acad. Sci.*, **81**, 4856–4859.

Lidicker, W. Z., Jr., 1965, Ecological observations on a feral house mouse population declining to extinction, *Ecol. Monogr.*, **36**, 27–50.

Lundelius, E. L., Jr., 1972, *Fossil Vertebrates from Late Pleistocene Ingleside Fauna, San Patricio County, Texas*, Bureau of Economic Geology Report of Investigations, No. 77, pp. 1–74.

Lundelius, E. L., Jr., 1983, Climatic implications of late Pleistocene and Holocene faunal associations in Australia, *Alcheringa*, **7**, 125–149.

Lundelius, E. L., Jr., Graham, R. W., Anderson, E., Guilday, J., Holman, J. A., Steadman, D., and Webb, S. D., 1983, Terrestrial vertebrate faunas, in Porter, S. C., ed., *Late-Quaternary Environments of the United States*, Vol. 1, *The Late Pleistocene*, Minneapolis, University of Minnesota Press, pp. 311–353.

Martin, L. D., and Neuner, A. M., 1978, The end of the Pleistocene in North America, *Trans. Nebr. Acad. Sci.*, **6**, 117–126.

Martin, P. J., 1982, Digestive and grazing strategies of animals in the arctic steppe, in Hopkins, D. M., Matthews, J. V., Jr., Schweger, C. E., and Young, S. B., eds., *Paleoecology of Beringia*, New York, Academic Press, pp. 259–266.

Martin, P. S., 1967, Prehistoric overkill, in Martin, P. S., and Wright, H. E., Jr., eds., *Pleistocene Extinctions: The Search for a Cause*, New Haven, Yale University Press, pp. 75–120.

Martin, P. S., 1973, The discovery of America, *Science*, **179**, 969–974.

Martin, P. S., 1984, Prehistoric overkill: the global model, in Martin, P. S., and Klein, R. G., eds., *Quaternary Extinctions: A Prehistoric Revolution*, Tucson, University of Arizona Press, pp. 354–403.

Martin, P. S., and Klein, R. G., eds., 1984, *Quaternary Extinctions: A Prehistoric Revolution*, Tucson, University of Arizona Press.

Martin, R. A., 1974, Fossil vertebrates from the Haile XIVA fauna, Alachua County, in Webb, S. D., ed., *Pleistocene Mammals of Florida*, Gainesville, The University Presses of Florida, pp. 100–113.

Matthiessen, P., 1981, *Sand Rivers*, New York, The Viking Press.

McNaughton, S. J., 1976, Serengeti migratory wildebeest: Facilitation of energy flow by grazing, *Science*, **191**, 92–94.

McNaughton, S. J., 1983, Serengeti grassland ecology: The role of composite environmental factors and contingency in community organization, *Ecol. Monogr.*, **53**, 291–320.

Mix, A. C., and Ruddiman, W. F., 1984, Oxygen isotope analyses and Pleistocene ice volumes, *Quat. Res.*, **21**, 1–20.

Moodie, K. B., and Van Devender, T. R., 1979, Extinction and extirpation in the herpetofauna of the southern High Plains with emphasis on *Geochelone wilsoni* (Testudinidae), *Herpetologica*, **35**, 198–206.

Mosimann, J. E., and Martin, P. S., 1975, Simulating overkill by Paleoindians, *Am. Sci.*, **63**, 304–313.

Murphy, D. A., and Crawford, H. S., 1970, *Wildlife Foods and Understory Vegetation in Missouri's National Forests*, Missouri Department of Conservation Technical Bulletin, No. 4, pp. 1–47.

Nowak, R. M., and Paradisio, J. L., 1983, *Mammals of the World*, Vol. 2, Baltimore, The Johns Hopkins University Press, pp. 1185–1187.

Péwé, T. L., 1975, *Quaternary Geology of Alaska*, U.S. Geological Survey Professional Paper, No. 835, pp. 1–145.

Rhodes, R. S., II, 1984, *Paleoecology and Regional Paleoclimatic Implications of the Farmdalian Craigmile and Woodfordian Waubonsie Mammalian Local Faunas, Southwestern Iowa*, Illinois State Museum Reports of Investigations, No. 40, pp. 1–51.

Rogers, K. L., 1976, Herpetofauna of the Beck Ranch local fauna (Upper Pliocene: Blancan) of Texas, *Pub. Mus. Mich. State Univ.*, **1**, 167–200.

Ruddiman, W. F., and McIntyre, A., 1982, Severity and speed of northern hemisphere glaciation pulses: The limiting case?, *Geol. Soc. Am. Bull.*, **93**, 1273–1279.

Schoener, T. W., 1982, The controversy over interspecific competition, *Am. Sci..*, **70**, 586–595.

Schoener, T. W., 1983, Field experiments on interspecific competition, *Am. Natural.*, **122**, 240–285.

Schweger, C. E., and Martin, J., 1976, *Grazing strategies of the Pleistocene Steppe-Tundra Fauna*, AMQUA Program and Abstracts, Tempe, Arizona State University, p. 157.

Shackleton, N. J., and Opdyke, N. D., 1973, Oxygen isotope and paleomagnetic stratigraphy of equatorial pacific core V28–238: Oxygen isotope temperatures and ice volumes on a 10^5 year and 10^6 year scale, *Quat. Res.*, **3**, 39–55.

Slaughter, B. H., 1966, The Moore Pit local fauna; Pleistocene of Texas, *J. Paleontol.*, **40**, 78–91.

Smith, G. I., and Street-Perrott, F. A., 1983, Pluvial lakes of the western United States, in Porter, S. C., ed., *Late-Quaternary Environments of the United States*, Vol. 1, *The Late Pleistocene*, Minneapolis, University of Minnesota Press, pp. 190–212.

Spaulding, W. G., Leopold, E. B., and Van Devender, T. H., 1983, Late Wisconsin Paleoecology of the American southwest, in Porter, S. C., ed., *Late-Quaternary Environments of the United States*, Vol. 1, *The Late Pleistocene*, Minneapolis, University of Minnesota Press, pp. 259–293.

Stuart, A. J., 1976, The history of the mammal fauna during the Ipswichian/last interglacial in England, *Philos. Trans. Roy. Soc. B*, **276**, 221–250.

Stuart, A. J., 1979, Pleistocene occurrences of the European pond tortoise (*Emys obicularis* L.) in Britain, *Boreas*, **8**, 87–97.

Stuart, A. J., 1982, *Pleistocene Vertebrates in the British Isles*, London, Longman.

Vereshchagin, N. K., and Baryshnikov, G. F., 1982, Paleoecology of the mammoth fauna in the Eurasian arctic, in Hopkins, D. M., Matthews, J. V., Jr., Schweger, C. E., and Young, S. B., eds., *Paleoecology of Beringia*, New York, Academic Press, pp. 267–279.

Vesey-Fitzgerald, D. F., 1960, Grazing succession among East African game animals, *J. Mammal.*, **41**, 161–172.

Walker, D. N., 1982, Early Holocene vertebrate fauna, in Frison, G. C., and Stanford, D., eds., *The Agate Basin Site: A Record of the Paleoindian Occupation of the Northwestern High Plains*, New York, Academic Press, pp. 274–308.

Watts, W. A., 1983, Vegetational history of the eastern United States 25,000 to 10,000 years ago, in Porter, S. C., ed., *Late Quaternary Environments of the United States*, Vol. 1, *The Late Pleistocene*, Minneapolis, University of Minnesota Press, pp. 294–310.

Webb, S. D., ed., 1974, *Pleistocene Mammals of Florida*, Gainesville, University Press of Florida.

Webb, T., III, and Bryson, R. A., 1972, Late- and postglacial climatic change in the northern Midwest, U.S.A.: Quantitative estimates derived from fossil pollen spectra by multivariate statistical analysis. *Quat. Res.*, **2**, 70–115.

Whitehead, D. R., Jackson, S. T., Sheehan, M. C., and Leyden, B. W., 1982, Late-glacial vegetation associated with caribou and mastodon in central Indiana, *Quat. Res.*, **17**, 241–257.

Wilson, J., 1973, Photosynthetic pathways and spatial heterogeneity on the North American Plains: Suggestion for the cause of the Pleistocene extinction, *Soc. Vert. Paleontol. Ann. Mtg. Abstr.*, p. 12.

Wright, H. E., Jr., 1981, Vegetation east of the Rocky Mountains 18,000 years ago, *Quat. Res.*, **15**, 113–125.

Wright, H. E., Jr., 1983, Introduction, in Wright, H. E., Jr., ed., *Late-Quaternary Environments of the United States*, Vol. 2, *The Holocene*, Minneapolis, University of Minnesota Press, pp. xi–xvii.

PART IV

MODERN EXTINCTIONS

8

EXTINCTION: WHAT IS HAPPENING NOW AND WHAT NEEDS TO BE DONE

PAUL R. EHRLICH

Department of Biological Sciences
Stanford University
Stanford, California

INTRODUCTION

The earth's biota now appears to be entering an era of extinctions that may rival or surpass in scale that which occurred at the end of the Cretaceous. As far as is known, for the first time in geologic history, a major extinction episode will be entrained by a global overshoot of carrying capacity by a single species—*Homo sapiens*. The episode, if it culminates as projected, will produce a crash in the population size of the species that caused it. Unfortunately, few laypeople are aware of the utter dependence of our species on the free services provided by natural ecosystems—and thus on other organisms that are key components of those systems (Holdren and Ehrlich, 1974; Bormann, 1976; Ehrlich et al., 1977; Westmann, 1977; Ehrlich and Mooney, 1983). Ironically, for the first time, a species engendering its own collapse has the knowledge necessary to avoid its fate but may not be able to disseminate that knowledge and act on it in time.

For population biologists, concern over the fate of humanity is only one sad aspect of the current situation. Another is the realization that the vast majority of

This paper is Contribution No. 1 from the Center for Conservation Biology, Department of Biological Sciences, Stanford University.

living species—our only *known* living companions in the universe—will be extinct long before scientists have even so much as identified and named them. Well under 2 million species have been named and described, while conservative estimates put the total in existence at between 5 and 10 million. Recently, however, studies of rain forest insects have led to the claim that there may be well over 30 million species inhabiting the Earth (Erwin, 1982). Since there is no accurate count—indeed not even an approximate count—of the number of extant species, direct estimation of rates of species extinction (as opposed to indirect estimates based on habitat destruction) is virtually impossible. But even ignoring its likely effect on human beings and not knowing exactly what is disappearing, the loss of a major portion of organic diversity must be viewed as a colossal tragedy (Ehrenfeld, 1978).

Furthermore, *species* extinction is only one part of the problem. At least in the temperate zones, it is the smaller part. There the disappearance of parts of species—subspecies, ecotypes, and genetically distinct populations—is much more threatening to the functioning of ecosystems. Numbers of such infraspecific units worldwide are even more difficult to estimate than those of species, but almost certainly there are billions.

WHAT IS HAPPENING NOW

A potential for comparing recent rates of extinction with those that occurred in the past exists only for two taxonomic groups, birds and mammals. In these groups, extinctions have been documented for a century or more, and living faunas are quite well known. Furthermore, the fossil record for the mammals can be used to make very rough estimates of the average life span of a species of higher vertebrate. Using this information, current extinction rates are estimated on the order of 4–40 times as high as those that persisted throughout most of geologic history and projected rates toward the end of the century would be 40–400 times as high (Ehrlich et al., 1977). Perhaps the best round number guesstimate is that species extinction in these groups is now or soon will be roughly 100 times the average level of the past 50 m.y.

Birds and mammals are, however, large, relatively well-studied organisms. It seems unlikely that in the last century or so many species of birds disappeared either unnoticed or before they became known to science. But this may not even be the case for small mammals, of which undescribed species may be vanishing along with tropical rain forests. The discovery in the 1960s of a small, endangered population of an undescribed species of cat on the island of Iriomote (near Okinawa) indicates that even relatively sizable mammals may still exist undescribed, especially if they are nocturnal.

When one turns to the insects, the group that accounts for the vast majority of described species diversity, extinction rates cannot be computed directly. The extant insect fauna is poorly known and even more poorly monitored. It is possible that only 5–10% of insect species have even been given scientific names. The upper bound on the total number of living species of all organisms given above is based on an estimate that there may be as many as 30 million species of tropical arthropods alone (Erwin, 1982).

A large group of insects for which there is some information on recent extinctions is the butterflies. The disappearance of several distinct taxa of butterflies, especially the satyrine *Cercyonis sthenele* and the lycaeninine *Glaucopsyche xerces*, both in San Francisco, has been documented. Most modern taxonomists, however, would consider both of these to be subspecies of species that still have extant populations. I know of no clearcut case in which a "full species" of butterfly has been shown to have disappeared from the Earth. This, however, does not mean that such an event has not occurred. Areas such as the Middle and High Atlas Mountains of Morocco, which might be expected to have a rather high level of endemism, were overgrazed long before their faunas were catalogued. Today they have butterfly faunas consisting mostly of species whose larvae feed on weeds and grasses. Also, a quite high rate of species extinction in butterfly faunas of tropical regions, especially of tropical moist forests, is both probable and also likely to go undetected.

Somewhat more can be said about the losses of populations of butterflies. Throughout North America and Europe, lepidopterists have watched the decimation of butterflies. Losses of some, such as populations of *Euphydryas editha* occupying patches of serpentine grassland near San Francisco, have been reasonably well documented as they have succumbed to development (Murphy and Ehrlich, 1981). In England, local extinctions have been closely studied (e.g., Muggleton and Benham, 1975). But mostly the evidence is anecdotal. Three years ago, Dennis Murphy and I traveled into the Pine Nut Mountains of Nevada to sample a population of *E. chalcedona* and found no butterflies and no foodplants—only a dusty mountain pass churned by the hooves of sheep. More than a decade ago, Wilfred Dowdeswell took me to see a major site for *Maniola jurtina* in England, and we arrived just as the last of it was being plowed under.

Everyone who has worked extensively with butterflies in the field can tell similar stories. One of the monuments to long-gone insects is a map of J. D. Gunder's butterfly collecting sites in the Los Angeles basin that hangs in the Department of Insects and Spiders of the American Museum of Natural History. Where butterflies thrived in the 1920s, there is now nothing but concrete. For those doing long-term research on butterfly populations, it is virtually impossible to find reasonably accessible sites that can be guaranteed to be free from disturbance for decades. Even relatively inaccessible sites are hardly secure—mineral exploration, for example, is now threatening our group's long-term study site for *E. gillettii* in the Gros Ventre Mountains of Wyoming, as well as a newly discovered population of *E. editha* in the Sweetwater Range of California and Nevada.

The situation with butterflies is, of course, just an indicator of the general problem of anthropogenic extinction. Butterflies are an excellent index of ecosystem integrity. First of all, they are the subject of considerable attention from naturalists, and the sizes of their populations are frequently noted. In addition, they are representatives of the most numerous terrestrial animals—herbivorous insects. And, since they are generally monophagous or oligophagous (Ehrlich and Raven, 1964), their diversity is closely related to plant diversity.

Because butterflies are almost never forced to extinction by overexploitation, the accelerating loss of butterfly populations in the temperate zones is an indicator of the more general loss of habitat—and a much more sensitive indicator than the disappearance of, say, organisms such as rhinoceri or cacti, which are being directly assaulted by human beings. Much habitat destruction in temperate zones is obvious even to the layperson, for example, the construction of buildings or freeways on relatively undisturbed land. Some of it is less obvious, as when diverse natural communities are overgrazed or replaced by golf courses or tree farms. And some is very subtle indeed, as exemplified by the unnoticed spread of pesticides and the increasing acidity of precipitation. All of these forms of habitat destruction, however, would be expected to lead to the loss of butterfly populations.

In the tropics, much less information is available on the status of any major group of organisms. Tropical rain forests occupy only about 6% of the earth's land surface; yet they probably contain two-thirds or more of the Earth's species. Furthermore, the biotas of the forests are very different from place to place—levels of endemism are high (e.g., Prance, 1982). Tropical forests are not uniform habitats in places such as the Amazon basin (which the uninitiated might assume to be a uniform stretch of "jungle"). The high diversity of species and the relatively restricted ranges of many of them make it safe to assume that the destruction of a hectare of tropical rain forest is much more likely to finish off an entire species than the destruction of a hectare of temperate zone habitat.

The rapid destruction of those crucial reservoirs of diversity has been an increasing focus of concern among professional biologists (e.g., Raven, 1976; Myers, 1980). Estimates of the rate of rain forest destruction vary (Ehrlich and Ehrlich, 1981), as do projections of how soon most of the biological diversity present in the forests will be lost. But most observers agree that unless dramatic changes occur, much of that diversity will vanish in a few decades. And most of the species that disappear will not even have been described and named.

The current extinction situation can best be summarized as follows:

1. With the exception of a mere handful of prominent, mostly endangered, mostly temperate zone organisms, monitoring of populations and species is utterly inadequate.
2. Direct comparisons of present species extinction rates with those of the past is impossible for most groups.
3. What evidence there is, however, that primarily comes from vertebrates, butterflies, and especially observations of wholesale assaults on habitat such as cutting of rain forests indicates that the Earth's biota is entering a period of rapid, anthopogenic extinctions that may rival or surpass in magnitude events such as the Cretaceous–Tertiary extinction episode some 65 m.y. ago.
4. In the temperate zones, extinctions are still largely of populations, rather than of entire species. In the tropical rain forests, however, species extinctions undoubtedly are already occurring at a high rate as well.

WHAT NEEDS TO BE DONE

The conservation community faces essential tasks. The first lies directly within the province of conservation biology and consists of vastly improving the monitoring of the earth's biota. The second, closely related to the first, is to design strategies to maximize its preservation.

Improved monitoring is crucial because biologists still do not know remotely enough about what lives where and hence what needs to be protected, especially in the tropics. Even in temperate zones, biological surveys tend to be woefully inadequate. The United States, with its immense wealth, does not have a published national flora, and only recently has the state of California begun to develop what promises to be an adequate inventory of its biological resources. Under those circumstances, it is hardly surprising that biologists have only a general notion of what is now disappearing from the poor nations of the tropics. And, sadly, there are still those, such as forest economists Roger A. Sedjo and Marion Clawson of Resources for the Near Future in Washington, D.C., who interpret this ignorance of detail as a justification for complacency (Boffey, 1983).

Obviously, even if essentially inexhaustible resources were put at the command of biologists today, comprehensive surveys of the Earth's biota could not be accomplished within the next crucial decade or so. Neither the time nor the trained personnel are available to do the job. But with some additional money and a lot more purposeful organization, much pertinent information could be gathered in a relatively short time.

I would suggest, for instance, that birds and butterflies be developed as key indicator groups in conservation biology and that efforts be made to develop techniques for monitoring them that will provide general information on rates of population extinction, habitat loss, and ecosystem degradation. In the temperate zones, there are large numbers of knowledgeable, willing amateurs who might be organized (partly on a volunteer basis, partly with financial assistance) to gather data on distribution and abundance according to carefully developed protocols and send it to processing centers. Christmas bird counts are an ancient tradition, and for more than three decades the Lepidopterists' Society has sponsored an annual "season summary" of butterflies collected in North America. In essence, people could become key elements in a widespread network of ecosystem monitoring while continuing to pursue their hobbies.

The problem in the tropics is both more pressing and more difficult. Monitoring of the rain forests themselves, both by satellite and on site, should be stepped up, and much more activity in these areas by professional biologists is certainly called for. But even here, potential contributions of nonprofessionals should not be ignored. Native peoples often have an excellent grasp of their local floras and faunas, and even in the tropics, enough birdwatchers and butterfly collecters are active to be organized into monitoring programs.

Research on the design of nature reserves has been escalating over the past few years, in large degree based on the theory of island biogeography (MacArthur and Wilson, 1967; Soule and Wilcox, 1980). The prototypical field experiment in this area is that of Lovejoy and his colleagues (1980, in press) on the loss of diversity from isolated remnant tracts of tropical rain forest. Our group is using a natural experiment to investigate the characteristics of reserves that will successfully support populations of butterflies and the plants they feed on. That experiment is the drying trend that, over the past 7000 yr, has isolated mesic biotas in desert mountain range "islands" in the western United States. These islands provide a wide variety of surrogate reserves with which to investigate questions such as the influence of size, shape, and distance from continuous habitat on the ability of a reserve to maintain diversity. Similar work has already been done with this system, (e.g., Brown, 1971), but our study goes a step beyond the others in looking simultaneously at two different groups, one of which is an essential resource for the other.

Many more such investigations should be underway to rapidly expand the empirical basis of island biogeographic theory—since nothing is clearer than that virtually all reserves in the near future will be newly isolated habitat islands. Such research can also help settle the persistent debate over the applicability of theory to reserve design (e.g., Simberloff and Abele, 1982; Wilcox and Murphy, 1984). That every problem of reserve design must be considered unique is both a counsel of despair and clearly not justified on the basis of present knowledge. But further development and testing of theory is clearly both possible and desirable. The problem of developing protocols for predicting what will occur as habitat is fragmented is a major challenge of population biology today.

All the investigatory activities outlined could be developed and carried out at a significant level with the expenditure of perhaps as little as $50 million/yr—an infinitesimal amount viewed in light of the risks to humanity posed by the depletion of the Earth's biota.

We cannot wait until the case for concern about the current extinction episode is strong enough to convince all decision makers. By the time that occurs, it will clearly be too late—most of the Earth's organic diversity will be gone. While attempts are being made to advance knowledge of rates and patterns of extinction and to develop principles of reserve design, enormous effort must be made to counter the forces of habitat destruction.

As has been detailed elsewhere (Ehrlich, 1980; Ehrlich and Ehrlich, 1981), this means attempting to end "development" of all relatively undisturbed habitat as rapidly as possible. It means taking steps to make areas already greatly disturbed by humanity more hospitable to other organisms. And it means dramatically curbing the spewing of poisons into the environment. These goals, so simple in outline, will of course be extremely difficult to reach. Great changes in the social–economic–political system will be required: changes in how people treat their environment and each other. But if people can be given an understanding that the very future of humanity depends on preserving organic diversity, then those changes will be seen as the only practical course of action—and humanity may pull itself together and make them.

If it has the time. In closing I must mention the greatest single threat of extinction hanging over our planet—large-scale thermonuclear war. Recent work on the ecological effects of such a conflict (e.g., Ehrlich, 1983; Ehrlich and Ehrlich, 1983) indicates that at the very least it would greatly accelerate the loss of organic diversity. If major atmospheric effects become global, as every analysis indicates is possible, then an extinction episode on the scale of that at the Cretaceous–Tertiary boundary could be entrained. Working to prevent World War III is a task that concerned ecologists should be involved in even as they deploy their special skills in other attempts to preserve the earth's living heritage.

ACKNOWLEDGMENTS

Anne H. Ehrlich, Richard W. Holm, Dennis D. Murphy, and B. A. Wilcox were kind enough to read and criticize this manuscript. This work is supported in part by a grant from the Koret Foundation in San Francisco.

REFERENCES

Boffey, P. M., 1983, Panel of experts disputes gloomy outlook for 2000, *New York Times*, May 30.

Bormann, F. H., 1976, An inseparable linkage: Conservation of natural ecosystems and conservation of fossil energy, *BioSci.*, **26**, 754–760.

Brown, J. H., 1971, Mammals on mountain tops: Nonequilibrium insular biogeography, *Am. Nat.*, **105**, 467–478.

Ehrenfeld, D., 1978, *The Arrogance of Humanism*, New York, Oxford University Press.

Ehrlich, A. H., and Ehrlich, P. R., 1983, After nuclear war, *Mother Earth News*, November/December.

Ehrlich, P. R., 1980, The strategy of conservation, 1980–2000, in Soule and Wilcox, eds., pp. 329–344.

Ehrlich, P. R., 1984, When the light is put away: Ecological effects of nuclear war, in Leaning, J., and Keyes, L., eds., *Counterfeit Ark: Crisis Relocation for Nuclear War*, Boston, Ballinger.

Ehrlich, P. R., and Ehrlich, A. H., 1981, *Extinction: The Causes and Consequences of the Disappearance of Species*, New York, Random House.

Ehrlich, P. R., and Mooney, H. A., 1983, Extinction, substitution, and ecosystem services, *Biosci.*, **33**, 248–254.

Ehrlich, P. R., and Raven, P. H., 1964, Butterflies and plants: A study in coevolution, *Evolution*, **18**, 586–608.

Ehrlich, P. R., Ehrlich, A. H., and Holdren, J. P., 1977, *Ecoscience: Population, Resources, Environment*, San Francisco, Freeman.

Erwin, T. L., 1982, Tropical forests: Their richness in Coleoptera and other arthropod species, *Coleopter. Bull.*, **36**, 74–75.

Holdren, J. P., and Ehrlich, P. R., 1974, Human population and the global environment, *Am. Sci.*, **62**, 282–292.

Lovejoy, T. E., Bierregaard, R. O., Rankin, J. M., and Schubart, H. O. R., 1983, Ecological dynamics of forest fragments, in Suttoon, S. L., Whitmore, T. C., and Chadwick, A. C., eds., *Tropical Rainforest: Ecology and Management*, Oxford, Blackwell, in press.

MacArthur, R. H., and Wilson, E. O., 1967, *The Theory of Island Biogeography*, Princeton, New Jersey, Princeton University Press.

Muggleton, J., and Benham, B., 1975, Isolation and decline of the large blue butterfly (*Maculinea arion*) in Great Britain, *Biol. Conserv.*, **7**, 119–128.

Murphy, D. D., and Ehrlich, P. R., 1981, Two California checkerspot butterflies: One new, one on the verge of extinction, *J. Lepidop. Soc.*, **34**, 316–320.

Myers, N., 1980, *Conversion of Tropical Moist Forests*, Washington, D.C., U.S. National Academy of Sciences.

Prance, G. T., ed., 1982, *Biological Diversification in the Tropics*, New York, Columbia University Press.

Raven, P. H., 1976, The destruction of the tropics, *Frontiers*, **40**, 22–23.

Simberloff, D., and Abele, L. G., 1982, Refuge design and island bigeographic theory effects and fragmentation, *Am. Nat.*, **120**, 41–50.

Soule, M. E., and Wilcox, B. A., eds., 1980, *Conservation Biology: An Evolutionary-Ecological Perspective*, Sunderland, Massachusetts, Sinauer Associates.

Westmann, W. E., 1977, How much are nature's services worth?, *Science*, **197**, 960–964.

Wilcox, B. A., and Murphy, D. D., 1984, Conservation strategy: The effects of fragmentation on extinction, submitted to *Am. Nat*.

9

ARE WE ON THE VERGE OF A MASS EXTINCTION IN TROPICAL RAIN FORESTS?

DANIEL SIMBERLOFF

Department of Biological Science
Florida State University
Tallahassee, Florida

INTRODUCTION

According to Myers (1981), destruction of moist tropical forests is proceeding so fast that they "may be reduced to degraded remnants by the end of the century, if they are not eliminated altogether. This will represent a biological debacle to surpass all others that have occurred since life first emerged 3.6 billion years ago." Is the situation really so dire?

I hope to bring to bear on this question recent approaches to biogeography (MacArthur and Wilson, 1967; Haffer, 1969) and to use real data. This is no mean task, as data on both the present tropical crisis and the mass extinctions of the geologic past are much less complete than those that ecologists are accustomed to using and render credible predictions extraordinarily difficult. This problem has been noted, and the contention advanced that we must act now even without strong evidence, since catastrophe is so imminent that conservation will be impossible by the time we have adequate evidence. "With the present rate of destruction of the tropical rain forests there is great danger of mass extinction . . . . Thousands of species could disappear before any aspect of their biology has been studied . . . . There is incomplete scientific evidence to prove this assertion, but if we wait for a generation to provide abundant evidence there will not be any rain forests left to prove it" (Gómez-Pompa et al., 1972). Soulé (1980) considers statistical confidence limits "luxuries that conservation biologists cannot now afford," but to me the urgency of the problem does not obviate statistical analysis.

HISTORICAL MASS EXTINCTIONS

Past mass extinctions (Newell, 1962, 1967) have been the most controversial phenomena to challenge paleobiologists (Schopf, 1974). During several relatively brief intervals, large fractions of the earth's biota have been eliminated (Table 1). No mass extinctions of plants are known, however (Newell, 1962, 1967). Taxonomic and stratigraphic uncertainty has beclouded identification of such episodes, but a recent systematic examination of fossil data confirms and focuses five marine mass extinctions (Raup and Sepkoski, 1982), while a spate of work on the Great American Interchange (Marshall et al., 1982) has elucidated a mass extinction of terrestrial mammals. A plethora of explanations have been proposed for mass extinctions. For marine mass extinctions, for example, meteorites, volcanic and metal poisons, extraterrestrial radiation, changes in temperature, salinity, and oxygen, and shortage of various resources or habitats have all been suggested causes (Valentine, 1973; Schopf, 1974).

Two related nomothetic principles, however, may explain at least part of many mass extinctions: area and provinciality effects (Moore, 1954; Flessa and Imbrie, 1973; Schopf, 1979). By *provinciality*, I mean how total area is divided into spatially separate units. In the evolutionary and paleontologic literature, this phenomenon is often called *endemism*, and ecologists sometimes term it *insularization*.

Perhaps ecology's oldest generalization is the species–area relationship (Watson, 1835; Connor and McCoy, 1979), that is, that large sites tend to have more species than small sites do, all other things being equal. This principle suggests that a decrease of area should be followed, with a time lag to be discussed below, by a decrease in biotic diversity. Of the six mass extinctions in Table 1, certainly four and perhaps five are accompanied by major marine areal decrease. The decrease was most severe for the most dramatic mass extinction, the Late Permian. Shallow seas decreased by 68%, and 52% of all families of marine animals disappeared.

New interest accrued to the relationship between area and biotic diversity in the wake of the equilibrium theory of island biogeography (MacArthur and Wilson, 1963, 1967), since the theory readily explains the relationship: larger sites have lower extinction rates and possibly higher immigration rates as well. However, there are other explanations for the species–area relationship (Simberloff, 1974a; Connor and McCoy, 1979), and most naturalists would agree that one of them–that larger areas tend to have more habitats, each with its own species complement—accounts for a large fraction of species–area relationships. Direct support for the equilibrium theory explanation is provided only by an experiment on arboreal arthropods of small mangrove islands (Simberloff, 1976). Although the deductive basis of the equilibrium theory area effect seems unexceptionable, the *speed* with

Table 1. Mass Extinctions of the Geological Past[a,b]

Period	Taxa	Area Change (%)	Maximum Length of Extinction (m.y.)	Provinciality	Extinction
Late Ordovician	Marine Animals	−38	5		−12% of families
Late Devonian	Marine animals	−11 to −21	25		−14% of families
	Amphibians		20		−99% of families
Late Permian	Marine animals	−68	22	14 down to 8	−52% of families
					−96% of species
	Amphibians		25		−78% of families
	Reptiles		25		−81% of families
Late Triassic	Marine animals	+10 to −27	5		−12% of families
	Amphibians		16.6		−99% of families
	Reptiles		16.6		−89% of families
Late Cretaceous	Marine animals	−27 to −43	6		−11% of families
	Reptiles		35		−57% of families
	Mammals		35		−32% of families
Pleistocene (Great American Interchange)	Mammals	0	3	2 down to 1	−27% of families
					−23% of genera

[a]Area change is for shallow seas.
[b]Data from Marshall et al. (1982), Newell (1967), Raup and Sepkoski (1982), Sepkoski (1976), Simberloff (1974), and D. Webb (personal communication, 1983).

which an areal decrease would be followed by a species diversity decrease by this route or any other is an empirical matter and might turn out to be so slow that conservationists would not concern themselves with species loss from this source.

Certainly, area changes cannot account for *all* mass extinctions. In Table 1, one can see that a number of mass extinctions of marine animals, accompanied by decrease in area of shallow seas, are contemporaneous, to the level of resolution afforded by the fossil record, with mass extinctions of terrestrial animals. As area of seas shrinks, area of land probably increases, so the terrestrial extinctions cannot be directly attributed to area decrease.

Part of this paradox may be resolved by a consideration of provinciality. That provinciality increases biotic diversity is another venerable rule. There appear to be two effects of provinciality on diversity. The first is evolutionary, the principle of "ecological vicars"—species that have evolved in different regions to perform similar ecological functions and that would not likely coexist in the same region. The same principle predicts a decrease in diversity when geological events decrease provinciality (Flessa and Imbrie, 1973; Schopf, 1979). The Late Permian marine extinction was accompanied not only by an areal decrease of 68% for shallow seas (Simberloff, 1974b) but by a decrease in number of marine provinces from 14 to 8 (Schopf, 1979). Theory provides no exact prediction of how much and how quickly diversity will decrease for a given decrease in provinciality. It is an empirical matter. For the Late Permian extinction, Schopf (1979) suggests that decrease in number of provinces was far more important than decrease in area, while the extinction of New World mammals in the Pleistocene Interchange was not accompanied by an areal decrease, but two provinces partially coalesced into one.

The second effect of provinciality on biotic diversity is much shorter term, and its study was inspired by the equilibrium theory. The conservation literature is confused about what the equilibrium theory predicts for the effect of number of provinces on biotic diversity. It is frequently asserted (references in Simberloff and Abele, 1976a) that single large refuges will, by the theory of island biogeography, preserve more species than will two or more small ones of equal total area. In fact, the theory makes no prediction (Simberloff and Abele, 1976a, 1982). This is an empirical matter and depends on the overlap of species sets in the separate refuges and the statistical form and parameters of the species–area relationship. For a variety of taxa and size ranges of sites, groups of small sites contain as many species as do single large ones and often more (Simberloff and Abele, 1982; Simberloff and Gotelli, 1984). A plausible explanation for these observations is that, on average, a single site embraces fewer habitats than does a group of distinct sites of equal total area, but there is no direct evidence that this is in fact the cause of the reported observations. In any event, no one claims that a single large site will contain fewer species than will a group of small ones of *smaller* total area, and this is the situation facing the tropics. If the group of small sites is but a subset of the single large one, it is also difficult to see how they would contain more habitats.

A final point must be noted about the historical mass extinctions. They may have occurred much more rapidly than Table 1 would suggest. As has often been noted (e.g., Newell, 1967; Hallam, 1973), the resolution of the fossil record is such that one can only place an upper bound on how long an extinction took, normally a matter of some few million years. The actual extinction, however, may have been much quicker. Recent interest in the impact hypothesis (Lewin, 1983) emphasizes the possibility that some historical mass extinctions may have been nearly instantaneous.

STATISTICAL PROTOCOLS

Any prediction of an imminent mass extinction in the tropics and comparison of what is likely to happen there with what happened during crises of the geological past must rest fundamentally on the species–area relationship, canonized as

$$S = cA^z \tag{1}$$

where S = number of species, A = area, and c and z are constants. Two other relationships between S and A fit many literature examples better than equation (1) does (Connor and McCoy, 1979). However, the qualitative results described below are valid if species and area are related by any of the relationships proposed by Connor and McCoy (1979). Similarly, $z = 0.25$; for real data, the exponent is usually around 0.25. Results are not substantially altered when the exponent to any value is set between 0.15 and 0.50.

Just as ecologists know that, on average, larger sites have more species than smaller ones do, so do they recognize that area is a very crude predictor of diversity. For a collection of 100 data sets studied by Connor and McCoy (1979), the average correlation coefficient between log area and log number of species [the transformation that linearizes equation (1)] is only 0.669. That is, only 44.8% of the variation in log S can be attributed to variation in log A. For three data sets, the correlation is negative. We will return to this problem.

The mass extinctions are almost all documented in terms of numbers of families. For purely statistical reasons, incompleteness of the fossil record is not nearly as debilitating at higher taxonomic levels as at the species level (Raup, 1972). This is a sampling problem. One is much less likely to miss all species in a genus or family than to miss any one of them. Ecological observations and theory, however, are at the species level. Empirically, one sees that family–area curves follow a similar form to species–area curves (Flessa, 1975), but with higher variance (personal observation). However, one can use rarefaction (Simberloff, 1978) to estimate how many families will be present if fewer species are present after an area reduction. One can estimate the new number of species from equation (1) and the new number of families by

$$E(F_B) = F_{\text{orig}} - \binom{S_{\text{orig}}}{S_{\text{new}}}^{-1} \cdot \sum_{i=1}^{F_{\text{orig}}} \binom{S_{\text{orig}} - S_i}{S_{\text{new}}} \tag{2}$$

A variance is easily calculated as well (Heck et al., 1975; Simberloff, 1978). Using the distribution of family sizes for living echinoids, Raup (1979) used equation (2) to estimate that 96% of all species went extinct in the Late Permian, given the observed extinction of 52% of all families.

A word is in order about rarefaction. It is used below because it seems the best statistical technique available for the questions raised, but it gives a *very* conservative estimate of decrease in number of families. This is because it assumes surviving species are a random sample with respect to family affiliation. However, confamilial species tend to be clumped both ecologically (with respect to both habitat and niche) and geographically. If a given species is extinguished by whatever means, the probability is increased that the *next* species to go extinct is in the same family. This is because the same forces that caused the extinction are more likely to affect confamilial than heterofamilial species. The effect of this nonrandomness on the rarefaction estimate of how many families will be left is to cause it to be too high; more families will be eliminated than estimated. Exactly how many more cannot be guessed without vastly more data on geography and ecology of tropical species of various taxa. But it will be the contention of this author that such data would allow a much more accurate assessment of the threat to the tropics than rarefaction permits. For now, the observation of the effects of this sort of contagion on random draws from a few distributions shows that the effect might well double the rarefaction estimate.

THE MODEL

I will attempt to fit tropical moist forest data to a model of increasing insularization. As tropical moist forest is destroyed or degraded, area decreases and insularization increases: a few large, continuous areas are converted into an archipelago of numerous, much smaller "islands." By the principles enunciated above, the area decrease should generate, with some lag time, a decrease in number of species. Conversely, if the component islands are large enough and last long enough, the increasing insularization might generate, also with some lag, an increase in number of species. We wish to predict the magnitude and speed of these two processes, and also how the change in number of species translates into change in number of families, for comparison to previous mass extinctions.

This model resembles the Pleistocene refuge theory of tropical diversity enunciated most explicitly by Haffer (1969) for neotropical birds, but even earlier in essence by Moreau (1963) for African birds. Haffer's model has also been adopted as an explanation for the great diversity of neotropical plants, butterflies, and several groups of vertebrates (references in Simpson and Haffer, 1978). His idea is that the Amazon rain forest, despite drastic recent anthropogenous insults, is near maximum extent, and that, as many as four times in the last 50,000 yr, moist forest has contracted to much smaller, isolated refugia during cooler, drier times. Intervening area became savanna, inhospitable to biota of moist forest. Populations of species of this biota are thus isolated and undergo allopatric speciation, increasing total number of species. Eventually, the climate becomes warmer and wetter again, the refugia expand, coalesce, and reconstitute a vast, continuous forest, and the new sister species invade one another's ranges or perhaps remain parapatric.

This model has been vigorously questioned by Benson (1982), Endler (1982), Strong (1982), and Beven et al. (1984). Much criticism centers around whether the posited Pleistocene climatic changes actually occurred and whether present biogeographic distributional data uniquely support the model. Its position seems established for now, however, by virtue of an approving review (Simpson and Haffer, 1978) and a massive symposium volume (Prance, 1982). Without arguing the merits of this debate, all parties would likely agree that severe insularization of a habitat, if maintained for long enough, would generate both a species loss from area decrease and a species gain from insularization if islands are sufficiently large and persistent. So I will assume that such insularization is now occurring and will be maintained and will proceed with my model.

Although different refuge theorists working with different taxa propose different numbers, sizes, and locations for refugia (cf. Simpson and Haffer, 1978), they agree that retrenchment in cooler times was severe. As an example (Figure 1), Pielou's rendition (1979) of Haffer's refugia (1969) will be used. The results do not change qualitatively when I use different schemes. Four regions of moist forest today are hypothesized to have been reduced to 10 refugia with total area just 16% of present (areas determined by planimetry). It is striking that few adherents of this view have suggested that this drastic areal decrease will cause extinctions; Fitzpatrick (1976) and Simpson and Haffer (1978) are exceptions. Rather, most have focused on opportunities for speciation that the increase in provinciality affords. Similarly, no refuge theorist proposes massive extinction once the refuges expand and coalesce. Instead, new sister species either coexist or remain parapatric.

One can easily rationalize these emphases in terms of time scales. For animals, at least, if the climatic change requires centuries or millennia, one would expect not immediate extinction, but movement to remain in the appropriate climatic regime (e.g., Coope, 1978). If the new climate does not persist long, one could hypothesize that extinction does not have time to occur, and the community trapped in the refuge never comes to a new equilibrium before area expands again. Of course, for diversity to increase, one must hypothesize that sufficient evolution occurs to generate new species in the same time span that is not long enough to allow extinctions.

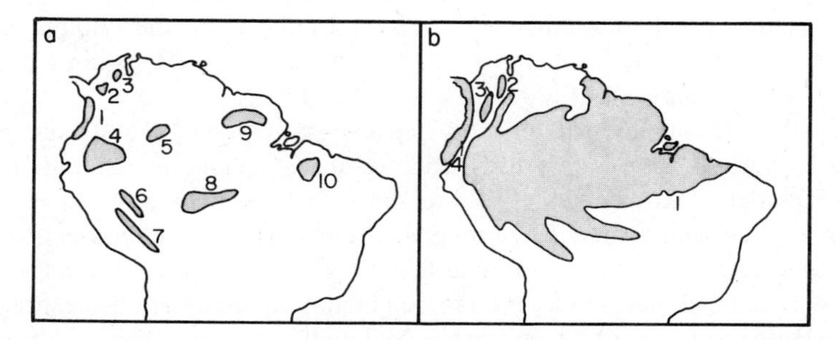

FIGURE 1. (*a*) The probable sites of patches of tropical rain forest that served as refugia for forest birds in dry periods in the Pleistocene (after Haffer, 1969). (*b*) The current extent of tropical rain forest in northern South America.

WHAT IS HAPPENING IN THE TROPICS

Biological data on tropical moist forest are almost nonexistent, and data on gross attributes like land cover are marginal. Myers (1980) made a heroic effort to collate, evaluate, and summarize the literature on how much forest there is and what is happening to it. The climatic range of tropical moist forest is ca. 16×10^6 km^2, of which the extent covered when Myers wrote was 9–11 $\times$ 10^6 km^2. The major previous statement (Sommer, 1976) had estimated 9.35×10^6 km^2 remaining. Sommer felt that, by 1976, 37% of Latin American forests, 41.6% of Asian forests, and 51.6% of African forests had already been destroyed, but since he had not considered edaphic or topographic factors, these estimates were likely inflated (Myers, 1980).

How quickly forest is being destroyed, where the process will be at the end of the century, and what the end point will be are even more matters for guesswork. Sommer (1976), using data from 13 nations encompassing 18% of all tropical moist forest, estimated a worldwide regression rate of 110,000 km^2/yr, which equals an astounding 21 ha/min. Other estimates (Myers, 1980) range from about half to about twice Sommer's estimate. Lanly (1982), using improved remote-sensing data on 76 nations comprising 97% of the tropics, estimates a current regression rate of 71,000 km^2/yr, while Myers (personal communication, 1983) suggests 92,000 km^2/yr. However, these figures refer only to deforestation, or complete removal of tree cover. If one includes other forms of conversion, like logging, that constitute sufficiently severe habitat change that one might expect extinction to result, a more likely estimate of regression rate is 119,000 km^2/yr (Lanly, 1982) to 200,000 km^2/yr (Myers, personal communication, 1983). For now I will use Sommer's estimate and err on the side of conservatism.

There is every reason to believe that this rate will increase, at least into the early years of the twenty-first century. The main reason for conversion is shifting agriculture (Lanly, 1982), which is very sensitive to population increase. The population of Africa is predicted to double in 23 yr, that of Latin America in 30 yr, and that of Asia in 36 yr (Kent, 1983). Pressure for food entailed by these increases is likely to be reflected in greatly increased shifting agriculture, with corresponding increased forest degradation. Indeed, the main brake on this degradation will probably be the increasing number of nations, like El Salvador, Madagascar, the Philippines, and Indonesia, in which forest conversion will be virtually complete within a decade if it is not already complete.

Where this destruction will end no one can say. Myers (1980) summarizes progress by various nations in setting aside refuges. Although some nations have made substantial commitments in this direction (particularly in light of the fact that most tropical moist forest resides in relatively underdeveloped countries), the total effort is woefully inadequate, with total area very small, protection limited, entire habitats not included, and so on. With rapid population increase, one cannot expect tropical nations to be willing to set aside land and to commit economic resources to long-term conservation.

From here on the discussion will be restricted to tropical moist forests of the neotropics. They comprise well over half of all tropical moist forest (Lanly, 1982). Neotropical forests are much less degraded than those of other regions. In Southeast Asia, for example, destruction is proceeding so quickly and there is so little forest left that primary lowland forest will be virtually extirpated by the year 2000. So focusing on American forests probably leads to an unduly optimistic picture. Nevertheless, availability of key biological data led this author to do so. Two taxa will be considered, land birds and flowering plants, for three reasons. First, these are the taxa for which most literature on neotropical conservation exists. Second, these taxa dominate Pleistocene refuge theory literature. Third, and most important, there is comprehensive data for these taxa on what species presently exist from massive compilations by Alwin Gentry for plants and J. V. Remsen for birds.

For plants, Raven (1976), in a seminal paper on the threat to the tropics, hazarded the guess that about 90,000 of the world's 240,000 species of flowering plants are found in tropical America. Gentry set out to compile all tropical American plants and found that Raven was remarkably accurate. From Gentry's list I find approximately 92,128 species in 239 families, although Gentry included nontropical species since his sources treated the neotropical biogeographic province as one entity. It is likely that fewer than 10,000 species are restricted to temperate parts of the province, so no great error will be introduced by using Gentry's list to represent tropical forest plants.

For area, Sommer (1976) suggests there has already been regression of 37%, but this is likely an overestimate since much that is in the climatic range of tropical moist forest probably was not tropical moist forest (see above). From that estimate, 10% was arbitrarily deducted, and it was assumed that in the recent past there has been regression of 27%, so that originally there were 6.93×10^6 km^2 of forest. It was also assumed that no plant species have yet been extinguished because of regression. Between now and the end of the century, Sommer's estimate of regression rate (1976) suggests 59,000 km^2 of forest destroyed per year in tropical America, or 952,470 km^2 by the end of the century. Myers's estimates (1980) are similar; he finds 5.6×10^6 km^2 now and, on a country-by-country basis, predicts a decrease of ca. 1,400,000 km^2 by the year 2000. Similarly, Lanly (1982) predicts a decrease in forested area of ca. 12.5% by the end of the century, but only for complete deforestation. He expects 3.9×10^6 km^2 of undisturbed productive forest to remain in the year 2000. In short, we are dealing with a projected areal decrease of 12–25% between now and the end of the century and ca. 40% between the Recent past and the end of the century. Finally, Gentry (1979) lists established national parks and equivalents for tropical America totaling ca. 96,700 km^2 of forest. This may be construed as a worst-case outcome of tropical forest destruction at some undefined future time.

Using Myers's predictions for amount of forest left at the end of the century and the worst-case estimate and assuming no extinction yet caused by recent regression, equilibrium numbers of species for those areas were calculated using equation (1) with $z = 0.25$ (Table 2). No substantial theory predicts how long it will take for these equilibria to be achieved. For plants, it may even take centuries because of their longevity. Indeed, since destruction will continue well beyond the end of

Table 2. Projected Plant Extinctions in New World Tropical Moist Forest

	Fraction of Original Area	Equilibrium Number of Species (N = 9)	Expected Number of Families $E(F)$
Originally	1.000	92,128	239
End of century	0.528	78,534	236.35
Worst-case end point	0.013	31,662	221.89

the century, the first equilibrium will never be achieved. Equation (2) gives rarefaction estimates of numbers of families expected for the predicted numbers of species. Confidence limits around these estimates are very small, but there is another reason, contagion, addressed above, why these predictions of decrease in numbers of families are likely gross underestimates. It is impossible to know just how low these estimates are, but they are almost certainly off by a factor of at least 2. It is also possible that predictions of numbers of species lost are low (see below).

Remsen's list of Amazon birds includes 704 species in 327 genera and 40 families. Consequences of three area reductions are considered: (1) If Amazonia were reduced by the end of the century exactly proportionally to Sommer's estimate for tropical forests generally (that is, to 0.593 of its extent before recent regression); (2) if Amazonia were reduced to refugia outlined by Haffer (Figure 1), a reduction to 0.160 of the original area; and (3) if Amazonia were reduced to national parks in Gentry's list in or near Amazonia, a reduction to 0.009 of the original area. For each reduction, number of species *at equilibrium* is computed by equation (1), and then the expected number of families and genera by equation (2). Results are in Table 3. The same caveats hold for rarefaction estimates of birds as for plants; it would not be surprising if actual decrease in generic and familial diversity turned out to be twice that predicted.

For neither taxon has provinciality been considered: not only will area be reduced, but it will be divided into a large number of mostly very small islands. To some

Table 3. Projected Extinctions of Land Birds of Amazonia

	Fraction of Original Area	Equilibrium Number of Species (N=9)	Expected Number of Genera $E(G)$	Expected Number of Families $E(F)$
Originally	1.000	704	327	40
End of century	0.593	618	304.63	39.72
Pleistocene refugia	0.160	445	251.80	38.73
Worst-case end point	0.009	217	155.26	34.77

extent, this is already true, and the effect will become even more extreme by the end of the century. Over the very long run, or maybe not so very long run, if refuge theorists are correct, increased provinciality would perhaps lead to new species. For example, the 84% decrease in area from the present to hypothesized Pleistocene refugia suggests, at equilibrium, a 37% decrease in number of species. The accompanying increase in provinciality from 4 to 10, given measured areas cited above, predicts an ultimate increase of 70% in number of species. This would be achieved once all allopatric populations had speciated.

I cannot be sanguine about this prospect for two reasons. First, though two species can arise from one very quickly if the right mutations occur, especially if selective pressures on two allopatric populations are very different, it does not seem plausible that this will happen often. Most newly allopatric populations will require at least a millennium or two to achieve species status, and I am not cheered by the thought of a thousand years of impoverishment.

Second, at some point, increases in provinciality without increases in area produce islands too small for long-term survival of some or all of the species initially stranded on them (Simberloff and Abele, 1976a, 1982; Simberloff and Gotelli, 1984). At exactly what point islands become so small that a substantial fraction of their biota will be extinguished must be determined empirically for each group, and this sort of research has just begun.

Shaffer (1981) contends that each species has a "critical population size" such that when populations fall below this point various stochastic forces, both genetic and ecological, are likely to extinguish it quickly. Above this point, extinction will be much slower. For tropical plants, there is no data that allow us to say anything about critical population sizes (and their associated critical areas). For birds, there are two hints that, at least for some species, they are quite large (the species require large areas). Barro Colorado Island in Lake Gatun was formed when the Chagres River was dammed during construction of the Panama Canal (Karr, 1982). Several bird species have since gone extinct. Since the area that is now the island was partly farmed and has since undergone secondary succession, habitat change alone would likely have produced some extinction (Simberloff and Abele, 1976a,b). But some extinctions are of forest species (Willis and Eisenmann, 1979) and cannot easily be attributed to succession. Several of these may be due to populations being reduced below their critical sizes. For example, some obligate army-ant-following species have disappeared, perhaps because a small (17-km^2) island simply does not have enough army ants to support them (Willis and Eisenmann, 1979).

Lovejoy and co-workers (1983) have recently begun an experiment in Brazil that should shed light on minimum population sizes and areas. They created islands of different size by felling surrounding forest and are monitoring avifaunal change. The two obligate ant followers have already disappeared from the smallest islands (1 and 10 ha). If many species have critical population sizes as large as the ant birds do, then the refuges set aside, even if guarded, may undergo substantial species loss before evolutionary changes occur. The data do not exist, but such problems will be greater for birds than for plants. Further, because of the longevity of plants, the time to extinction even after area becomes too small may be centuries.

Another problem, in addition to lack of data, pervades this effort to assess minimum areas. Area alone cannot adequately predict suitability for any particular species, just as it is an insufficient predictor of species richness (Boecklen and Gotelli, 1984). Predictions of faunal collapse based on species-area relationships alone typically have confidence intervals so large as to render them ludicrous. For example, the 95% simultaneous prediction interval for the estimate by Soulé et al. (1979) of the extinction coefficient for the Nairobi National Park spans 10 orders of magnitude (Boecklen and Gotelli, 1984). Their forecast in effect is that the Nairobi reserve will lose between 0.5 and 99.5% of its species in 5000 yr with a 95% level of confidence!

We surely can do better by looking carefully at habitats and biogeographic ranges of species. Kitchener et al. (1980), Game and Peterken (1983), and Simberloff and Abele (1982), among others, have all emphasized that by judiciously selecting sites to encompass a large variety of habitats, it may be possible to conserve as many species in a smaller total area as might have been predicted from species–area relationships to require a very large single area. Ashton (1981) has noted that many tropical tree species have very specialized edaphic requirements, but if these requirements are met, they may persist in small populations, sometimes fewer than 100 individuals.

Just as a detailed assessment of habitats should aid conservation prospects, so might ecological considerations enable us to predict more accurately the outcome of any particular action. Two sorts of studies are particularly needed. First, obligatory linkages between species are complex and numerous in the tropics, so that extinction of one species will likely have cascading effects that lead to other extinctions (Janzen, 1974). For example, Lovejoy (1973) notes that a single bee species cross-pollinates the Brazil nut tree, and the tree, while in flower, is the bee's major food. Similarly, Howe (1977) has shown that some frugivorous birds are seasonally each restricted to fruits of a single tree species, which in turn is dispersed almost exclusively by the bird that feeds on it.

Second, felling the forest over large areas will likely affect hydrology and other climatic factors (Myers, 1980, 1981; Ashton, 1981), which in turn could lead to extinctions as habitat changes. For example, Myers (1980) sketches a scenario in which reduced evapotranspiration of a reduced Amazonia leads to a drier forest in the entire region, and Salati (cited in Webster, 1983) suggests in addition major increases in flooding and river flow rates. On a small scale, Lovejoy et al. (1983) have observed increased wind-caused tree falls in their isolated 10-ha Brazilian forest and suspect a number of other tree deaths are caused by microhabitat change. They suggest that replacements for such deaths caused by habitat change are likely to be by different species, so that eventually proportions of different species in the forest will change. Local extinctions may result from such habitat modification.

Finally, sites set aside for conservation in Amazonia might not contain as many species as species–area relationships would have predicted. Prance (1977) and Gentry (1979) have argued that these sites do not encompass some major centers of plant diversity (Pleistocene refugia?) nor some edaphic and other ecologic conditions that make Amazonia so species-rich.

A MASS EXTINCTION?

Are we, then, on the verge of a mass extinction in the tropics? This author knows of no data to show that any plant or bird species has yet gone extinct in the New World tropics, but because these two taxa fossilize so poorly, we know of few such extinctions from the late Pleistocene, even though we are sure that extinctions occurred then. Perhaps the best circumstantial evidence that neotropical bird species are already extinct from forest destruction is provided by Hilty (1984), who has examined biogeographic ranges of Colombian birds. Several forest species' known ranges are completely obliterated and the birds have not been seen since, even though some are large, obvious birds that could be used for food. For example, the Cauca Guan, *Penelope perspicax*, is the size of a turkey and was restricted to the upper reaches of the Cauca Valley, an area in which forest is virtually all gone. For plants, range data are not nearly as reliable, so such inferences are more difficult.

Comparing data from Tables 2 and 3 to those of Table 1, we find that, at least in the New World, even with an increase in the rate of destruction, there is not likely to be a mass extinction by the end of the century comparable to those of the geological past. Doubling the rarefaction prediction of numbers of families lost (for reasons stated above), a loss of about 15% of all plant species and 2% of all plant families would be expected if forest regression proceeded as predicted until 2000 and then stopped completely (Table 2). How long it would take after that for these new equilibria to be achieved I cannot guess. For Amazon birds, the comparable figures are a 12% species loss and a 1% family loss (Table 3). As horrible as it is to contemplate losses of 15,000 plant species and 100 bird species, they are not comparable to the major geological mass extinctions.

However, sometime in the *next* century, and probably sooner rather than later if Myers's summary (1980) of forest regression rates is accurate and if there are no major changes in the way forests are treated, things may get much worse. If tropical forests in the New World were reduced to those currently projected as parks and refuges, by the end of the century about 66% of all plant species and 14% of families would disappear, after equilibration occurs (a few families may persist outside the New World). For the Amazon birds, 69% of all species and 26% of families would disappear before a new equilibrium was reached (a few families may persist outside Amazonia). Perhaps captive propagation, habitat manipulation, and other conservation techniques will ameliorate this situation somewhat, and it may be that the extinctions that establish the new equilibrium will be spread out over several millennia. However, the estimates throughout this chapter have been conservative, and the region (the neotropics) that is in the best condition now and is likely to persist longer has been treated. All told, then, it is clear that Myers was remarkably close to the mark. The imminent catastrophe in tropical forests *is* commensurate with all the great mass extinctions except for that at the end of the Permian.

ACKNOWLEDGMENTS

I thank Alwin Gentry, J. V. Remsen, Gary Graves, Steve Hilty, Norman Myers, Peter Raven, Dave Raup, and Dave Webb for laboriously gathered data and thoughtful advice.

REFERENCES

Ashton, P. S., 1981, Techniques for the identification and conservation of threatened species in tropical forests, in Synge, H., ed., *The Biological Aspects of Rare Plant Conservation*, New York, Wiley, pp. 155–164.

Benson, W. W., 1982, Alternative models for infrageneric diversification in the humid tropics: Tests with passion vine butterflies, in Prance, G. T., ed., *Biological Diversification in the Tropics*, New York, Columbia University Press, pp. 608–640.

Beven, S., Connor, E. F., and Beven, K., 1984, Avian biogeography in the Amazon Basin and the biological model of diversification, *J. Biogeogr*, **11**, 383–399.

Boecklen, W. J., and Gotelli, N., 1984, Island biogeographic theory and conservation practice: Species–area or specious–area relationships, *Biol. Conserv.*, **29**, 63–80.

Connor, E. F., and McCoy, E. D., 1979, The statistics and biology of the species–area relationship, *Am. Nat.*, **113**, 791–833.

Coope, G. R., 1978, Constancy of insect species versus inconstancy of Quaternary environments, in Mound, L. A., and Waloff, N., eds., *Diversity of Insect Faunas*, Oxford, Blackwell, pp. 176–187.

Endler, J. A., 1982, Pleistocene forest refuges: Fact or fancy? in Prance, G. T., ed., *Biological Diversification in the Tropics*, New York, Columbia University Press, pp. 179–200.

Fitzpatrick, J. W., 1976, Systematics and biogeography of the tyrannid genus *Todirostrum* and related genera (Aves), *Bull. Mus. Compar. Zool.*, **147**, 435–463.

Flessa, K. W., 1975, Area, continental drift and mammalian diversity, *Paleobiology*, 1, 189–194.

Flessa, K. W., and Imbrie, J., 1973, Evolutionary pulsations: Evidence from Phanerozoic diversity patterns, in Tarling, D. H., and Runcorn S. K., eds., *Implications of Continental Drift for the Earth Sciences*, Vol. 1, London, Academic Press, pp. 247–285.

Game, M., and Peterken, G. F., 1983, Nature reserve selection in central Lincolnshire woodlands.

Gentry, A., 1979, Extinction and conservation of plant species in tropical America: A phytogeographical perspective, in Hedberg, I., ed., *Systematic Botany, Plant Utilization and Biosphere Conservation*, Uppsala, Sweden, Almqvist and Wiksell, pp. 115–126.

Gómez-Pompa, A., Vazquez-Yanes, C., and Guevara, S., 1972, The tropical rainforest: A nonrenewable resource, *Science*, **177**, 762–765.

Haffer, J., 1969, Speciation in Amazonian forest birds, *Science*, **165**, 131–137.

Hallam, A., 1973, Provinciality, diversity, and extinction of Mesozoic marine invertebrates in relation to plate movements, in Tarling, D. H., and Runcorn, S. K., eds., *Implications of Continental Drift for the Earth Sciences*, Vol. 1, London, Academic Press, pp. 287–294.

Heck, K. L, Jr., van Belle, G., and Simberloff, D., 1975, Explicit calculation of the rarefaction diversity measurement and the determination of sufficient sample size, *Ecology*, **56**, 1459–1461.

Hilty, S. L., 1985, Zoogeographic changes in Colombian avifauna: A preliminary blue list.

Howe, H. F., 1977, Bird activity and seed dispersal of a tropical wet forest tree, *Ecology*, **58**, 539–550.

Janzen, D. H., 1974, The deflowering of Central America, *Nat. Hist.*, **4**, 48–53.

Karr, J. R., 1982, Avian extinction on Barro Colorado Island, Panama: A reassessment, *Am. Nat.*, **119**, 220–234.

Kent, M. M., 1983, 1983 world population data sheet of the Population Reference Bureau, Inc., Washington, D.C., Population Reference Bureau.

Kitchener, D. J., Chapman, A., Dell, J., Muir, B. G., and Palmer, M., 1980, Lizard assemblage and reserve size and structure in the Western Australian wheatbelt: Some implications for conservation, *Biol. Conserv.*, **17**, 25–62.

Lanly, J.-P., 1982, *Tropical Forest Resources (F.A.O. Forestry Paper 30)*, Rome, Food and Agriculture Organization of the United Nations, 106 pp.

Lewin, R., 1983, Extinctions and the history of life, *Science*, **221**, 935–937.

Lovejoy, T. E., 1973, The Transamazonica: Highway to extinction?, *Frontiers*, **38**, 18–23.

Lovejoy, T. E., Bierregaard, R. O., Rankin, J. M., and Schubart, H. O. R., 1983, Ecological dynamics of forest fragments, in Sutton, S. L., Whitmore, T. C., and Chadwick, A. C., eds., *Tropical Rain Forest: Ecology and Management*, Oxford, Blackwell, pp. 377–384.

MacArthur, R. H., and Wilson, E. O., 1963, An equilibrium theory of insular zoogeography, *Evolution*, **17**, 373–387.

MacArthur, R. H., and Wilson, E. O., 1967, *The Theory of Island Biogeography*, Princeton, New Jersey, Princeton University Press, 203 pp.

Marshall, L. G., Webb, S. D., Sepkoski, J. J., Jr., and Raup, D. M., 1982, Mammalian evolution and the Great American Interchange, *Science*, **215**, 1351–1357.

Moore, R. C., 1954, Evolution of late Paleozoic invertebrates in response to major oscillations of shallow seas, *Bull. Mus. Compar. Zool.*, **112**, 259–286.

Moreau, R. E., 1963, Vicissitudes of the African biomes in the late Pleistocene, *Proc. Zool. Soc. Lond.*, **141**, 395–421.

Myers, M., 1980, *Conversion of Moist Tropical Forests*, Washington, D.C., National Academy of Sciences, 205 pp.

Myers, N., 1981, Conservation needs and opportunities in tropical moist forests, in Synge, H., ed., *The Biological Aspects of Rare Plant Conservation*, New York, Wiley, pp. 141–154.

Newell, N. D., 1962, Paleontological gaps and geochronology, *J. Paleontol.*, **36**, 592–610.

Newell, N. D., 1967, Revolutions in the history of life, *Geol. Soc. Am. Spec. Pap.*, **89**, 63–91.

Pielou, E. C., 1979, *Biogeography*, New York, Wiley, 351 pp.

Prance, G. T., 1977, The phytogeographic subdivisions of Amazonia and their influence on the selection of biological reserves, in Prance, G. T., and Elias, T. S., eds., *Extinction is Forever*, New York, New York Botanical Garden, pp. 195–213.

Prance, G. T., 1982, Forest refuges: Evidence from woody angiosperms, in Prance, G. T., ed., *Biological Diversification in the Tropics*, New York, Columbia University Press, pp. 137–159.

Raup, D. M., 1972, Taxonomic diversity during the Phanerozoic, *Science*, **177**, 1065–1071.

Raup, D. M., 1979, Size of the Permo-Triassic bottleneck and its evolutionary implications, *Science*, **206**, 217–218.

Raup, D. M., and Sepkoski, J. J., Jr., 1982, Mass extinctions in the marine fossil record, *Science*, **215**, 1501–1503.

Raven, P. H., 1976, Ethics and attitudes, in Simmons, J. B., et al., eds., *Conservation of Threatened Plants*, New York and London, Plenum Press, pp. 155–179.

Schopf, T. J. M., 1974, Permo-Triassic extinctions: Relation to sea-floor spreading, *J. Geol.*, **82**, 129–143.

Schopf, T. J. M., 1979, The role of biogeographic provinces in regulating marine faunal diversity through geologic time, in Gray, J., and Boucot, A. J., eds., *Historical Biogeography, Plate Tectonics, and the Changing Environment*, Corvallis, Oregon, Oregon State University Press, pp. 449–457.

Sepkoski, J. J., Jr., 1976, Species diversity in the Phanerozoic: Species–area effects, *Paleobiology*, **2**, 298–303.

Shaffer, M. L., 1981, Minimum population sizes for species conservation, *BioSci.*, **31**, 131–134.

Simberloff, D., 1974a, Equilibrium theory of island biogeography and ecology, *Ann. Rev. Ecol. System.*, **5**, 161–182.

Simberloff, D., 1974b, Permo-Triassic extinctions: Effects of area on biotic equilibrium, *J. Geol.*, **82**, 267–274.

Simberloff, D., 1976, Experimental zoogeography of islands: Effects of island size, *Ecology*, **57**, 629–648.

Simberloff, D., 1978, Use of rarefaction and related methods in ecology, in Cairns, J., Jr., Dickson, K. L., and Livingston, R. J., eds., *Biological Data in Water Pollution Assessment: Quantitative and Statistical Analyses*, ASTM STP 652, Philadelphia, American Society for Testing and Materials, pp. 150–165.

Simberloff, D., and Abele, L. G., 1976a, Island biogeography theory and conservation practice, *Science*, **191**, 285–286.

Simberloff, D., and Abele, L. G., 1976b, Island biogeography and conservation: Strategy and limitations, *Science*, **193**, 1032.

Simberloff, D., and Abele, L. G., 1982, Refuge design and island biogeographic theory: Effects of fragmentation, *Am. Nat.*, **120**, 41–50.

Simberloff, D., and Gotelli, N., 1984, Effects of insularisation on plant species richness in the prairie-forest ecotone, *Biol. Conserv.*, **29**, 27–46.

Simpson, B. B., and Haffer, J., 1978, Speciation patterns in the Amazonian forest biota, *Ann. Rev. Ecol. System.*, **9**, 497–518.

Sommer, A., 1976, Attempt at an assessment of the world's tropical moist forests, *Unasylva*, **28**, 5–24.

Soulé, M. E., 1980, Thresholds for survival: Maintaining fitness and evolutionary potential, in Soulé, M. E., and Wilcox, B. A., eds., *Conservation Biology: An Evolutionary-Ecological Perspective*, Sunderland, Massachusetts, Sinauer, pp. 151–169.

Soulé, M. E., Wilcox, B. A., and Holtby, C., 1979, Benign neglect: A model of faunal collapse in the game reserves of East Africa, *Biol. Conserv.*, **15**, 259–272.

Strong, D. R., Jr., 1982, Comment, in Prance, G. T., ed., *Biological Diversification in the Tropics*, New York, Columbia University Press, pp. 157–158.

Valentine, J. W., 1973, *Evolutionary Paleoecology of the Marine Biosphere*, Englewood Cliffs, New Jersey, Prentice-Hall, 511 pp.

Watson, H. C., 1835, Remarks on the geographical distribution of British plants: n.p., London.

Webster, B., 1983, Forest's role in weather documented in Amazon, *New York Times*, July 5, p. 13.

Willis, E. O., and Eisenmann, E., 1979, A revised list of birds on Barro Colorado Island, Panama, *Smithson. Contrib. Zool.*, **291**, 1–31.

PART V

MODELING EXTINCTION EVENTS

10

CAUSES AND CONSEQUENCES OF MASS EXTINCTIONS: A COMPARATIVE APPROACH

DAVID JABLONSKI

Department of the Geophysical Sciences
University of Chicago
Chicago, Illinois

INTRODUCTION

Mass extinctions have been recognized in the fossil record for well over a century. Sudden changes of rock types and their contained fossils were part of Cuvier's theory of the Earth (1817) long before Darwin provided a noncatastrophic theory that allowed for evolutionary change in the intervals between faunal revolutions as well. However, despite the tremendous expansion of our descriptive knowledge of the fossil record, we still know remarkably little about the mechanisms that underlie mass extinctions. The recent upsurge of interest in mass extinctions, largely kindled by the Alvarez hypothesis for an extraterrestrial impact at the end of the Cretaceous Period, is generating a wealth of new data and (especially) hypotheses bearing on the causes of major faunal turnovers and of the role of such events in shaping the global biota. Unfortunately, attention has been focused primarily on the end-Cretaceous event, and the other major extinctions have been largely neglected as sources of information on the behavior of communities and clades under conditions of marked perturbation.

Comparisons among mass extinctions can shed light on causal mechanisms:

shared patterns in such factors as timing, ecologic or taxonomic distribution, or geography would provide useful constraints on extinction hypotheses. There is no reason to assume a priori that all mass extinctions share a common cause, but common features may imply similar mechanisms, whereas a lack of common patterns would indicate a series of unique causal events. A comparative approach requires additional biological data beyond global tallies at high taxonomic levels, as well as a shift of attention from the present bias toward the victims of extinctions to an assessment of both victims and survivors (Flessa and Jablonski, 1984).

In this chapter the major mass extinctions of the Phanerozoic will be compared with respect to hypothesized biotic patterns, such as coeval marine/terrestrial and high/low latitude effects, and hypothesized causal mechanisms, such as extraterrestrial impacts and marine regressions. Suitable data are still sparse, especially above the often-deceptive scale of local sections (see Newell, 1967, 1982; Sepkoski, 1982a; for discussion), but new information and new analyses of patterns in marine invertebrates at the generic and familial level can be brought to bear, along with a review of previous geological and paleontological work. Patterns of extinction and survival will then be compared between mass extinction events and the intervening times of background levels of extinction in order to assess the role of mass extinctions in shaping large-scale evolutionary patterns.

My preliminary analyses suggest that there are in fact large-scale patterns shared by all of the five major extinction events and that some of the proposed general mechanisms are more plausible than others. For example, of all the environmental upheavals proposed to have been responsible for the major extinctions, marine regression remains the most consistent correlate, although the causal link is probably indirect and related to such factors as climatic change and habitat destruction. At the same time, the similarities in patterns of extinction and survival between the iridium-laden end-Cretaceous extinction and the other major events, for which there is as yet no geochemical evidence for an extraterrestrial impact, may raise doubts about the uniqueness of the Cretaceous event and therefore about its hypothesized extraterrestrial trigger. On the other hand, if apparent 26- (or 28-) m.y. periodicities in cratering intensities (Alvarez and Muller, 1984) and extinction rates (Raup and Sepkoski, 1984; Sepkoski and Raup, this volume) reflect a cause-and-effect relationship, comparative extinction studies will constrain models for physical and biological effects of periodic impacts. As defined in this chapter, true mass extinctions occur less frequently than 26 m.y., and comparative studies may reveal initial conditions that set the stage for a particularly severe biotic response to extraterrestrial impacts.

The data presented here also suggest that mass extinctions are not simply intensifications of background extinction processes: although there is some biological selectivity to mass extinctions, patterns of extinction and survival are qualitatively as well as quantitatively different from those of the background intervals. Mass extinctions, therefore, must be treated as a separate class of macroevolutionary phenomena, shaping biotas in ways that are predictable neither from microevolutionary processes nor from the macroevolutionary processes that operate when background levels of extinction prevail. Again, these distinctions are particularly important if the perturbations (extraterrestrial or otherwise) that trigger mass extinctions occur with some frequency and periodicity.

A WORKING DEFINITION

Extinction at all taxonomic levels is one of the harsh realities of the history of life. As data on Phanerozoic extinction rates become increasingly refined, it is important to clarify what is, and is not, a mass extinction. Criteria are needed to pinpoint those time intervals that stand out above background levels as deserving particular attention (see e.g., Newell, 1967, 1982; Sepkoski, 1982a; Raup and Sepkoski, 1982).

Mass extinctions should satisfy criteria of timing, breadth, and magnitude. First, a mass extinction should be an event that is brief relative to the average duration of the taxa involved rather than a prolonged interval. For marine invertebrate families, this would require heightened extinction rates over a time period of 1–10 m.y. or less [comparable to Sepkoski's (1982a) criterion of one to two stratigraphic stages]. One remarkable outcome of the ongoing accumulation of Phanerozoic diversity data (Sepkoski, 1982b) has been the persistence of the abrupt nature of extinction events at the family level, despite improved stratigraphic resolution (Raup and Sepkoski, 1982, 1984) (see Figure 1).

To qualify as mass extinctions, these episodes should also have considerable taxonomic—and thus, presumably, ecologic and biogeographic—breadth, affecting a variety of higher taxa. For example, the end-Cretaceous event encompassed

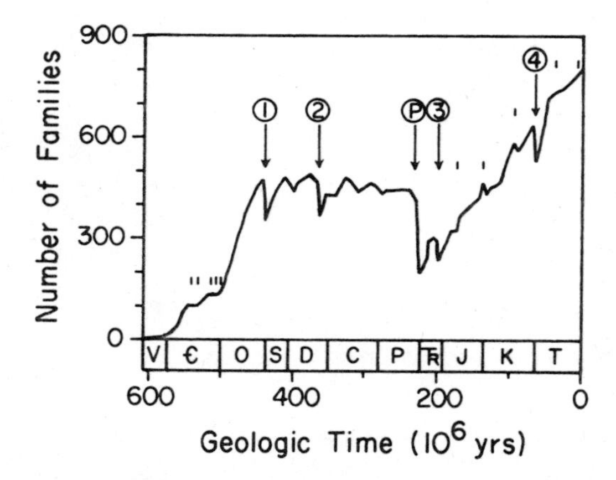

FIGURE 1. Standing diversity of families of skeletonized marine animals through the Phanerozoic, showing the five major mass extinction events. Extinctions: 1, end-Ordovician (Ashgillian); 2, Late Devonian (Frasnian); P, end-Permian (Guadalupian—Dzhulfian); 3, Late Triassic (Norian); 4, end-Cretaceous (Maestrichtian). Small vertical lines above the diversity curve mark the "lesser mass extinctions" recognized by Sepkoski: five Cambrian biomere events, the Toarcian event in the Early Jurassic, the Cenomanian event in the mid-Cretaceous, the end-Eocene event, and the Late Pliocene event. After Sepkoski (1982a); data from Sepkoski (1982b).

benthic bivalves as well as planktic Foraminifera and marine ammonites as well as terrestrial dinosaurs. An extinction peak generated by ammonites alone, although deserving careful study, should not be designated a mass extinction. Similarly, regionally restricted extinctions probably do not belong in the mass extinctions category.

All of the Major and Intermediate extinction events recognized by Sepkoski (1982a) clearly exhibit the timing and breadth required for mass extinctions. Of those, the Late Triassic (Norian) event is least securely within that category when viewed at the family level: thirty-one families of cephalopods become extinct, but other marine invertebrate groups lose only six or fewer families apiece. When evaluated at the generic level, however, the Norian event retains the breadth of a true mass extinction (see Hallam, 1981a). Of Sepkoski's Lesser Extinctions, the Cambrian trilobite extinctions that define biomere boundaries (see also Palmer, 1965, 1979) would not qualify as mass extinctions in the sense used here unless they prove to be accompanied by significant turnovers in other groups as well (which may be the case; see Rowell and Brady, 1976, on inarticulate brachiopods).

Certainly the most distinctive signature of mass extinctions is their magnitude. However, significance levels for elevated extinction rates are difficult to evaluate, particularly with respect to overall Phanerozoic diversity patterns. There may be so much variance in so-called background extinction levels that a single statistic for the global marine biota will be of little use, and the possibility of a continuum of extinction magnitudes arises (see Raup and Sepkoski, 1982; Quinn, 1983). This possibility is underscored by the spectrum of extinction events identified by Sepkoski (1982a) and the range of extinction peaks detected by Raup and Sepkoski (1984). One solution would be a shift in emphasis to taxon-specific changes in extinction rates. Because different groups and different habitats are typified by markedly different background extinction rates (Bretsky, 1969; Van Valen, 1973; Levinton, 1974; Stanley, 1979, 1982a; Kauffman, 1978; Jablonski, 1980, 1982; and many others), global tallies of total losses and gains at high taxonomic levels can be deceptive. For example, for most benthic invertebrate groups, a catastrophic doubling of generic or familial extinction rates would not bring them to equal on an absolute scale the breakneck background rates typical of many ammonoid clades. On the other hand, a modest elevation of already high ammonoid extinction rates could mask the normal turnover rates of less hurried groups and generate an apparent peak in global extinction. Extinction magnitudes would be more appropriately measured with respect to a given taxon's "normal" turnover rate than in absolute terms.

Mass extinctions, then, should be recognized on the basis of relatively sudden, simultaneous elevation of taxonomic extinction rates over background levels (by a factor of at least 2, following Sepkoski, 1982a) within each of a number of disparate higher taxa. Not included among these criteria is a required decline in standing diversity. Even severe extinction might be accompanied by origination rates—and thus rebounds—sufficiently rapid to mask such declines at the available level of stratigraphic resolution (see also McLaren, this volume). Simple counts of standing diversity and global extinction rates have been extremely useful and in fact are the basis of most of the calculations presented in this chapter, but contemporaneous, extreme elevations in extinction rates above background levels for a number of higher taxa would be the preferred measure.

COMPARISONS AMONG MASS EXTINCTIONS

The five largest mass extinctions of the Phanerozoic are the end-Ordovician (Ashgillian), Late Devonian (Frasnian), end-Permian (Guadalupian-Dzulfian), Late Triassic (Norian), and end-Cretaceous (Maestrichtian) events (Figure 1). Of these five, the end-Permian event is by far the most severe, removing approximately 50% of marine invertebrate families (Sepkoski, 1982a) and ending the reign of the Paleozoic Evolutionary Fauna (Sepkoski, 1981). Estimates of species-level extinction required to account for such losses range from 77% (Valentine et al., 1978) to a staggering 96% (Raup, 1979); at the generic level, the end-Permian extinction is equivalent to some 85 m.y. of background extinction (Raup, 1978). The other large extinction events vary in absolute magnitude, but all four involve the extinction of 15–22% of the standing diversity of skeletonized marine families, as tabulated by Sepkoski (1982a,b). Sepkoski also recognizes 8–10 "small events of mass extinction," and Raup and Sepkoski (1984; Sepkoski and Raup, this volume) document extinction peaks of even lesser magnitude, but these will not be considered here.

A comparative examination of the five major mass extinctions emphasizing shared environmental and biological changes rather than relative magnitudes may provide insights into causal mechanisms. Attempts to attribute all mass extinctions to a single driving mechanism are legion, but most do not withstand this comparative approach. Perhaps this is only to be expected: there is no guarantee that a common cause underlies these events, even when shared extinction patterns are observed. It is possible that global biotic responses to a variety of perturbations are similar once a threshold in severity is crossed. Also, as McLaren (1983; and this volume) points out, the same proximate cause may arise from a number of ultimate causes. For example, an extinction-inducing climatic deterioration might result from changes in land/sea configurations, a decline in the solar constant, extraterrestrial impacts, explosive volcanic activity, or a biogenic decline in atmospheric CO_2. Despite these problems, comparisons among extinctions can falsify otherwise intractable hypotheses and, by corroborating others, can suggest avenues of further research.

Plankton and Benthos

All of the large mass extinctions strongly affect both planktic and benthic marine organisms (Newell, 1967; Sepkoski, 1982a; Tappan, 1980, 1982). This pattern alone is sufficient to rule out many hypotheses, most notably those invoking influxes of cosmic rays or high-energy radiation derived from supernovas or other astronomical sources (e.g., Schindewolf, 1963; Reid et al., 1978; Hunt, 1978). Such events might have devastating consequences for the terrestrial biota and for organisms occupying the first few meters of the water column but would have little impact on the pelagic or benthic species enjoying the shielding effects of greater depths (Black, 1967; Waddington, 1967; Worrest, 1982). Also excluded by combined plankton—benthos extinction are most hypotheses based on biotic interactions. For example, evolution of a particularly effective durophagous predator might account for suddenly heightened extinction in shelled benthos but could not explain the contemporaneous extinction of oceanic plankton; similarly, origination of a superior competitor for essential nutrients among the plankton would have relatively little impact on the world benthos. Reversals or disruptions of competitive hierarchies (and other biotic interactions) might well be the mechanism of final extinction for many individual taxa, but the simultaneous, widespread occurrence of such biotically mediated extinctions appears likely only in the wake of major environmental perturbations.

Marine and Terrestrial

Simultaneous extinction among marine and terrestrial organisms permits a similar set of arguments, but the data are not as clear cut. The end-Cretaceous extinctions are apparently synchronous between land and sea within the limits of available stratigraphic resolution (i.e., on the scale of tens to hundreds of thousands of years) now that revision of the magnetostratigraphy of the Cretaceous—Tertiary boundary in New Mexico indicates correlation with the marine event (Butler and Lindsay, 1983). Such close synchroneity argues against a wealth of dinosaur-specific hypotheses for the end-Cretaceous event, from the fortuitous radiation of egg-eating mammals to the evolution of noxious defensive compounds in angiosperms (see Swain, 1976) or of competitively superior caterpillars (Flanders, 1962). Simultaneous extinction of dinosaurs, rudist bivalves, and calcareous nannoplankton demands a more far-reaching explanation.

Unfortunately, the vagaries of the terrestrial fossil record and the difficulties in correlation between marine and nonmarine sequences make definite conclusions difficult for the other extinction events. The end-Ordovician terrestrial record is essentially nonexistent. The Late Devonian extinction apparently had little effect on freshwater fishes (McGhee, 1982) or on land plants (Chaloner and Sheerin, 1979; Knoll, 1984; but see Scheckler, 1984, and note that there are Frasnian extinction maxima in Knoll's ordinal and familial plots, although total numbers are low). There were major shifts in the distribution and composition of terrestrial plant and animal communities during the Late Permian and through the Triassic. However, the link between these demonstrably diachonous changes on land and the marine extinctions, which also were a relatively prolonged process, is not clear (see Olson, 1982; Benton, 1983; Knoll, 1984). As Pitrat (1973) points out, the abruptness and magnitude of the terrestrial faunal turnover near the Permo-Triassic boundary may be exaggerated by changes in sedimentation in the extraordinarily rich Karroo deposits of South Africa. Nevertheless, the record does indicate that the Late Permian and Triassic were not uneventful for the terrestrial biota.

The Late Triassic was also a time of terrestrial vertebrate turnover, although again the abruptness of the event may be exaggerated by incomplete sampling and biostratigraphic imprecision (Olsen and Galton, 1977). However, the extinction of the widespread Gondwanan *Dicroidium* flora and of the rhynchosaurian reptiles does appear to have been close to, and possibly overlapping in time with, the marine Norian extinctions (see Bakker, 1977; Schopf and Askin, 1980; Hallam, 1981a, 1984; Benton, 1983; see Charig, 1984, for the competition-based scenario). Simultaneously heightened marine and terrestrial extinction rates, therefore, are at least plausible for three of the large mass extinctions but are apparently absent for the Late Devonian and not expected for the end-Ordovician.

Global Refrigeration

Climatic extremes, particularly major global cooling episodes, have often been invoked to explain mass extinctions, and Stanley (1984) has recently revived global refrigeration as a general mechanism. Climatic changes do roughly coincide with most extinction events, but simple global refrigeration may be insufficient to produce the observed patterns of extinction and survival. Glaciations accompany extinctions in the Late Precambrian, in which planktonic acritarchs suffer heavily; however, the radiation and extinction of the Ediacaran biota—assuming Seilacher's (1984) interpretation is correct—clearly postdates this glaciation (see Frakes, 1979; Vidal and Knoll, 1982; Cloud and Glaessner, 1982; Crowley, 1983). The end-Ordovician extinction is clearly associated with glaciation (Berry and Boucot, 1973; Sheehan, 1973; Brenchley and Newall, 1984), and the Late Devonian event may have been accompanied by glaciation as well (Copper, 1977; Frakes, 1979; Crowell, 1982).

After the Devonian, the glaciation–extinction correlation breaks down (Figure 2). The prolonged 90-m.y. Permo-Carboniferous glaciation (Crowell, 1978, 1982;

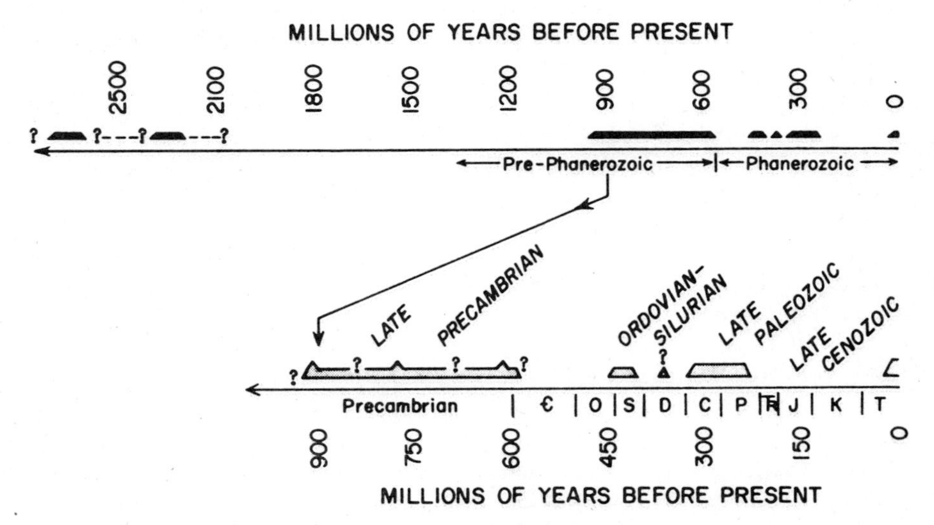

FIGURE 2. Timing of ice ages through geologic time, plotted linearly to two scales. Note poor correspondence between these glacial episodes and the extinction events plotted in Figure 1. After Crowell (1982).

Frakes, 1979) lacks extreme extinction rates, even at maximum glaciation near the Carboniferous–Permian boundary. The locus of continental glaciation shifted as Gondwana drifted across the South Pole, and the ice cap(s) waxed and waned repeatedly as evidenced by pervasive cyclic sedimentation during this interval, so it would be difficult to argue for the evolution of a uniquely extinction-resistant, glacially acclimated world biota.

The great end-Permian extinction has an even more perplexing relationship to the Late Paleozoic Ice Age. Fossiliferous glacial-marine sediments extend into the Artinskian of Australia (ca. 265 m.y. according to Harland et al., 1983) and the glacial dropstone facies may extend into the Kazanian (ca. 256 m.y.) (Crowell, 1978) or even the latest Permian Tatarian (roughly equal to the Dzhulfian) (Hambrey and Harland, 1981). Probable glaciation thus appears to overlap or be only slightly older than the onset of high end-Permian extinction rates (i.e., Guadalupian times, roughly 253–258 m.y.), and the continued high extinction rates—and marine regression—recorded by Schopf (1974) and Sepkoski (1982a) for the Dzhulfian thus actually correspond to deglaciation and the 5–8 m.y. (at most) of unglaciated Permian time remaining before the close of the Paleozoic.

After the Late Paleozoic Ice Age, there were no detectable continental glaciations for over 200 m.y. (e.g., Frakes, 1979; Crowley, 1983). The Late Triassic and end-Cretaceous extinctions, therefore, took place in the absence of major global refrigeration (although, as discussed below, some climatic changes probably did occur). Of the 8–10 "lesser mass extinctions" (Sepkoski, 1982a), the Late Eocene event appears to a marked global cooling, including the initiation to deep-sea cold-water circulation (reviews by Arthur, 1979; Kennett, 1983), and the late Pliocene event appears to correspond to the onset of the current Ice Age (e.g., Crowley, 1983). The late Pliocene event is particularly mild relative to most of the other mass extinctions and indeed may not qualify at all under the working definition used here. Few higher taxa are lost, with the most severe extinctions being at the species level among the plankton (Tappan and Loeblich, 1971, 1973). The roughly contemporaneous molluscan extinction was regionally restricted, a purely circum-North Atlantic event rather than global in extent (Stanley and Campbell, 1981; Stanley, 1982b, 1984), and it is not clear how this event should be compared to the large extinctions; for example, can this pattern and magnitude of extinction be regarded as a reflection of the same mechanism that triggered the massive, also glaciation-associated, end-Ordovician event? There is, it seems, no simple relationship between global refrigeration and mass extinction.

High and Low Latitudes

Despite a lack of correspondence between glacial episodes and mass extinctions, latitudinal variations in extinction rates are common to all the large mass extinctions. Among marine organisms, low-latitude taxa, particularly in reef communities, were invariably more severely affected than temperate, polar, and cosmopolitan taxa. This biogeographic extinction pattern has been reported for the end-Ordovician (Skevington, 1974; Sheehan, 1979; Berry, 1979), the Late Devonian (Copper,

1977), the end-Permian (Newell, 1971; Valentine, 1972, 1973; Valentine and Moores, 1973; Bretsky, 1973; Waterhouse, 1973; Waterhouse and Bonham-Carter, 1976), the Late Triassic (Fabricius, 1966; Hallam, 1981a), and the end-Cretaceous (Kauffman, 1979, 1984; Thierstein, 1981; Hallock, 1982) (see also Kauffman, 1984). For each of these extinctions, the disruption of the reef-building community was so profound or the destruction of reef-building habitats so prolonged that there was a lag of several millions years before carbonate buildups reappeared in the fossil record (e.g., Newell, 1971; Heckel, 1974; Boucot, 1983; Copper, 1974; Stanley, 1981). For example, after the collapse of the Cretaceous rudist-buildup community, there was a 10-m.y. hiatus before the resurgence of scleractinian coral reefs (Newell, 1971).

Reef communities aside, the generalization that tropical communities are more severely affected by mass extinctions requires further testing. Given the anecdotal nature of the evidence, the proposed pattern might simply be a reflection of the global diversity gradient: the tropics are more diverse, so they have more taxa to lose. However, a preliminary examination of the geographic affinities of some of the victims and survivors of the end-Permian and end-Cretaceous events does suggest that tropical families were affected disproportionately. Stratigraphic ranges for Permo-Triassic articulate brachiopod families were compiled from Sepkoski (1982b), with additions and corrections from the literature, and biogeographic affinities were assigned on the basis of Waterhouse and Bonham-Carter (1975), Hoover (1979), and Cooper and Grant (1972–1976). Extinction was high in both categories, but 75% of the exclusively tropical families became extinct, whereas only 56% of the extratropical families were lost (Figure 3).

Cretaceous–Tertiary bivalves and gastropods also show disproportionate extinction in the tropics. Family ranges again were based primarily on Sepkoski (1982b), with biogeographic data compiled from a variety of sources (e.g., Sohl, 1960, 1964, 1971; Kollmann, 1976–1982, 1979; Kauffman, 1973). Among the bivalves, 86% of the exclusively tropical families became extinct, versus only 17% of the extra-

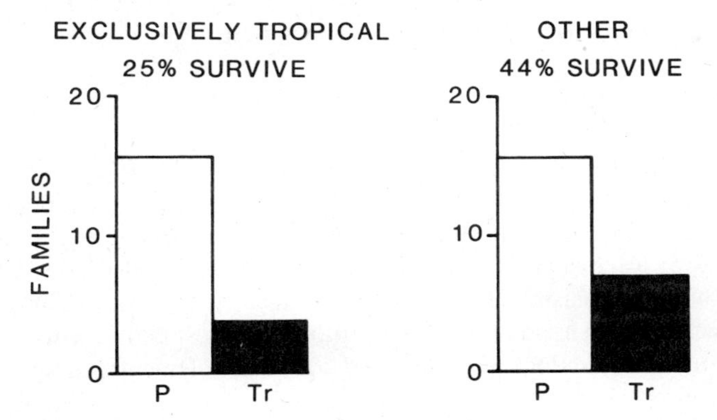

FIGURE 3. Extinction in Permo-Triassic articulate brachiopod families with respect to latitudinal distribution. Significantly fewer tropical families survive the end-Permian mass extinction than families that range into, or are restricted to, extratropical regions. See text for sources.

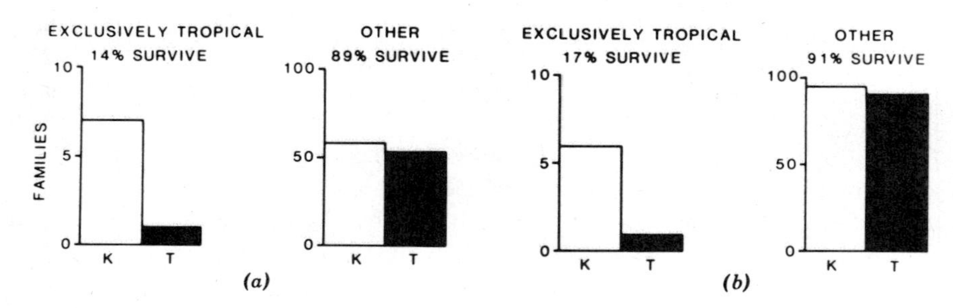

FIGURE 4. Extinction in (*a*) Cretaceous–Tertiary bivalves and (*b*) with respect to latitudinal distribution. Significantly fewer tropical families survive the end-Cretaceous mass extinction than families that range into, or are restricted to, extratropical regions. See text for sources.

tropical families (Figure 4*a*). Similarly, 80% of the gastropod families restricted to the Cretaceous tropics became extinct, whereas only 9% of the extratropical families were lost (Figure 4*b*). High survivorship among high-latitude mollusks at lower taxonomic levels is suggested by the similarity between latest Cretaceous and earliest Tertiary faunas in northern Europe (Rosenkrantz, 1960; Heinberg, 1979a,b; Voigt, 1981; Birkelund and Hakansson, 1982), in California (Saul, 1983), in the Dakotas (Cvancara, 1966), and in northern Alaska (L. Marincovich, personal communication). Further analyses are needed to confirm that this pattern occurs among other taxa and during the other extinctions, but these results do support the suggestion that the tropical biota suffer more heavily than those of other regions during mass extinction events, even in the absence of major/severe global refrigeration. It is possible that the vulnerability of the tropics reflects its biogeographic structure, with high proportions of endemic taxa in both marine (Valentine, 1973) and terrestrial (McCoy and Connor, 1980; Brown and Gibson, 1983; Simberloff, this volume) environments rather than a climatic pattern common to the mass extinctions. As discussed below, endemic clades appear too vulnerable to mass extinctions regardless of their latitudinal situation. More data are needed before these not entirely mutually exclusive hypotheses can be tested.

Extraterrestrial Impacts

Collision with extraterrestrial objects or other astronomical mishaps such as nearby supernovas have been suggested repeatedly as causes for mass extinctions (e.g., de Laubenfels, 1956; Schindewolf, 1963; Russell and Tucker, 1971; Urey, 1973; Lemcke, 1975; Reid et al., 1978; Russell, 1979; Napier and Clube, 1979), often seemingly because plausible earthbound mechanisms had been exhausted. Only in the past 5 yr have such hypotheses entered the realm of testability, with the discovery of anomalous concentrations of iridium and other trace elements, as well as sanidine spherules and shock-metamorphosed quartz, at the Cretaceous–Tertiary boundary all over the world (Alvarez et al., 1980, 1982, 1984a,b; Smit and Hertogen, 1980; Kyte, et al., 1980; Ganapathy, 1980; Ganapathy et al., 1981; Smit and Klaver, 1981; Orth et al., 1981, 1982; Smit and ten Kate, 1982; DePaola et al., 1983;

Kyte, 1983; Montanari et al., 1983; Gilmore et al., 1984; Bohor et al., 1984; Smit and Kyte, 1984). The coincidence of such apparent impact signatures with the extinction level at over 50 localities on land and sea is extremely impressive, but problems remain.

Doubts regarding the hypothesized extraterrestrial impact at the end of the Cretaceous are both geochemical and paleontologic. Little is known about the geochemistry of most trace elements, particularly their diagenetic behavior. Platinum group metals such as iridium and osmium, for example, may be concentrated in sediments under anaerobic conditions (Goldschmidt, 1954; Rucklidge et al., 1982; Keith, 1982; but see Finkelman and Aruscavage, 1981), and many Cretaceous–Tertiary boundary clays apparently record reducing microenvironments rich in pyrite or, in terrestrial sections, are associated with coals. Iridium concentrations similar to or higher than those reported for the Cretaceous–Tertiary boundary are recorded from deep-sea manganese nodules (Harriss et al., 1968), carboniferous coals [Block and Dams, 1975; Botoman and Stith, 1981; although these concentrations may be detrital rather than diagenetic in origin (Finkelman and Aruscavage, 1981) and thus less of a problem for the impact hypothesis], and kerogen in a Permian limestone (Kucha, 1981). Reports of iridium enrichment in a mid-Cretaceous bituminous shale (Wezel, et al., 1981) are apparently attributable to contamination, however (Alvarez et al., 1984a), and in any event it remains to be clearly worked out whether these diagenetic processes are really incompatible with end-Cretaceous impact hypotheses (Van Valen, 1984). Magnetic spherules in Lower Cretaceous hardground of Italy (Castellarin et al., 1973) deserve further investigation as well.

Particulates from the Hawaiian volcano Kilauea contain much iridium as well as several of the other elements (selenium, arsenic) that are also anomalously enriched at the Cretaceous–Tertiary boundary (Zoller et al., 1983). Kilauean elemental ratios differ from the boundary measurements, however, and the one end-Cretaceous volcanic eruption sufficiently large to blanket the Earth with the appropriate particulates, the Deccan Trap basalts, extended over too long an interval to produce the relatively well-defined spike observed in the boundary clays (see Wensink, 1973; Mahoney et al., 1982). Osmium isotope ratios are consistent with meteoritic origin for the end-Cretaceous boundary material, but the significantly different ratios from Denmark and Colorado would require contamination from crustal sources or the simultaneous impact of at least two meteorites of differing composition (Luck and Turekian, 1983). (Even this seeming inconsistency can be accommodated ad hoc if the impact episodes occur as comet showers rather than as single, stray asteroids.) The mineralogy of the iridium-rich fine fraction of the Cretaceous–Tertiary boundary clay differs among localities but is similar to clays stratigraphically above and below it (Rampino and Reynolds, 1983; but see Bohor, 1984, and Alvarez et al., 1984a). None of these data conclusively falsify the impact hypothesis for the source of the end-Cretaceous iridium, but taken together they raise stubborn questions about iridium anomalies as unique signatures of extraterrestrial impacts. Along with these problems, which may be resolved by the time this chapter is published, remain among-site isotopic variations and conflicting results regarding land versus sea impact sites, indicating that the impact evidence is not quite as monolithic as its most zealous advocates would suggest (see also Grieve, 1984, and especially Van Valen, 1984).

Paleobiological data are also still somewhat equivocal. Although there is a remarkable stratigraphic correspondence between the iridium anomaly and the biological extinction horizon at most localities, Officer and Drake (1983) note several deep-sea cores in which the correspondence is not precise. A number of these imprecisions are clearly the result of bioturbation or of disturbance when the core was taken (Alvarez et al., 1984a), but at least two cores may still provide reliable data. At DSDP Site 524 in the eastern South Atlantic, the anomaly appears approximately 40 cm below the final dwindling of Cretaceous nannofossils; judging by estimated sedimentation rates for that locality, the final extinction would be approximately 80,000 yr too late. The iridium anomaly is smudged out over some tens of centimeters in the core, which might be expected owing to bioturbation, but mathematical models of biogenic mixing processes indicate that this would not be sufficient to shift the peak of the anomaly from the level of the final nannofossil disappearances to its present position. On the other hand, at DSDP Site 465 in the northwest Pacific, the iridium anomaly appears 30 cm above the biostratigraphic boundary and thus is about 100,000 yr too late to be the signature of an event responsible for the end-Cretaceous extinction. Clearly, further microstratigraphic studies of the boundary interval are needed to corroborate or revise these important results; for example, the critical core at Site 465 appears to have been disturbed during drilling (cf. Kyte, 1983). If the initial findings are correct, two interpretations are possible. If the biological extinction was globally synchronous, these data would suggest that the iridium anomaly is diagenetic or sedimentologic in origin and its presence depends on special conditions that varied in age slightly from locality to locality. If the iridium anomaly is a true time horizon, as now appears to be more likely, then the oceanic plankton extinction extended over the order of 50,000–100,000 yr, which would falsify many of the present models for the physical and biological effects of the impact (see Percival and Fischer, 1977; Hsü et al., 1982a,b; Perch-Nielsen et al., 1982, for further discussion of noninstantaneous plankton extinctions).

The magnitude of the end-Cretaceous extinctions may also be inconsistent with the more extreme models for an extraterrestrial impact. Toon et al. (1982; Pollack et al., 1983) indicate that light levels would have been too low for photosynthesis for 2–12 months and that continental temperatures would drop below freezing for 4–24 months. Add to this the hypothesized heat shock, cyanide poisoning, ozone depletion, and acid rain (Emiliani et al., 1981; Hsü et al., 1982a,b,; Lewis et al., 1982; O'Keefe and Ahrens, 1982), and extinction rates and patterns begin to appear too mild, particularly outside the tropics, to be consistent with the available impact scenarios (see above and also Kauffman, 1979, 1984; Buffetaut, 1980, 1984; Van Valen, 1984). More plausible models are needed for the biotic effects of extraterrestrial impact, particularly in light of evidence that final plankton extinction may have postdated the iridium anomaly by some tens of thousands of years (Hsü et al, 1982a,b; Perch-Nielsen et al., 1982).

The other large mass extinctions apparently lack geochemical anomalies. The negative results for the Late Devonian (McGhee et al., 1984), the Permo-Triassic (Asaro et al., 1982), and the Cambrian biomeres (Orth et al., 1984) have been regarded by some impact enthusiasts as evidence that a comet impact was the extinction-triggering event! On the other hand, iridium anomalies, along with microtektites providing clear evidence for an extraterrestrial impact, are present near the Eocene–Oligocene boundary (O'Keefe, 1980; Glass, 1982; Ganapathy, 1982; Alvarez et al., 1982) and in the late Pliocene (ca. 2.3 m.y. ago; Kyte et al., 1981). The former is associated with a "lesser mass extinction" and a pronounced cooling event in terrestrial, shallow-water and deep-sea environments (Sepkoski, 1982a; reviews in Kennett, 1983; Savage and Russell, 1983), whereas the latter event appears to have had no significant global effects [the Neogene molluscan extinctions discussed by Stanley and Campbell (1981; Stanley, 1982b, 1984; Vermeij, this volume) were apparently resticted to the North Atlantic and Caribbean and were underway by ca. 3.0 m.y. ago]. The role of the Late Eocene impact in the extinctions is still unclear. At least three microtektite horizons are known from the early Late Eocene to mid-Oligocene (39.4–31.0 m.y.) and none appear to correspond exactly to significant marine or terrestrial extinctions (Keller et al., 1983), and changes in plate tectonic configurations and in deep-sea circulation patterns may be sufficient to explain the observed climatic changes and faunal turnovers (Keigwin, 1980; Van Couvering et al., 1981; Burns and Nelson, 1981; Corliss, 1981; Cavelier et al., 1981; Haq, 1981, 1982; Miller and Curry, 1982; Norris, 1982; Crowley, 1983; Kennett, 1983; Keller, 1983a,b; Snyder et al., 1984).

Too little is known about the physical effects of impacts to decide whether the end-Cretaceous, Late Eocene, and Late Pliocene biotic patterns can all be accommodated by impact models. This problem becomes particularly severe if periodic comet showers are invoked to trigger all of the post-Paleozoic extinction peaks of Raup and Sepkoski (1984) (Davis et al., 1984; Whitmire and Jackson, 1984; Rampino and Stothers, 1984; see Jablonski, 1984b). In addition to the problem of relative magnitudes of extinctions, it is curious that end-Cretaceous deep-sea benthic Foraminifera are virtually unaffected by the extinction, but Late Eocene deep-sea Foraminifera show a graded but pronounced turnover (see Douglas and Woodruff, 1981; Kennett, 1983; Snyder et al., 1984); that deep-sea ostracodes show an "unexpectedly catastrophic" end-Cretaceous extinction (ca. 25% of the genera) but a diversification in the Late Eocene (Benson et al., 1984); and that dinoflagellate plankton extinctions may actually have been more profound at the end of the Eocene than at the end of the Cretaceous (cf. Tappan and Loeblich, 1971, 1973; Bujak and Williams, 1979; Hansen, 1979; McLean, 1981; De Coninck and Smit, 1982; Thierstein, 1982). A comparison among extinctions, then, is ambiguous: not all extinctions have geochemical evidence for extraterrestrial impacts, impacts do not always bring large mass extinctions, and biotic responses are different for the two iridium-associated extinction events. The impact hypothesis would more comfortably fit the Late Eocene and end-Cretaceous paleontological data if it could be demonstrated that the most severe global effects of extraterrestrial impacts would entail climatic perturbations on time scales of thousands to tens or hundreds of thousands of years.

Paleoceanographic Changes

Changes in global ocean chemistry, salinity, or circulation have also been hypothesized as general mechanisms for the large mass extinctions. However, timing and magnitude of Phanerozoic chemical or salinity excursions do not appear to be in step with extinction events. For example, only one of the oceanic anoxic events of the mid-Cretaceous (reviewed by Arthur, 1979; Jenkyns, 1980; Schlanger and Cita, 1982; Kennett, 1983) coincides with the "lesser mass extinction" at the end of the Cenomanian (Sepkoski, 1982a; Kauffman, 1984). The world ocean was well oxygenated during the last 20 m.y. of the Cretaceous, and during the Late Permian and the early Paleozoic, black shales closest in age to the end-Ordovician event were deposited during the Early Silurian, postextinction, transgression (Leggett, 1978, 1980). Black shale facies do expand near the time of the Late Devonian event, at least in western Europe (McLaren, 1982; Eder and Franke, 1982), but the Late Triassic event probably represents the strongest association between anoxic conditions and mass extinction on a potentially global scale (Hallam, 1981a,b; Ager, 1981).

Proposed salinity extremes (Fischer, 1964; Holser, 1977; Stevens, 1977; Keith, 1982) or catastrophic injections of brackish water into the world ocean (Gartner and Keany, 1978; Thierstein and Berger, 1978; Berger and Thierstein, 1979; Gartner and McGuirk, 1980) have not been supported for the end-Cretaceous by the subsequent biostratigraphic or isotopic analyses (Boersma and Shackleton, 1979; Buchardt and Jørgensen, 1979; Perch-Nielsen et al., 1979; Clark and Kitchell, 1979, 1981; Watts et al., 1980; Thierstein, 1980; Berger et al., 1981; Kitchell and Clark, 1982; Kauffman, 1984), nor are they consistent with the latitudinal patterns of extinction discussed above. Although oceanic circulation rates and oxygenation levels undeniably have varied significantly through the Phanerozoic, it appears that these variations have not played a primary role in mass extinctions. Instead, they may be a byproduct of climatic or sea-level perturbations (which in turn may have several possible underlying causes, as mentioned above) that under certain conditions also trigger extinction events (e.g., Fischer and Arthur, 1977; Fischer, 1981, 1984).

Marine Regression

Extensive withdrawals of the sea have been hypothesized to cause mass extinctions through species–area effects (Schopf, 1974; Simberloff, 1974), habitat destruction (Newell, 1967), and climatic change (Haq, 1973). Global marine regressions do indeed coincide with the end-Ordovician (Berry and Boucot, 1973; Sheehan, 1973; Lenz, 1976; McKerrow, 1979; Chen, 1984; Brenchley and Newall, 1984), the end-Permian (Newell, 1967; Schopf, 1974; Hallam, 1984), the end-Triassic (Hallam, 1981a, 1984), the end-Cretaceous (Matsumoto, 1980; Kauffman, 1984), and the Late Devonian (although evidence here is weakest; for a diversity of opinions, see McLaren, 1970, 1982, 1983; Johnson, 1974; House, 1975a,b; Schlager, 1981; Burchette, 1981; Hallam, 1984; Johnson et al., 1985), but the association is troublesome for at least two reasons. Not all regressions are accompanied by mass

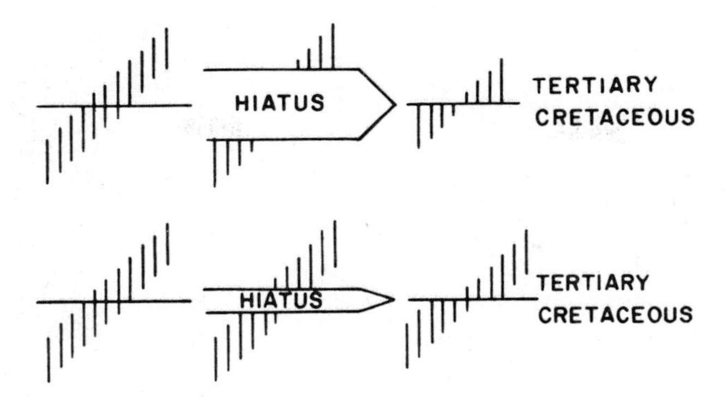

FIGURE 5. The truncation effect. A stratigraphic hiatus generated by marine regression will truncate stratigraphic ranges and thus artificially enhance the apparent abruptness of an extinction event. After Birkelund and Hakansson (1982).

extinctions, and regressions distort the fossil record so that a literal interpretation of biostratigraphic evidence can be misleading.

Regression of the sea interrupts marine sedimentation and permits erosion and reworking of older deposits during subaerial exposure and subsequent transgression. Such processes can seriously distort patterns of extinction in the fossil record in two different ways. First, regression can enhance the apparent magnitude and abruptness of extinction (Figure 5). Time is missing from the stratigraphic record, so that species' ranges appear to terminate at a single horizon; this might be termed the truncation effect (see, for example, Birkelund and Hakansson, 1982; Kauffman, 1984). On the other hand, progressive loss of sampling area with the retreat of the sea can result in an artificially gradual extinction. If there is a progressive decline in sampling or preservational quality approaching an extinction boundary—or simply if last sampled occurrence of each taxon is randomly distributed relative to its true disappearance—then even a razor-sharp, instantaneous extinction event will appear as a progressive decline (see Signor and Lipps, 1982).

The Lazarus Effect

It is difficult to correct for the truncation effect, but artificially gradual extinction can be assessed by examining the observed stratigraphic record of taxa known to cross the extinction boundary. The proportion of those taxa that must have survived through a given time interval (because they appear on either side of it) yet are not recorded within it will give a rough indication of completeness of the fossil record for the interval in question. This disappearance and apparent extinction of taxa that later reappear unscathed can be termed the *Lazarus effect* (Figure 6). The magnitude of the Lazarus effect is an indication of the distortion suffered by the fossil record in that time interval. The sediments of the Gulf and Atlantic Coastal Plain of North America contain some of the most diverse and well-preserved Late Cretaceous molluscan faunas in the world (Sohl, 1960, 1964, 1971). In this region, the progressive decline of bivalve and gastropod generic diversity to 90% of Campanian

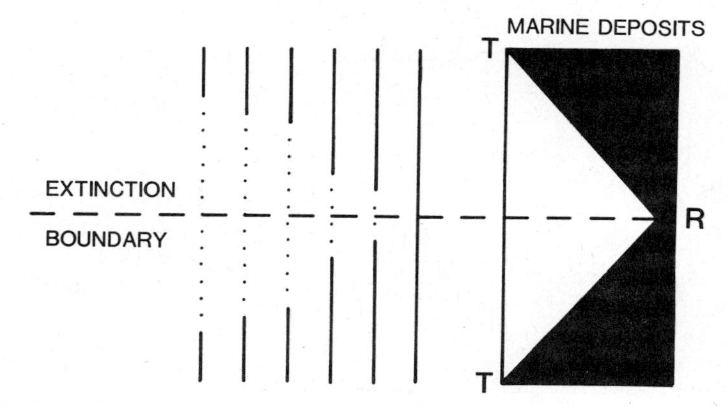

FIGURE 6. The Lazarus effect. If the quality of the stratigraphic record declines during a regression, an extinction event can be made to appear artificially gradual. This can be controlled for by examining the sampling record for taxa known to cross the boundary. The proportion of taxa that vanish during the regression but reappear unscathed sometime later (the Lazarus effect) provides an estimate of sampling bias.

levels in the early Maestrichtian, then to 75% in the mid-Maestrichtian, with 54% surviving into the Cenozoic might be regarded as strong evidence for a gradual, rather than catastrophic, faunal change in the latest Cretaceous. This can be tested by evaluating the record of the Coastal Plain Cretaceous genera known to persist into the Cenozoic. As shown in Figure 7, the 120 Coastal Plain genera known to

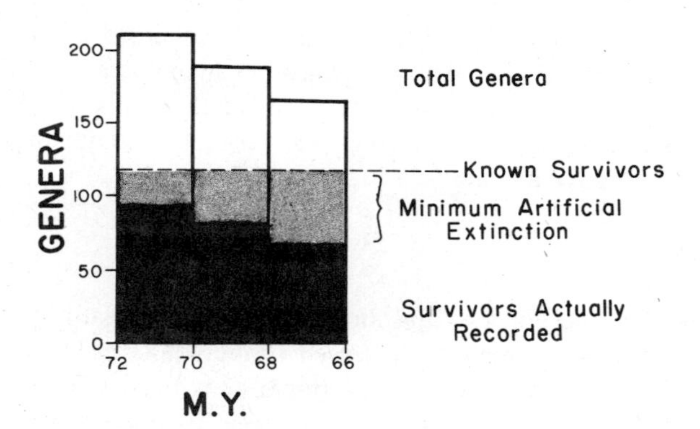

FIGURE 7. Artificial extinction in Late Cretaceous bivalves and gastropods. In the Gulf and Atlantic coastal plain, 114 Late Cretaceous genera survive into the Tertiary. There is a progressive decline in the total number of genera present in the last 6 m.y. of the Cretaceous (depicted in the total height of each histogram), which might suggest a gradual end-Cretaceous extinction. However, there is also a decline over that time interval in the number of genera known to survive the extinction that are actually recorded in the fossil record of the region (black histograms). Since 114 genera are known to cross the Cretaceous–Tertiary boundary, the stippled part of the histograms represents the minimum artificial extinction for each interval, that is, the proportion of taxa that exhibit the Lazarus effect, seeming to become extinct but returning unscathed at some time after the extinction.

survive the end-Cretaceous event appear to dwindle to 72%, then to 56%, of their Late Campanian numbers during the last 6 m.y. of the Cretaceous. This pattern of apparent extinction and subsequent return is an artifact of changes in depositional environments and stratigraphic incompleteness near the extinction boundary. The artificial decline in diversity provides a basis for evaluating apparently gradual declines in other groups that do suffer final extinction at the boundary—the 56% Lazarus effect in the Coastal Plain bivalves and gastropods, for example, is remarkably similar to the overall gradual decline observed for those groups in this region. This suggests that a hypothesis of abrupt extinction cannot be rejected for these taxa at the rather coarse scale of less than 2 m.y., the duration of the last time interval examined.

The Lazarus effect is apparent for most extinction events, with the most impressive examples occurring at the Permo–Triassic boundary. Preservational biases near this boundary are so strong, for example, that Batten (1973) found more Paleozoic families and genera of gastropods in mid-Triassic strata than in the Upper Permian! Similar patterns, on a less massive scale, are also recorded for genera and families of Permo–Triassic bivalves and articulate brachiopods (Nakazawa and Runnegar, 1973; Waterhouse and Bonham-Carter, 1976) and perhaps among higher taxa of bryozoans (Boardman, 1984). These data suggest that the prolonged nature of the Guadalupian–Dzulfian extinction peak detected by Raup and Sepkoski (1982) and Sepkoski (1982a) is at least in part an artifact.

The end-Cretaceous event exhibits the Lazarus effect in other regions as well. For example, in the classic boundary sections in Denmark, the Cenozoic bryozoan fauna appears immediately above the Cretaceous–Tertiary unconformity, but the Cretaceous survivors vanish from the record for the duration of an entire foraminiferal zone before they are recorded again (Birkelund and Hakansson, 1982); in the same sections, all articulate brachiopods disappear, with Cretaceous holdovers and new Danian species reappearing at roughly the same level as the bryozoans (cf. Surlyk and Johansen, 1984). Similarly, following the Late Devonian event, the order Stromatoporoidea, comprising the major reef builders of the mid-Paleozoic, disappears completely from the record for the first substage of the Famennian and returns in the last substage of the Devonian (C. W. Stearn, in McLaren, 1982; see also Bogoyavlenskaya, 1982). Finally, of the eight families of cystoid echinoderms known to survive the end-Ordovician event, only one is actually recorded from the first series of the Silurian (Paul, 1982).

The Lazarus effect shows how the fine structure of the record of biotic change may be distorted or concealed, but there is no denying that mass extinctions are real phenomena: the biologic discontinuites are too profound and lasting. Neither dinosaurs nor ammonites returned in the Cenozoic. Furthermore, the observed declines of some taxa are too prolonged and encompass too many transgressive–regressive pulses to be accounted for purely in terms of sampling biases. Such prolonged declines include the Late Cretaceous records of the inoceramid bivalves (Dhondt, 1983; Alvarez et al., 1984b) and ammonites (Hancock, 1967; Wiedmann, 1973; Kennedy, 1977). Sampling biases may preclude a similar conclusion for the dinosaurs (Russell, 1975, 1984), but this does not invalidate the shorter-term latest-

Cretaceous decline inferred by Van Valen and Sloan (1977; Van Valen, 1984) and Archibald and Clemens (1982, 1984; Clemens, 1982, and this volume), because as Signor and Lipps (1982) point out, the pattern of gradual replacement of dinosaurs by Tertiary mammals—rather than the dwindling of a single group—could not be generated by a catastrophic event (the data have, however, recently been challenged by Smit and van der Kaars, 1984). The same conclusions could be drawn for the early appearance of Tertiary forms in the high-latitude Upper Cretaceous of North America (Hickey et al., 1983) and—if bioturbation can be ruled out, as suggested by the isotopic data of Perch-Nielsen et al. (1982)—for the rapid but progressive post-iridium replacement of Cretaceous nannofloras by Tertiary forms (Percival and Fischer, 1977; Thierstein, 1982; Hsü et al., 1982a,b; Perch-Nielsen et al., 1982).

It should again be emphasized that the existence of a pronounced Lazarus effect does not demonstrate abrupt extinction: it is simply a method for estimating uncertainties in stratigraphic ranges around a particular event. This uncertainty can be a double-edged sword. For example, the fact that the surviving Cretaceous brachiopods disappear at the same horizon as the exterminated species and fail to reappear throughout the basal Danian *eugubina* zone (Surlyk and Johansen, 1984) undercuts the claim for an instantaneous extinction precisely at the boundary clay: any pattern of decline could have occurred during the interval in which all brachiopods, victims and survivors alike, were unrecorded.

Species–Area Effects Are Not Enough

One of the most elegant explanations for the association between mass extinctions and marine regression has been the extension of the equilibrium theory of island biogeography to high taxonomic levels and geologic time scales. Schopf (1974) and Simberloff (1974) argued that the end-Permian extinction could be explained in terms of species–area relationships during the Late Permian regression. Following MacArthur and Wilson (1967; see also Preston, 1962; Simberloff, 1972, 1974), these authors concluded that the end-Permian extinction could be explained simply by decreased habitable shelf area leading to decreased population sizes, which increased species' vulnerability to stochastic extinction processes. Extinction rates remained elevated above origination rates until an equilibrium diversity was reached for the new shelf area.

These extinction hypotheses, based on the biogeographic effects of areal reduction of the continental shelf, do not take into account the extensive shallow-water habitats around oceanic islands. This is not a trivial problem, because conical islands are not subject to reduction in shallow-water area during regression (Stanley, 1979, 1984), and overall habitat diversity can be maintained during a major drop in sea level through the emergence of drowned sea mounts to replace those that have suffered subaerial exposure (Jablonski, 1984a). Unfortunately, the biotas of ancient oceanic islands rarely can be directly assessed, because their records are obscured and ultimately destroyed by plate-tectonic processes (Ordovician volcaniclastic deposits may contain some intriguing exceptions; Neuman, 1972, 1976; Lockley, 1984).

The potential role of island refuges in sheltering shallow marine diversity from species–area effects during regression can be estimated by analyzing the proportion of present-day benthic marine families represented on oceanic islands (Jablonski, 1984a; Jablonski and Flessa, 1984). It turns out that oceanic islands harbor a remarkably large proportion of the shallow-water families of bivalves, gastropods, asteroids, ophiuroids, echinoids, and scleractinian corals (Figure 8), despite today's highly provincial biogeographic structure (e.g., Campbell and Valentine, 1977). Of the 276 families considered, 239 (87%) have species reported from one or more of the 22 oceanic islands for which data were obtained; 220 families (78%) occur on two or more of the islands. These results suggest that reduction in continental shelf area during marine regression is not sufficient in itself to explain mass extinctions, because even a marine regression that totally eliminates the shelf biota would cause at most a 13% extinction of Recent families—a figure that does not compare well with estimated 52% familial extinction at the end of the Permian.

Differences in species–family ratios between mainland and island faunas add a complicating factor (Jablonski and Flessa, 1984). There are significantly fewer species per family on islands relative to continental shelves for echinoids (with the relationship between island and mainland differences in species–family ratios being about 1 : 1.3) and bivalves (a relative difference of 1 : 1.5), but not for gastropods.

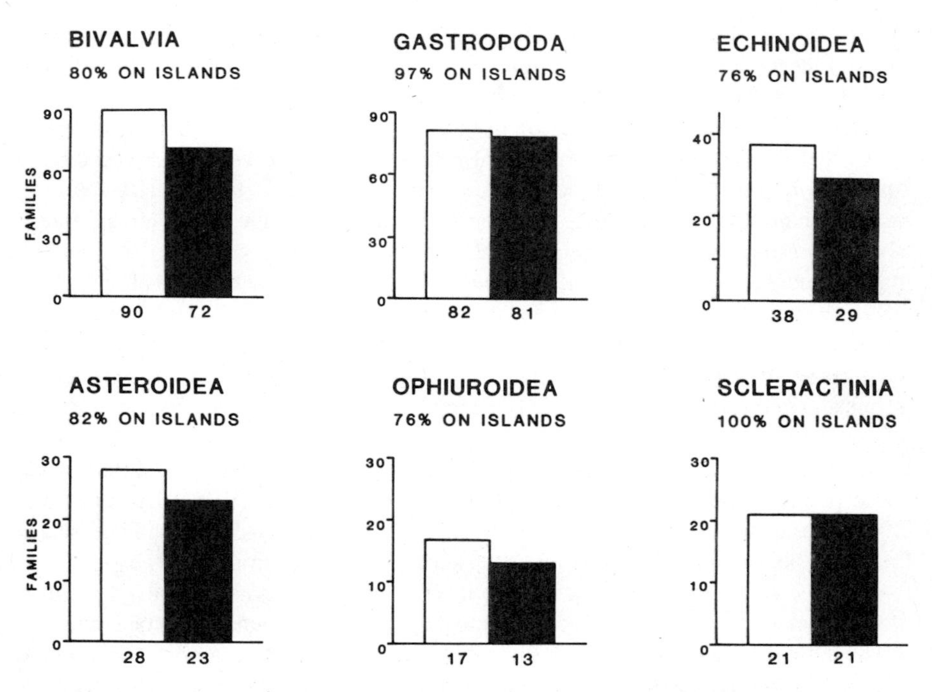

FIGURE 8. Most of the shallow-water families in the Recent world biota (white bars) are represented on 22 oceanic islands (black bars), suggesting that even an extreme regression on the world's continental shelves would not be sufficient to produce a major mass extinction through area effects. Representation of modern families on islands among the six classes range from 76 to 100%. After Jablonski (1984a).

This raises the possibility that, at least for some taxa, families on islands may be more extinction prone than they are on the mainland. However, it is not clear that these relatively small differences in island versus mainland species–family ratios are sufficient to offset the advantages of widespread dispersion at the family level among oceanic islands. As discussed below, taxonomic structure (including species–family ratios) appears to be of secondary importance in determining survival during mass extinctions because the biological attributes that give rise to species-rich clades tend to make those clades extinction prone. The Lazarus effect observed at most large extinction events is in itself evidence for the existence of refugia effective at the familial level for millions of years.

If species–area effects are insufficient to explain the association between mass extinctions and marine regression, other causal mechanisms must be sought. There is a sizable literature on this subject, but few critical tests have been attempted, so I will undertake only a brief discussion of two of the most plausible mechanisms. First, regression of extensive seas would destroy unique habitats and biogeographic provinces. A number of authors have underscored the unusual communities and endemic centers characteristic of many epeiric settings (e.g., Kauffman, 1973; Johnson, 1974; Bambach, 1977; Sheehan, 1982). These biogeographic entities could be lost through habitat destruction or disrupted due to climatic changes accompanying regression. For example, the end-Cretaceous regression destroyed the shallow-water chalk habitat of Cretaceous epeiric seas; chalk deposition at shelf depths was never widespread again after this time, and many of the taxa characteristic of these environments apparently suffered accordingly (Jablonski and Bottjer, 1983).

The most far-reaching effects of marine regression on the global biota would most likely have been climatic in origin. Extensive shallow-water seas would have an amelioriating effect on global climate, particularly when latitudinally trending seaways facilitate heat exchange between high and low latitudes. In contrast, extensive regression will lead to heighted albedo and increased land area in mid and high latitudes, and thus overall global cooling, as well as an increase in continentality and thus of seasonality (Donn and Shaw, 1977; Barron et al., 1980, 1981; Thompson and Barron, 1981; Brass et al., 1982; Burett, 1982; Barron, 1983). These climatic changes would be accentuated as increased continental areas are exposed to weathering and reduced atmospheric CO_2 content (Berner et al., 1983); a number of climatic and geochemical models suggest or even require elevated atmospheric CO_2 levels during the warm Cretaceous and attribute climatic changes of shifts of this atmospheric composition toward present-day values (Barron, 1983; Berner et al., 1983; see also Tappan, 1968; Worsley, 1974; Fischer, 1981, 1984). Time scales for atmospheric and climatic response remain poorly constrained, however.

There are few hard data on the biotic effects of the climatic changes held to result from extensive regression, but some plausible consequences can be recounted in the hopes of stimulating further research. In the Cretaceous example, the climatic changes resulting from regression would have decreased terrestrial productivity, with adverse effects most marked among large herbivores such as dinosaurs (Russell, 1966; Axelrod and Bailey, 1968; Sloan, 1976; Archibald and Clemens, 1982); as

Schopf (1982) points out, other changes in the terrestrial environment during regression, such as dissection of floodplains by rivers cutting to a new base level, would have further disrupted the habitats of large land-dwellers. Regression from a highstand might also trigger considerable vegetational changes by lowering the groundwater table (Mörner, 1978, 1984).

Regarding patterns of extinction in the tropics at the end of the Cretaceous, Hickey's (1981, 1984) data for the lack of severe extinction among low-latitude terrestrial floras actually appear to document high variance in the extent of floral turnover among tropical localities. This pattern is more suggestive of the mosaic pattern of local extinction and persistence in low latitudes during the Pleistocene, with many species becoming restricted to relatively small refugia within a fragmented tropics during glacial extremes (see Simberloff, this volume). While such waxing and waning of tropical forests might induce little global extinction in plants, birds, and insects, it would elevate extinction rates of organisms requiring more extensive home ranges, such as dinosaurs and other large vertebrates. Extinction in tropical marine biotas would also be expected to be high during regression: contraction or loss of tropical provinces and increased seasonality would inevitably eliminate some taxa and reduce growth rates of reef builders, and lowered base level would disrupt carbonate platforms due to erosion and terrigenous influx.

The role of extensive epicontinental seaways in the amelioration of global climate helps to explain the inexact correlation between regression and mass extinction. The initial state of the land–sea distribution will determine the magnitude of the climatic and biotic response to regression; eustatic sea-level drops that entail relatively little lateral movement of the strandline will have a less profound effect on global climate and biogeography than those drops that occur from a situation of maximum continental inundation. For example, the glacio-eustatic fluctuations of the late Cenozoic were vertically as impressive as those of the Cretaceous–Tertiary or the Ordovician–Silurian, but even maximum interglacial transgression did not create shelf seas comparable to those lost at the close of the Paleozoic or Mesozoic or at the end of the Ordovician. Consequently, the late Cenozoic extinctions are far more modest than those of the glacial-free end-Permian or end-Cretaceous events or of the glaciated end-Ordovician event (see Sheehan, 1982). [It should be noted that the Vail et al. (1977) curve, which has been used to estimate regression magnitudes, is really a measure of onlap/offlap histories and thus not a reliable comparator of global transgression or regression events (Watts, 1982; Hallam, 1984; Steckler, 1984).]

This scenario of regression-related, climatically mediated extinction is consistent in a qualitative way with the available paleontologic and stratigraphic data and with current climatic models, but it is exceedingly difficult to test critically. Data bearing on this scenario would include further verification of latitudinal patterns of extinction, examination of the behavior of geographic ranges of higher taxa during extinctions, stable isotope paleotemperature data indicating increased seasonality during regressions, and characterization of the biology of victims and survivors across extinction events of various magnitudes.

Many of the biogeographic and paleoclimatic arguments presented here would also be consistent with an extraterrestrial impact if the primary biotic consequences of such impacts (at least at higher taxonomic levels) are driven by climatic perturbations extending over thousands to tens of thousands of years, instead of the 1–10 yr propounded by Alvarez et al. (1984b). A synergistic effect between regression and impact events in which the state of the biosphere just prior to an impact plays a crucial part may be in the final analysis the most plausible hypothesis of all. The most severe extinctions would occur when an impact coincides with regression of extensive epicontinental seas. This might explain the great variance in amplitude of the 26-m.y. extinction periodicity documented by Raup and Sepkoski (1984) as well as the apparent failure of major extinctions to accompany every extensive regression or to occur at sufficiently high frequencies to agree with calculations of impact probabilities over the Phanerozoic (e.g., Toon, 1984). In addition, current models of impact-triggered climate change show harshest effects in continental interiors, with coastal and marine environments buffered by the ocean's thermal inertia (Toon, 1984); again, high sea level would mitigate the effects of impacts, low sea levels would exacerbate them. Even if extraterrestrial impacts are the pacemakers of mass extinctions, the terrestrially dictated initial state of the global system probably determines magnitude and pattern of the biotic response.

COMPARISONS OF BACKGROUND AND MASS EXTINCTION

If mass extinctions are sufficiently large or the perturbations that trigger them sufficiently severe, a taxon's survival may be a matter of chance and the postextinction fauna would be simply an impoverished, random sample of the preextinction biota. If the end-Permian extinction really did remove 96% of the marine species, higher taxa might easily have been lost through sampling effects, regardless of quality of adaptation of individual taxa—what Raup (1979) referred to as an evolutionary founder effect. On the other hand, the latitudinal and other biotic patterns common to the mass extinctions suggests the operation of some form of selectivity. The bases of this selectivity do appear to be distinct from the traits that confer extinction resistance during times of background extinction, bearing out Raup's suggestion that the survivors of mass extinctions would not necessarily comprise an advance in the overall adaptive level of the biota as measured during background times. There are as yet few data suitable for close comparisons of background and mass extinctions, but patterns of survival and extinction among Late Cretaceous molluscan genera illustrate some of the differences between macroevolutionary processes during mass extinctions relative to those during times of background extinction levels.

The Biology of Lazarus Taxa

Although Lazarus taxa in the Late Cretaceous of the Gulf and Atlantic Coastal Plain appear to be drawn from most high taxa and ecological groupings, a few generalizations can be made about the biological attributes of these clades that apparently

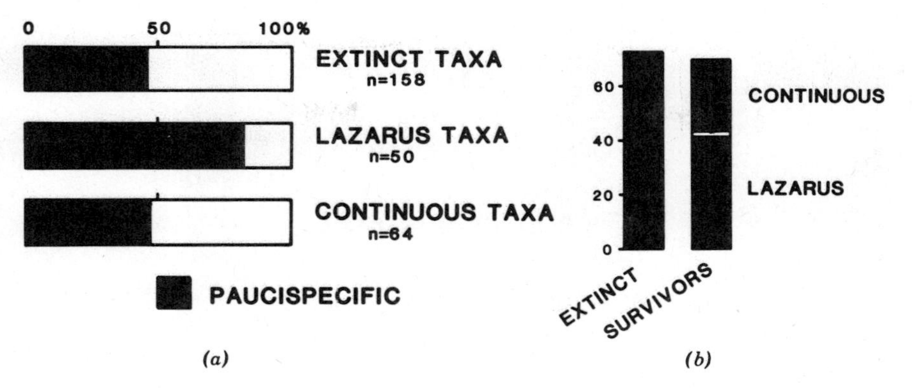

FIGURE 9. Lazarus taxa (taxa that vanish from the record during mass extinctions but return at a later time) are not a random sample of the biota. Among Cretaceous–Tertiary bivalves and gastropods of the Gulf and Atlantic coastal plain, (*a*) 82% of the Lazarus taxa are species poor (paucispecific), while 46% of the genera that genuinely become extinct at the boundary, and 47% of the genera that are continuously recorded in the sediments underlying the boundary, are species poor. (*b*) Of the 143 species-poor genera considered, 73 are lost at the end of the Cretaceous, and 70 survive, with about 40 of the survivors exhibiting the Lazarus effect and about 30 being continuously recorded in the latest Cretaceous sediments of the region.

go extinct only to reappear after the mass extinction event. First, 82% of the Lazarus clades tend to be species poor, arbitrarily defined in this example as comprising only one or two species in the Coastal Plain in the last 6 m.y. of the Cretaceous (Figure 9*a*). This result makes sense in light of sampling theory: all other factors being equal, a clade's probability of being detected within a given stratigraphic interval will be proportional to the number of species within the clade, particularly in an interval of declining sampling quality. Both the truly extinct taxa and the taxa with continuous stratigraphic records comprise about 46% of the species-poor taxa. Viewed another way, (Figure 9*b*), about half (73) of the 143 species-poor genera are lost and half survive, with the survivors approximately equally apportioned between genera with continuous records and Lazarus genera.

Second, Lazarus taxa tend to be widespread. Only 12% of the Lazarus genera among the Late Cretaceous mollusks of the Gulf and Atlantic Coastal Plain are geographically restricted to that region (Figure 10*a*). This also makes sense from the standpoint of sampling effects: widespread taxa are less likely to be driven into final extinction by local environmental changes. In contrast to the Lazarus taxa, 42% of the extinct taxa were endemic to the Coastal Plain. This distinction is seen more strikingly in Figure 10*b*; of the 77 endemic genera, 87% became extinct at the end of the Cretaceous, but only 7% of the endemics exhibit the Lazarus effect and 6% are continuously recorded.

Because Lazarus taxa are not a random sample of all clades but tend to be species poor and widespread, the Lazarus effect must be employed with some caution as a null hypothesis for patterns of decline before extinction boundaries. Species-rich endemics might be undergoing genuine extinction even as Lazarus taxa are temporarily dropping out of the stratigraphic record. On the other hand, analysis of

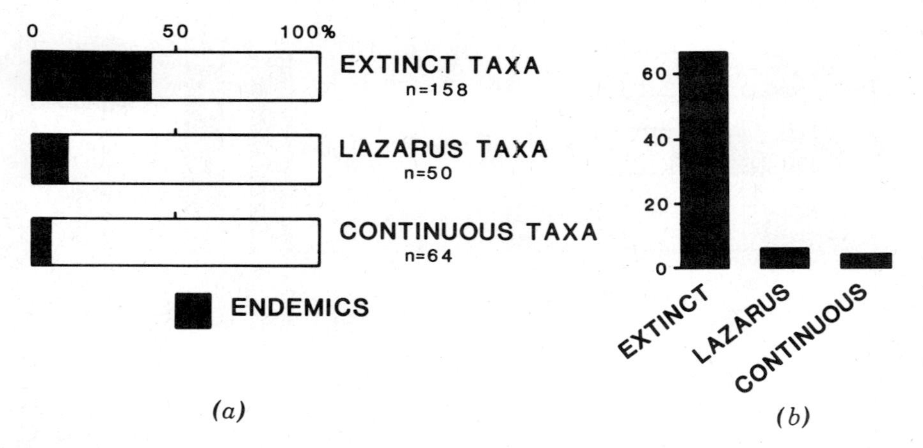

FIGURE 10. Lazarus taxa also tend to be geographically widespread. (*a*) Only 10% of the Lazarus taxa among Gulf and Atlantic Coastal Plain bivalve and gastropod genera are restricted to that region (endemic). (*b*) Of the 77 genera endemic to the Gulf and Atlantic Coastal Plain, only 6 are Lazarus taxa and only 4 are continuously recorded; the remaining 67 are lost at the end of the Cretaceous.

victims, survivors, and Lazarus taxa reveals some among-clade patterns of extinction and persistence that contrast with patterns during background times and merit further investigation.

Background and Mass Extinctions

Relatively little is known about the biotic factors that govern background extinction rates at the generic and familial level. One clade characteristic that appears to impart extinction resistance is taxonomic richness: species-poor clades have a greater probability of terminal extinction owing to random events than species-rich clades (e.g., Gould et al., 1977; Stanley, 1979; Hansen, 1980; Raup, 1981; Gould, 1982a,b; Strathmann and Slatkin, 1983). There are, of course, exceptions to this generalization even during times of background extinction, but the most striking violation of this probabilistic approach to extinction occurs during mass extinction events. Turning again to the Late Cretaceous molluscan data from the Gulf and Atlantic Coastal Plain, Figure 11 shows that species richness did not improve a clade's chance of surviving the Cretaceous–Tertiary boundary and that species-poor genera did not suffer disproportionate extinction. In fact, for the gastropods, species-rich genera constituted about 50% of the victims but only about 33% of the survivors, so that species-rich taxa were actually underrepresented among the survivors. These data appear to falsify the probabilistic assumption that species richness determines clade survival during mass extinction.

It is probably not species richness per se that is a liability during mass extinctions, but the covariation of biological attributes that determine a clade's extinction rate with traits that affect its speciation rate. For example, among bivalves and gastropods, low larval dispersal capability tends to heighten speciation rates and is generally accompanied by restricted geographic ranges and narrow environmental

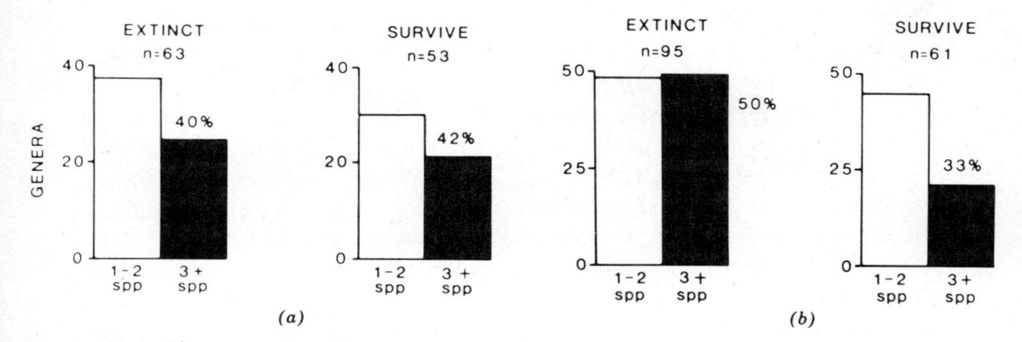

FIGURE 11. During mass extinctions, species-rich clades are not at an advantage. (*a*) Among bivalves of the Gulf and Atlantic Coastal Plain, the species-rich genera (here recognized as containing three or more species in the region) are equally represented among victims and survivors of the end-Cretaceous extinction. (*b*) Among gastropods, species-rich genera are disproportionately represented among the extinct taxa.

tolerances, imparting high extinction rates as well. Conversely, species with high dispersal capability, and thus low speciation rates, tend to have broader geographic ranges and environmental tolerances, which in turn impart extinction resistance to those species (see Jackson, 1974; Scheltema, 1977, 1978; Jablonski, 1980, 1982; Hansen, 1980; Jablonski and Lutz, 1983; Scheltema and Williams, 1983; Jablonski and Flessa, 1984). Thus, high species richness can contribute to the persistence of species-rich groups during background extinctions, when overall extinction rates are relatively low. Under these circumstances, the volatility of such clades will be less likely to result in terminal extinction. However, the species-rich groups will fare poorly during episodes of mass extinction, presumably because the individual species tend to be stenotopic and geographically restricted. More work is needed to determine how many of these species-rich clades represent predominantly intra-provincial diversifications, which would have a low probability of surviving severe and widespread environmental perturbations.

Unlike most bivalves and gastropods, the Mesozoic ammonites consist of short-lived diverse families and long-lived families containing few, long-lived species and genera (Ward and Signor, 1983), so that it is unclear whether genus or species richness ever contributes to clade survival, even during times in which other taxa exhibit background extinction levels. However, this group is so volatile (see Stanley, 1979) that it is difficult to separate the effects of background and mass extinction over the course of its boom-and-bust history. A number of short-lived, diverse ammonite clades do terminate together at minor mass extinctions recognized by Raup and Sepkoski (1984), suggesting a low threshold for the switch to a mass extinction regime for this group (a circular argument if, as discussed above, ammonites are actually the sole source of some of those extinction peaks). But as also observed for bivalves and gastropods, ammonite clades consisting of few long-lived taxa tend to survive or be the last holdouts at mass extinction boundaries (Ward and Signor, 1983), and these same clades tend to be geographically widespread rather than endemics (J. Wiedmann, in Kauffman, 1984).

The biologic characteristics of Lazarus taxa also suggest that a clade's biogeographic distribution may be more important than its taxonomic structure in determining survival or termination during mass extinctions. For the Coastal Plain bivalves, 33% of the extinct genera were restricted to that region versus only 3% of the survivors; among the gastropods, 48% of the victims were endemics versus only 11% of the survivors (Figure 12). Vermeij (this volume) also found that endemic taxa exhibited disproportionate extinction, regardless of species richness, relative to widespread taxa during late Pliocene molluscan extinctions (see Bretsky, 1973; Boucot, 1975; Hoffman and Szubzda-Studencka, 1982; Martinell and Hoffman, 1983; for additional examples).

Anstey (1978) also detected qualitative differences between background and mass extinctions in Paleozoic Bryozoa. Morphologically simple genera (interpreted as ecological generalists) exhibit fairly steady extinction rates throughout the Paleozoic, but losses among morphologically complex genera (inferred specialists) peak during mass extinctions. If the complex, specialist genera are more species rich and geographically restricted than the simple, generalist species, this would translate into the same pattern of extinction observed for Late Cretaceous mollusks. The fact that complex taxa have lower extinction rates than simple taxa when the effects of mass extinction are removed (Anstey, 1978, p. 415) reinforces similarities with the end-Cretaceous pattern.

The marked disruption of low-latitude communities relative to high-latitude ones during mass extinctions (discussed above) would enhance the lack of correlation between species richness and survivorship. Clades conforming to the latitudinal diversity gradient, and thus attaining peak species richness in the tropics, would be more severely affected than clades in which diversity is evenly distributed with respect to latitude or is highest outside the tropics. Clades restricted to the tropics, which are often among the most species rich of all, consequently would be among the most vulnerable to mass extinctions. This is, as discussed above (Figures 3–

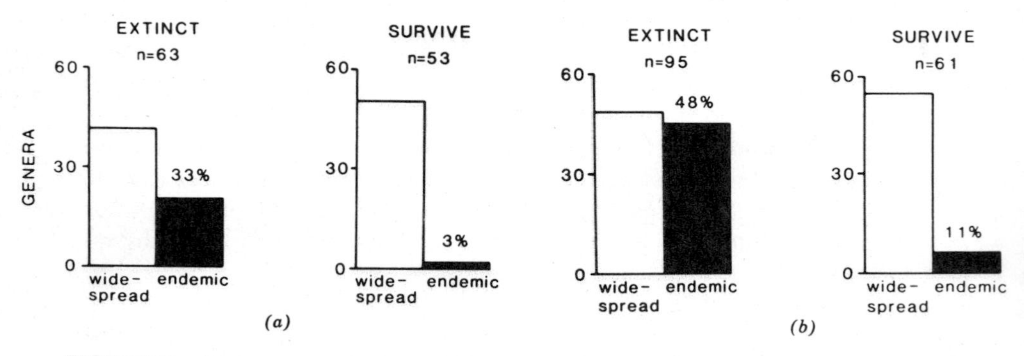

FIGURE 12. During mass extinctions, clades with broad geographic distributions tend to survive, and endemic clades tend to become extinct. (*a*) Among bivalves of the Gulf and Atlantic Coastal Plain, 33% of the genera that became extinct at the end of the Cretaceous were restricted to the region, but only 3% of the survivors were endemics. (*b*) Among the gastropods, 48% of the victims were endemics, but only 11% of the survivors were.

5), the pattern observed in the fossil record, with tropical clades being most severely affected during mass extinctions.

Taken together, these data suggest that a theory of evolution above the species level must include an alternation between two macroevolutionary regimes. Mass extinctions are selective rather than random in their removal of taxa, but extinction or survival depends on patterns of geographic distribution and other traits that apparently are less decisive during background times, when clades having high speciation rates tend to be among the most persistent, and conventional selection at the individual level is more effective. Clades capture new adaptations during background times, when selection at the level of individuals and species affords an opportunity for traits to originate and persist beyond the perilous species-poor initial stages in the history of a new clade. Selectivity will be indifferent to many of these new adaptations during mass extinctions, and traits will be lost not because they are disadvantageous but because they occur in clades that lack the environmental tolerances or geographic distribution necessary for survival during those times. Both macroevolutionary and microevolutionary selective processes are disrupted. For example, Fürsich and Jablonski (1984) suggested that the drilling habit in naticid gastropods was lost during the Norian extinction event, soon after its initial appearance in the Late Triassic of northern Italy, despite the undoubted expansion of available resources that this innovation entailed.

Survivors of mass extinctions, therefore, represent a biased sample of the preextinction biota; but the bias will rarely coincide with the patterns of clade expansion and contraction before the event. Dwindling, "endangered" taxa may, of course, suffer final extinction during the perturbation triggering the mass extinction, but species-poor clades with the appropriately broad geographic distribution will survive; even expanding clades will be lost if they lack such distributions—particularly if their affinities are tropical. Caution is needed in interpreting selectivity patterns because taxa characterized by a particular reproductive mode, body size, or membership in a particular community may have survived not because these traits were directly selected for, but because these traits were characteristic of a province (or latitude) that suffered relatively little disturbance. Strathmann (1978) and Valentine and Jablonski (1983) suggested that at least some Paleozoic crinoids and articulate brachiopods may have had planktotrophic larvae, with today's nonplanktotrophic development in these groups having become fixed at the end of the Paleozoic. While it might be argued that planktotrophy was directly selected against (see also Valentine, 1983), it is also plausible that extinction of tropical and subtropical clades was so severe (see Figure 3) that the planktotrophic mode so typical of those latitudes today was lost for these taxa.

The role of background processes in shaping the composition of the world biota becomes even more difficult to assess if the alternation of macroevolutionary regimes takes place with a 26-m.y. periodicity. Because all but the most prolific speciators among marine benthos exhibit origination rates of one or fewer species per species per million years (see Raup, 1978; Stanley, 1979; Jablonski, 1982), a maximum interval of 26 m.y. before the rules of survival and extinction change may not be sufficient for the process of species selection to get very far. It may be that most evolutionary trends are sufficiently short-lived to fall between mass extinction events, and the long-term evolutionary trends that we observe in the fossil record are a biased sample, perceived only because they occurred in clades that also possess the properties required to persist through mass extinction events. Stabilizing species selection (Gilinsky, 1981; Gould, 1982b) would be a special case, potentially mediated by mass rather than background processes by eliminating outlying morphologies.

On the other hand, not all extinction peaks are of equal magnitude; for example, the mid-Miocene extinction event of Raup and Sepkoski (1984) represents the loss of a mere 5–10 families from a standing diversity of over 900 (contra Alvarez and Muller, 1984, who misinterpreted the Raup–Sepkoski diagram). There may be a threshold of perturbation required to effect a shift from background to mass extinction macroevolutionary regimes, so that some of the smaller peaks of Raup and Sepkoski (1984) really could represent simple intensifications of background processes (cf. Quinn, 1983), which would permit species selection to operate uninterruptedly over more than one 26-m.y. interval. These would probably include at least some of the events that fail to meet the mass extinction criteria suggested herein, for example, the Bajocian, Hauterivian, and middle Miocene extinction peaks, the first two of which have now been rejected by Sepkoski and Raup (this volume). Furthermore, the threshold is probably taxon specific, so that events of intermediate severity may effect a shift in macroevolutionary regime for some clades, such as the volatile ammonoids, while other clades, such as the imperturbable gastropods, remain in the background regime.

Threshold effects are probably insufficient to explain the large-scale faunal changes that appear to proceed unimpeded by mass extinctions (e.g., Jablonski et al., 1983; Sepkoski, 1984; Sepkoski and Miller, 1984). Some of these trends may even be reinforced or accelerated by the mass extinction regime. For example, near-shore taxa tend to be widespread and eurytopic—and thus have a relatively high probability of survival during mass extinctions—and prolonged onshore–offshore expansions of major innovations have typified the evolution of shelf communities throughout the Phanerozoic (Jablonski et al., 1983, and references therein). Recognition of these kinds of dynamic interactions between regimes will permit development of a macroevolutionary theory that can encompass patterns of differential extinction and survival during normal times and during times of mass extinction.

SUMMARY AND CONCLUSIONS

Mass extinctions play an important part in shaping the world's biota, but the driving mechanisms are still poorly understood. Comparisons among mass extinctions and times of background extinction represent a relatively unexplored approach to the mass extinction problem. However, now that as many as 12 extinction peaks have been documented for the past 250 m.y. (reduced to 8 statistically significant peaks by Sepkoski and Raup, this volume), applying consistent criteria for the classification of mass and background extinction becomes increasingly important. One approach is to test for statistically significant outliers above background levels, but a virtual continuum of extinction events seems to emerge from the Raup and Sepkoski (1984) analysis. A different kind of working definition is proposed here: an event should be considered a true mass extinction only if it comprises the elevation—by at least a factor of two—of extinction rates over background levels for a large number of ecologically disparate higher taxa over a period that is brief relative to the average duration of the taxa involved. These criteria will be most useful if they employ taxon-specific background rates. Otherwise, elevations above background levels by groups typified by continually high turnover rates can override patterns exhibited by a majority of less volatile taxa.

Comparisons among mass extinctions are fraught with problems, but the generalizations that can be drawn permit the exclusion of a number of hypothesized extinction mechanisms. For the five largest mass extinctions of the Phanerozoic (end-Ordovician, Late Devonian, end-Permian, Late Triassic, and end-Cretaceous), both planktic and benthic organisms are affected; both marine and non-marine groups also appear to be affected, although relative severities differ and precise land–sea synchroneity is difficult to demonstrate. Excursions in sulfur, carbon, and oxygen isotope ratios are associated with certain extinction boundaries, but no convincing evidence has emerged for paleoceanographic mechanisms for mass extinctions. The mass extinctions do not correlate in any clearcut way with global refrigerations, because the end-Permian, Late Triassic, and end-Cretaceous lack glaciations and the Permo–Carboniferous glaciation lacks a mass extinction. Some kind of climatic perturbation may still be implicated by the preferential extinction of tropical marine organisms and the relatively high survivorships of high-latitude forms during mass extinctions. However, alternative explanations for this biogeographic pattern emerge from the data presented here: species richness and endemic distribution, both attributes of many tropical taxa, heighten clade vulnerabilities during mass extinction events.

With the possible exception of the Late Devonian event, marine regression remains the best-documented associate of the major mass extinctions, although causal linkages remain uncertain. Species–area effects during regression are unlikely to produce familial extinctions of the appropriate magnitude, because even in today's highly provincialized biota, total eradication of the continental shelf biota would remove only 13% of the marine families of bivalves, gastropods, echinoids, asteroids, ophiuroids, and scleractinian corals. The great majority of the families would persist around oceanic islands, which would suffer little net reduction in shallow-water area during regressions. The persistence of shallow-water refugia during extinction-associated regressions is indicated by the large proportion of taxa that appear to become extinct during the event but reappear unscathed at some later time. This phenomenon, the Lazarus effect, can be used to assess patterns of extinction near mass extinction boundaries: apparent gradual declines in taxonomic diversity leading to the extinction event can only be accepted as genuine if they exceed the magnitude of the Lazarus effect.

Not all marine regressions bring mass extinctions, probably because climatic and biogeographic effects of regressions depend on the state of land–sea relationships immediately prior to the regression. Even small eustatic sea-level drops can expose large continental areas if the initial state included extensive epicontinental seas. Conversely, even large eustatic drops that began with strandlines close to the edge of the continental margins would be expected to have little effect on the shallow-water biota; that is probably at least part of the explanation for the lack of major extinctions during the major glacioeustatic excursions of the late Neogene.

Marine regressions are not necessarily the sole driving mechanism for mass extinctions either, in light of increasing evidence for extraterrestrial impacts at some or all of these events. At present, of the five major mass extinctions, only the end-Cretaceous event has direct geochemical evidence for an extraterrestrial impact, but the discovery of a possible 26-m.y. extinction periodicity in phase with a similar periodicity in terrestrial cratering certainly raises the possibility that all of the extinction events had extraterrestrial triggers. Statistical, geochemical, and biological problems remain with the extraterrestrial impact hypothesis, but thus far it has withstood all attempts at direct falsification. From a paleobiological viewpoint, perhaps the most significant difficulty is that the postulated physical effects of the impact are more severe than can be accommodated by patterns of extinction and survival near extinction boundaries. In addition, not all extinction events appear to be abrupt and not all groups that vanished near the end of the Cretaceous underwent geologically instantaneous extinctions. A synergistic effect between regressions and extraterrestrial impacts, in which impacts fail to trigger severe mass extinctions unless they coincide with major regressions, would account both for the patterns of extinction and survival at individual events and for the variation in extinction magnitudes among peaks.

Regardless of the forcing mechanism, mass extinctions represent the operation of a macroevolutionary regime that is distinct from the one that operates during

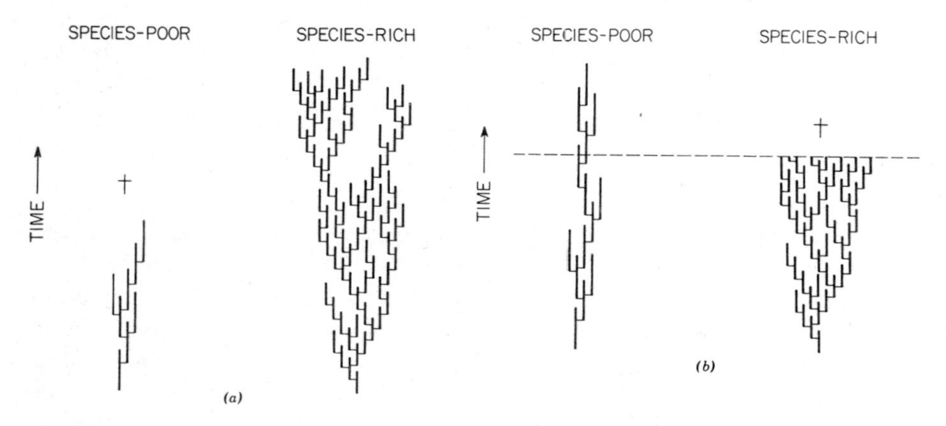

FIGURE 13. Alternative macroevolutionary regimes. (*a*) During times of background levels of extinction, species-rich clades tend to be buffered against extinction by the sheer numbers of their constituent species, while species-poor clades at more vulnerable to stochastic, background extinction. (*b*) During mass extinctions, species richness is no longer an advantage and for some taxa is even correlated with extinction probability, while species-poor clades tend to survive if they are geographically widespread.

times of background extinction (Figure 13). In background times, species-rich clades are generally buffered from extinction by their large number of component species, whereas species-poor clades tend to be at risk. During mass extinctions, however, geographically widespread clades tend to be the survivors, regardless of their species richness. For example, species richness failed to enhance the probability of survival of bivalve and gastropod clades during the end-Cretaceous event; for the gastropods, species richness was if anything a liability. Species richness in itself is not a disadvantage, but the biological attributes that heighten speciation rates covary with other traits that increase species' vulnerability to extinction (e.g., low dispersal capability, narrow environmental tolerance). This covariation is not precise and is relatively unimportant during the less extreme conditions of background extinction, but it evidently becomes critical during mass extinction.

The alternation of macroevolutionary regimes in which different traits enhance survival during background and mass extinction episodes indicates that mass extinctions will shape the biota in unexpected ways. During mass extinctions, quality of adaptation or fitness values—at any hierarchical level, from microevolution to species selection—are far less important than membership in the particular communities, provinces, or distributional categories that suffer minimal disturbance during mass extinction events. These qualitative differences between background and mass extinction suggest a threshold effect rather than a continuum in the switch between macroevolutionary regimes; the frequency of alternation between the two regimes—which may vary among environments and clades—will play a crucial role in the rate and pattern of macroevolutionary change.

NOTE ADDED IN PROOF

The case for an impact at the Cretaceous-Tertiary boundary continues to gain corroboration (e.g., Bonté et al., 1984; Brooks et al., 1985; Smit and Romein, 1985), although detractors point out that volcanicity has not been fully ruled out as a source of geochemical anomalies and environmental perturbations (e.g., Officer and Drake, 1985). The detection of what appears to be soot at several end-Cretaceous localities (Wolbach et al., 1985) may be another line of evidence for a catastrophic close to the Mesozoic, but it is not really conclusive on the trigger of the inferred conflagration, and suffers from a paucity of control samples. Iridium anomalies have now been reported for the late Devonian and end-Permian mass extinctions (Playford et al., 1984; Sun et al., 1984; Xu et al., 1985), as predicted by a scenario of repeated impact-forced events. However, each of the additional anomalies has its unique ambiguities and problems in interpretation—including the spectre of biological concentration. The situation is further complicated by an anomaly near—but not precisely coinciding with—the base of the Cambrian (Hsü et al., 1985); no mass extinction is present, and the Cambrian explosion of skeletonized organisms was evidently well underway. The Raup-Sepkoski extinction periodicity hypothesis is still viable as of this writing (e.g., Kitchell and Pena, 1984); many of Hoffman's (1985) objections are answered by Sepkoski and Raup (this volume). Astronomical mechanisms proposed for the reported biotic periodicity have not fared so well—according to recent analyses, interstellar molecular clouds are insufficiently dense, and orbits for small solar companions may be too unstable, to accomplish the job (e.g., Clube and Napier, 1984; Thaddeus and Chanan, 1985; Bahcall and Bahcall, 1985). If the ongoing search for a solar companion is inconclusive, the compounded uncertainties of the revised astronomical models may render them virtually untestable (Jablonski, 1984b). Weissman (1985) maintains that the composition of melt material in known craters argues against comets as impactors. Potential impact-triggered extinction mechanisms are still poorly understood, but data from the end-Cretaceous (Smit and Romein, 1985) and end-Eocene (Corliss et al., 1985) events further document some temporal structure to the extinctions, rather than a strictly instantaneous event. See Jablonski (1985, in press) for a more detailed look at the qualitative differences between background and mass extinction.

ACKNOWLEDGMENTS

I am grateful to Karl W. Flessa, J. John Sepkoski, Jr., and especially Susan M. Kidwell for an exceptionally helpful set of reviews and to Norman D. Newell for initial inspiration. Thanks also to Walter Alvarez, Nancy Beckvar, Frank Grober, David M. Raup, and Sherman Suter for fruitful discussions. Research supported in part by NSF Grant EAR 81-21212 and EAR 84-17011.

REFERENCES

Ager, D. V., 1981, Major marine cycles in the Mesozoic, *J. Geol. Soc. Lond.*, **138**, 159–166.

Alvarez, W., and Muller, R. A., 1984, Evidence in crater ages for periodic impacts on the Earth, *Nature*, **308**, 718–720.

Alvarez, L. W., Alvarez, W., Asaro, F., and Michel, H. V., 1980, Extraterrestrial cause for the Cretaceous-Tertiary extinction, *Science*, **208**, 1095–1108.

Alvarez, W., Alvarez, L. W., Asaro, F., and Michel, H. V., 1982, Current status of the impact theory for the terminal Cretaceous extinction, *Geological Society of America Special Paper 190*, pp. 305–328.

Alvarez, W., Alvarez, L. W., Asaro F., and Michel, H. V., 1984a, The end of the Cretaceous: Sharp boundary or gradual transition?, *Science*, **223**, 1183–1186.

Alvarez, W., Kauffman, E. G., Surlyk, F., Alvarez, L. W., Asaro, F., and Michel, H. V., 1984b. Impact theory of mass extinctions and the invertebrate fossil record, *Science*, **223**, 1135–1141.

Anstey, R. L., 1978, Taxonomic survivorship and morphologic complexity in Paleozoic bryozoan genera, *Paleobiology*, **4**, 407–418.

Archibald, J. D., and Clemens, W. A., 1982, Late Cretaceous extinctions, *Am. Sci.*, **70**, 377–385.

Archibald, J. D., and Clemens, W. A., 1984, *Mammal Evolution Near the Cretaceous-Tertiary Boundary*, in W. A. Berggren and J. A. Van Couvering, eds., *Catastrophes and Earth History*, Princeton, Princeton University Press, 339–371.

Arthur, M. A., 1979, Paleoceanographic events—Recognition, resolution and reconsideration, *Rev. Geophys. Space Phys.*, **17,** 1474–1494.

Asaro, F., Alvarez, L. W., Alvarez, W., and Michel, H., 1982, Geochemical anomalies near the Eocene/Oligocene and Permian/Triassic boundaries, *Geological Society of America Special Paper* **190,** 517–528.

Axelrod, D. I., and Bailey, H. P., 1968, Cretaceous dinosaur extinction, *Evolution*, **22**, 595–611.

Bahcall, J. N., and Bahcall, S., 1985, The Sun's motion perpendicular to the galactic plane, *Nature,* **316,** 706–708.

Bakker, R. T., 1977, Tetrapod mass extinctions; A model of the regulation of speciation rates and immigration by cycles of topographic diversity, in Hallam, A., ed., *Patterns of Evolution*, Amsterdam, Elsevier, pp. 439–468.

Bambach, R. K., 1977, Species richness in marine benthic habitats through the Phanerozoic, *Paleobiology,* **3**, 152–167.

Barron, E. J., 1983, A warm, equable Cretaceous: The nature of the problem, *Earth Sci. Rev.*, **19**, 305–338.

Barron, E. J., Sloan II, J. L., Harrison, C. G. A., 1980, Potential significance of land-sea distribution and surface albedo variations as a climatic forcing factor: 180 m.y. to the present, *Palaeogeogr., Palaeoclimatol., Palaeoecol.*, **30**, 17–40.

Batten, R. L., 1973, The vicissitudes of the gastropods during the interval of Guadalupian-Ladinian time, in Logan, A., and Hills, L. V., eds., *The Permian and Triassic Systems and Their Mutual Boundary,* Can. Soc. Petrol. Geol. Mem. 2, pp. 596–607.

Benson, R. H., Chapman, R. E., and Deck, L. T., 1984, Paleoceanographic events and deep-sea ostracodes, *Science,* **224**, 1334–1336.

Benton, M. J., 1983, Dinosaur success in the Triassic: A noncompetitive ecological model, *Quart. Rev. Biol.*, **58**, 29–55.

Berger, W. H., and Thierstien, H. R., 1979, On Phanerozoic mass extinctions, *Naturwissenschaften,* **66,** 46–47.

Berger, W. H., Vincent E., and Thierstein, H. R., 1981, The deep-sea record: Major steps in Cenozoic ocean evolution, *Soc. Econ. Paleont. Min. Spec. Pub.* **32,** 489–504.

Berner, R. A., Lasaga, A. C., and Garrels, R. M., 1983, The carbonate-silicate geochemical cycle and its effect on atmospheric carbon dioxide over the past 100 million years, *Am. J. Sci.*, **283,** 641–683.

Berry, W. B. N., 1979, Graptolite biogeography: A biogeography of some Lower Paleozoic plankton, in Gray, J., and Boucot, A. J., eds., *Historical Biogeography, Plate Tectonics, and the Changing Environment*, Corvallis, Oregon, Oregon State University Press, 105–115.

Berry, W. B. N., and Boucot, A. J., 1973, Glacio-eustatic control of Upper Ordovician-Early Silurian platform sedimentation and faunal changes, *Geol. Soc. Am. Bull.*, **84**, 275–284.

Birkelund, T., and Hakansson, E., 1982, The terminal Cretaceous extinction in Boreal shelf seas: A multicausal event, Geological Society of America Special Paper 190, pp. 373–384.

Black, D. I., 1967. Cosmic ray effects and faunal extinctions at geomagnetic reversals, *Earth Planet. Sci. Lett.*, **3**, 225–236.

Block, C., and Dams, R., 1975, Inorganic composition of Belgian coals and coal ashes, *Envir. Sci. Technol.*, **9**, 146–150.

Boardman, R. S., 1984, Origin of the post-Triassic Stenolaemata (Bryozoa): A taxonomic oversight, *J. Paleontol.*, **58**, 19–39.

Boersma, A., and Shackleton, N. J., 1979, Some oxygen and carbon isotope variations across the Cretaceous/Tertiary boundary in the Atlantic Ocean, in Christiansen, W. K., and Birkelund, T., eds., *Cretaceous-Tertiary Boundary Events Symposium,* Vol. 2, University of Copenhagen, pp. 50–53.

Bogoyavlenskaya, O. V., 1982, Late Devonian to Early Carboniferous stromatoporoids, *Paleontol. J.*, **16**, 29–36.

Bohor, B. F., 1984, Analysis of the Cretaceous–Tertiary boundary clay: Methodology questioned, *Science*, **223**, 190–191.

Bohor, B. F., Foord, E. E., Modreski, P. J., and Triplehorn, D. M., 1984, Mineralogic evidence for an impact event at the Cretaceous–Tertiary boundary, *Science*, **224**, 867–868.

Bonté, P., Delacotte, O., Renard, M., Laj, C., Boclet, D., Jehanna, C., and Racchia, R., 1984, An Iridium rich layer at the Cretaceous/Tertiary boundary in the Bidart section (southern France), *Geophys. Res. Lett.*, **11**, 473–476.

Botoman, G., and Stith, D. A., 1981, Analysis of Ohio coals, 1977–1978, *Ohio Dept. Nat. Res. Dir. Geol. Surv., Inf. Circ.*, **50**, 54 pp.

Boucot, A. J., 1975, *Evolution and Extinction Rate Controls*, Amsterdam, Elsevier, 427 pp.

Boucot, A. J., 1983. *Does evolution take place in an ecological vacuum? II*, J. Paleontol, **57**, 1–30.

Brass, G. W., Saltzman, E., Sloan II, J. L., Southam, J. R., Hay, W. W., Holser, W. T., and Peterson, W. H., 1982, Ocean circulation, plate tectonics, and climate, in NRC Geophysics Study Committee, *Climate In Earth History*, Washington, D.C., National Academy Press, pp. 83–89.

Brenchley, P. J., and Newall, G., 1984, Late Ordovician environmental changes and their effects on faunas, in D. L. Bruton, ed., *Aspects of the Ordovician System*, Oslo, Universitetsforlaget, pp. 65–79.

Bretsky, P. W., 1969, Evolution of Paleozoic benthic marine invertebrate communities, *Palaeogeogr., Palaeoclimatol., Palaeoecol.*, **6**, 45–59.

Bretsky, P. W., 1973, Evolutionary patterns in the Paleozoic Bivalvia: Documentation and some theoretical considerations, *Geol. Soc. Am. Bull.*, **84**, 2079–2096.

Brooks, R. R., Reeves, R. D., Yang, X.-H. Ryan, D. E., Holzbecher, J., Collen, J. D., Neall, V. E., and Lee, J., 1984, Elemental anomalies at the Cretaceous-Tertiary boundary, Woodside Creek, New Zealand, *Science*, **226**, 539–541.

Brown, J. H., and Gibson, A. C., 1983, *Biogeography*, St. Louis, Mosby, 643 pp.

Buchardt, B., and Jørgensen, N. O., 1979, Stable isotope variations at the Cretaceous/Tertiary boundary in Denmark, in Christensen, W. K., and Birkelund, T., eds., *Cretaceous–Tertiary Boundary Events Symposium*, Vol. 2, University of Copenhagen, pp. 54–63.

Buffetaut, E., 1980, Determination de la nature des événements de la transition Crétacé-Tertiare: La contribution de l'étude des Crocodiliens, *Mém. Soc. Géol. France N.S.*, **139**, 47–52.

Buffetaut, E., 1984, Selective extinction and terminal Cretaceous events, *Nature*, **310**, 276.

Bujak, J. P., and Williams, G. L., 1979, Dinoflagellate diversity through time, *Mar. Micropaleontol.*, **4**, 1–12.

Burchette, T. P., 1981, European Devonian reefs: A review of current concepts and models, *Soc. Econ. Paleontol. Mineral Spec. Pub.*, **30**, 85–142.

Burett, C. F., 1982, Phanerozoic land-sea and albedo variations as climate controls, *Nature*, **296**, 54–56.

Burns, D. A., and Nelson, C. S., 1981, Oxygen isotopic paleotemperatures across the Runangan-Whaingaroan (Eocene-Oligocene) boundary in a New Zealand shelf sequence, *N. Z. J. Geol. Geophys.*, **24**, 529–538.

Butler, R. F., and Lindsay, E. H., 1983, Magnetic mineralogy of continental deposits, San Juan Basin, New Mexico, *EOS, Trans. Am. Geophys. Union*, **64**, 683 (Abstract).

Campbell, C. A., and Valentine, J. W., 1977, Comparability of modern and ancient faunal provinces, *Paleobiology*, **3**, 49–57.

Castellarin, A., Del Monte, M., and Frascari R. S., F., 1973. Cosmic fallout in the "hard grounds" of the Venetian region (southern Alps), *Gior. Geol. Bologna*, **39**, 333–346.

Cavelier, C., Chateauneuf, J.-J., Pomerol, C., Robussier, D., Renard, M., and Vergnaud-Grazzini, C., 1981, The geological events at the Eocene/Oligocene boundary, *Palaeogeogr. Palaeoclimatol., Palaeoecol.*, **36,** 223–248.

Chaloner, W. G., and Sheerin, A., 1979, Devonian macrofloras, in House, M. R., Scrutton, C. T., and Bassett, M. G., eds., *The Devonian System*, Special Papers in Palaeontology 23, pp. 145–161.

Charig, A. J., 1984, Competition between therapsids and archosaurs during the Triassic Period: A review and synthesis of current theories, *Symp. Zool. Soc. Lond.*, **52**, 80–87.

Chen Xu, 1984, Influence of the Late Ordovician glaciation on basin configuration of the Yangtze Platform in China, *Lethaia*, **17**, 51–59.

Clark, D. L., and Kitchell, J. A., 1979, Injection events in ocean history, *Nature*, **278**, 669.

Clark, D. L., and Kitchell, J. A., 1981, Terminal Cretaceous extinctions and the Arctic spillover model, *Science*, **212**, 577.

Clemens, W. A., 1982, Patterns of extinction and survival of the terrestrial biota during the Cretaceous/Tertiary transition, *Geological Society of America Special Paper* **190,** pp. 407–413.

Cloud, P., and Glaessner, M. F., 1982, The Ediacarian Period and System: Metazoa inherit the Earth, *Science*, **218**, 783–792.

Clube, S. V. M., and Napier, W. M., 1984, Terrestrial catrastrophism: Nemesis or galaxy? *Nature*, **311**, 635–636.

Cooper, G. A., and Grant, R., 1972–1976, Permian brachiopods of west Texas, I–V, *Smithsonian Contrib. Paleobiol.*, 14, 15, 19, 21, 24, 3159 pp.

Copper, P., 1974. Structure and development of early Paleozoic reefs, *Proc. 2nd Internatl. Symp. Coral Reefs*, I, 365–386.

Copper, P., 1977, Paleolatitudes in the Devonian of Brazil and the Frasnian–Famennian mass extinction, *Palaeogeogr., Palaeoclimatol., Palaeoecol.*, **21**, 165–207.

Corliss, B. H., 1981, Deep-sea benthonic foraminiteral faunal turnover near the Eocene/Oligocene boundary, *Mar. Micropaleont.*, **6,** 367–384.

Corliss, B. H., Aubrey, M.-P., Berggren, W. A., Fenner, J. M., Keigwin, L. D., Jr., and Keller, G., 1984, The Eocene/Oligocene boundary event in the deep sea, *Science*, **226,** 806–810.

Crowell, J. C., 1978, Gondwanan glaciation, cyclothems, continental positioning, and climate change, *Am. J. Sci.*, **278**, 1345–1372.

Crowell, J. C., 1982, Continental glaciation through geolog time, in NRC Geophysics Study Committee, *Climate in Earth History*, Washington, D.C., National Academy Press, pp. 77–82.

Crowley, T. J., 1983. The geologic record of climatic change, *Rev. Geophys. Space Phys.*, **21**, 828–875.

Cuvier, G., 1817, *Essay on the Theory of the Earth*, Edinburgh, W. Blackwood, 348 pp. (Reprinted 1978, Arno Press.)

Cvancara, A. M., 1966, Revision of the fauna of the Cannonball Formation (Paleocene) of North and South Dakota. Part I. Bivalvia, *Contrib. Univ. Michigan Mus. Paleontol.*, **20**, 277–374.

Davis, M., Hut, P., and Muller, R. A., 1984, Extinction of species by periodic comet showers, *Nature*, **308**, 715–717.

De Coninck, J., and Smit, J., 1982, Marine organic-walled microfossils at the Cretaceous-Tertiary boundary in the Barranco del Gredero (S.E. Spain), *Geol. Mijn.*, **61**, 173–178.

de Laubenfels, M. W., 1956, Dinosaur extinction: One more hypothesis, *J. Paleontol.*, **30**, 207–212.

DePaolo, D. J., Kyte, F. T., Marshall, B. D., O'Niel, J. R., and Smit, J., 1983, Rb–Sr, Sm–Nd, K–Ca, O, and H isotopic study of Cretaceous–Tertiary boundary sediments, Caravaca, Spain: Evidence for an oceanic impact site, *Earth Planet. Sci. Lett.*, **64**, 356–373.

Dhondt, A. V., 1983, Campanian and Maastrichtian inoceramids: A review, *Zitteliana*, **10**, 689–701.

Donn, W., and Shaw, D., 1977, Model of climate evolution based on continental drift and polar wandering, *Geol. Soc. Am. Bull.*, **88**, 390–396.

Douglas, R. G., and Woodruff, F., 1981, Deep-sea benthic Foraminifera, in Emiliani, C., ed., *The Sea*, Vol. 7, Wiley-Interscience, New York, pp. 1233–1327.

Eder, W., and Franke, W., 1982, Death of Devonian reefs, *N. Jb. Geol. Paläont. Abh.*, **163**, 241–243.

Emiliani, C., Kraus, E. B., and Shoemaker, E. M., 1981, Sudden death at the end of the Mesozoic, *Earth Planet. Sci. Lett.*, 55, 317–334.

Fabricius, F. H., 1966, *Beckensedimentation und Riffbildung an der Wende Trias/Jura in den Bayerisch-Tiroler Kalkalpinen*, Leiden, E. J. Brill, 143 pp.

Finkelman, R. B., and Aruscavage, P. J., 1981, Concentration of some platinum-group metals in coal, *Internat. J. Coal Geol.*, **1**, 95–99.

Fisher, A. G., 1964, Brackish oceans as a cause of the Permo-Triassic marine faunal crisis, in Nairn, A. E. M., ed., *Problems in Palaeoclimatology*, Wiley-Interscience, London, pp. 566–577.

Fischer, A. G., 1981, Climatic oscillations in the biosphere, in Nitecki, M. H., ed., *Biotic Crises in Ecological and Evolutionary Time*, New York, Academic Press, pp. 103–131.

Fischer, A. G., 1984, The two Phanerozoic supercycles, in Berggren, W. A., and Van Couvering, J. A., eds., *Catastrophes and Earth History*, Princeton, Princeton University Press, pp. 129–150.

Fischer, A. G., and Arthur, M. A., 1977, Secular variations in the pelagic realm, *Soc. Econ. Paleont. Mineral. Spec. Pub.*, **25**, 19–50.

Flanders, S. E., 1962, Did the caterpillar exterminate the giant reptile?, *J. Res. Lepidopt.*, 1, 85–88.

Flessa, K. W., and Jablonski, D., 1984, Extinction is here to stay, *Paleobiology*, **9**, 315–321.

Frakes, L. A., 1979, *Climates Throughout Geologic Time*, Amsterdam, Elsevier, 310 pp.

Fürsich, F. T., and Jablonski, D., 1984, Late Triassic naticid drillholes: Carnivorous gastropods gain a major adaptation but fail to radiate, *Science*, **224,** 78–80.

Ganapathy, R., 1980, A major meteorite impact on the Earth 65 million years ago: Evidence from the Cretaceous–Tertiary boundary clay, *Science*, **209**, 921–923.

Ganapathy, R., 1982, Evidence for a major meteorite impact on the Earth 34 million years ago: Implications for Eocene extinctions, *Science*, **216**, 885–886.

Ganapathy, R., Gartner, S., and Jiang, M.-J., 1981, Iridium anomaly at the Cretaceous–Tertiary boundary in Texas, *Earth Planet. Sci. Lett.*, **54**, 393–396.

Gartner, S., and Keany, J., 1978, The terminal Cretaceous event: A geologic problem with an oceanographic solution, *Geology*, **6**, 708–712.

Gartner, S., and McGuirk, J. P., 1980, Terminal Cretaceous extinction: Scenario for a catastrophe, *Science*, **206**, 1272–1276.

Gilinsky, N. L., 1981, Stabilizing species selection in the Archaeogastropoda, *Paleobiology*, **7**, 316–331.

Gilmore, J. S., Knight, J. D., Orth, C. J., Pillmore, C. L., and Tschudy, R. H., 1984, Trace element patterns at a non-marine Cretaceous–Tertiary boundary, *Nature*, **307**, 224–228.

Glass, B. P., 1982, Possible correlations between tektite events and climatic changes? *Geological Society of America Special Paper 190*, 251–256.

Goldschmidt, V. M., 1954, *Geochemistry*, Oxford, Clarendon Press.

Gould, S. J., 1982a, Darwinism and the expansion of evolutionary theory, *Science*, **216**, 380–387.

Gould, S. J., 1982b, The meaning of punctuated equilibrium and its role in validating a hierarchical approach to macroevolution, in R. Milkman, ed., *Perspectives on Evolution*, Sunderland, Massachusetts, Sinauer, pp. 83–104.

Gould, S. J., Raup, D. M., Sepkoski, J. J., Jr., Schopf, T. J. M., and Simberloff, D. S., 1977, The shape of evolution: A comparison of real and random clades, *Paleobiology*, **3**, 23–40.

Grieve, R. A. F., 1984, Physical evidence of impact, *Nature*, **310**, 370.

Hallam, A., 1981a, The end-Triassic bivalve extinction event, *Palaeogeogr., Palaeoclimatol., Palaeoecol.*, **35**, 1–44.

Hallam, A., 1981b, *Facies Interpretation and the Stratigraphic Record*, San Francisco, W. H. Freeman, 291 pp.

Hallam, A., 1984, Pre-Quaternary sea-level changes, *Ann. Rev. Earth Planet. Sci.*, **12**, 205–243.

Hallock, P., 1982, Evolution and extinction in larger Foraminifera, *Proc. 3rd N. Am. Paleontol. Conv.*, **1**, 221–225.

Hambrey, M. J., and Harland, W. B., eds., 1981, *Earth's Pre-Pleistocene Glacial Record*, Cambridge, Cambridge University Press, 1004 pp.

Hancock, J. M., 1967, Some Cretaceous-Tertiary faunal changes, in Harland, W. B., et al., eds., *The Fossil Record*, London, Geological Society of London, pp. 91–104.

Hansen, J. M., 1979, Dinoflagellate zonation around the Boundary, in Birkelund, T., and Bromley, R. G., eds., *Cretaceous–Tertiary Boundary Events Symposium*, Vol. 1, University of Copenhagen, pp. 136–141.

Hansen, T. A., 1980, Influence of larval dispersal and geographic distribution on species longevity in neogastropods, *Paleobiology*, **6**, 193–207.

Haq, B. U., 1973, Transgression, climatic change and the diversity of calcareous nannoplankton, *Mar. Geol.*, **15**, M25–M30.

Haq, B. U., 1981, Paleogene paleoceanography: Early Cenozoic oceans revisited, *Oceanol. Acta.*, **4** (Suppl.), 71–82.

Haq, B. U., 1982, Climatic acme events in the sea and on land, in NRC Geophysics Study Committee, *Climate in Earth History*, National Academy Press, Washington, D.C., pp. 126–132.

Harland, W. B., Cox, A. V., Llewellyn, P. G., Pickton, C. A. G., Smith, A. G., and Walters, R., 1983, *A Geologic Time Scale*, Cambridge University Press, 131 pp.

Harriss, R. C., Crocket, C. H., and Stainton, M., 1968, Pd., Ir and Au in deep-sea manganese nodules, *Geochim. Cosmochim. Acta*, **32**, 1049–1056.

Heckel, P. H., 1974, Carbonate buildups in the geologic record: A review, *Soc. Econ. Paleontol. Mineral. Spec. Pub.*, **18**, 90–154.

Heinberg, C., 1979a, Bivalves from the latest Maastrichtian of Stevns Klint and their stratigraphic affinities, in Birkelund, T., and Bromley, R. G., eds., *Cretaceous–Tertiary Boundary Events Symposium*, Vol. 1, University of Copenhagen, pp. 58–64.

Heinberg, C., 1979b, Evolutionary ecology of nine sympatric species of the pelecypod Limopsis in Cretaceous chalk, *Lethaia*, **12**, 325–340.

Hickey, L. J., 1981, Land plant evidence compatible with gradual, not catastrophic, change at the end of the Cretaceous, *Nature*, **282**, 529–531.

Hickey, L. J., 1984, Changes in the angiosperm flora across the Cretaceous-Tertiary boundary, in Berggren, W. A., and Van Couvering, J. A., eds., *Catastrophes and Earth History*, Princeton, Princeton University Press, pp. 279–313.

Hickey, L. J., West, R. M., Dawson, M. R., and Choi, D. K., 1983, Arctic terrestrial biota: Paleomagnetic evidence of age disparity with mid-northern latitudes during the Late Cretaceous and Early Tertiary, *Science*, **221**, 1153–1156.

Hoffman, A., 1985, Patterns of family extinction depend on definition and geological timescale, *Nature*, **315**, 659–662.

Hoffman, A., and Szubdza-Studencka, B., 1982, Bivalve species duration and ecologic characteristics in the Badenian (Miocene) marine sandy facies of Poland, *N. Jb. Geol. Paläont. Abh.*, **163**, 122–135.

Holser, W. T., 1977, Catastrophic chemical events in the history of the ocean, *Nature*, **267**, 403–408.

Hoover, P. R., 1979, *Early Triassic Terebratulid Brachiopods from the Western Interior of the United States*, U.S. Geological Survey Professional Paper 1057, 21 pp.

House, M. R., 1975a, Facies and time in Devonian tropical areas, *Proc. Yorkshire Geol. Soc.*, **40**, 233–288.

House, M. R., 1975b, Faunas and time in the marine Devonian, *Proc. Yorkshire Geol. Soc.*, **40**, 459–490.

Hsü, K. J., He, O., McKenzie, J. A., Weissert, H., Perch-Nielsen, K., Oberhänsli, H., Kelts, K., LaBrecque, J., Tauxe, L., Krähenbühl, U., Percival, S. F., Jr., Wright, R., Karpoff, A. M., Petersen, N., Tucker, P., Poore, R. Z., Gombos, A. M., Pisciotto, K., Carman, H. F., Jr., and Schreiber, E., 1982a, Mass mortality and its environmental and evolutionary consequences, *Science*, 216, 249–256.

Hsü, K. J., McKenzie, J. A., and He, Q. X., 1982b, Terminal Cretaceous environmental and evolutionary changes, *Geological Society of America Special Paper* **190,** pp. 317–328.

Hsü, K. J., Oberhänsli, H., Gao, J. Y., Sun, S., Chen, H., and Krähenbuhl, U., 1985, "Strangelove ocean" before the Cambrian explosion, *Nature*, **316,** 809–811.

Hunt, G. E., 1978, Possible climatic and biological impact of nearby supernovae, *Nature*, 271, 430–431.

Jablonski, D., 1980, Apparent versus real biotic effects of transgression and regression, *Paleobiology*, 6, 398–407.

Jablonski, D., 1982, Evolutionary rates and modes in Late Cretaceous gastropods, *Proc. 3rd N. Am. Paleont. Conv.*, 1, 257–262.

Jablonski, D., 1984a (1985), Marine regressions and mass extinctions: a test using the modern biota, in Valentine, J. W., ed., *Phanerozoic Diversity Patterns: Profiles in Macroevolution*, Princeton, Princeton University Press.

Jablonski, D., 1984b, Keeping time with mass extinctions, *Paleobiology*, 10, 139–145.

Jablonski, D., 1985, Background and mass extinctions: The alternation of macroevolutionary regimes, *Science*, in press.

Jablonski, D., in press, Evolutionary consequences of mass extinctions, in Raup, D. M., and Jablonski, D., eds., *Pattern and Process in the History of Life*, Berlin, Springer-Verlag.

Jablonski, D., and Bottjer, D. J., 1983, Soft-bottom epifaunal suspension-feeding assemblages in the Late Cretaceous: Implications for the evolution of benthic paleocommunities, in Tevesz, M. J. S., and McCall, P. L., eds., *Biotic Interactions in Recent and Fossil Benthic Communities*, New York, Plenum Press, pp. 747–812.

Jablonski, D., and Flessa, K. W., 1984 (1985), The taxonomic structure of shallow-water marine faunas: Implications for Phanerozoic extensions, *Malacologia*, **27,** 43–66.

Jablonski, D., and Lutz, R. A., 1983, Larval ecology of marine benthic invertebrates: Paleobiological implications, *Biol. Rev.*, 58, 21–89.

Jablonski, D., Sepkoski, J. J., Jr., Bottjer, D. J., and Sheehan, P. M., 1983, Onshore–offshore patterns in the evolution of Phanerozoic shelf communities, *Science*, 222, 1123–1125.

Jackson, J. B. C., 1974, Biogeographic consequences of eurytopy and stenotopy among marine bivalves and their evolutionary significance, *Am. Nat.*, 108, 541–560.

Jenkyns, H. C., 1980, Cretaceous anoxic events: From continents to oceans, *J. Geol. Soc. Lond.*, 137, 171–188.

Johnson, J. G., 1974, Extinction of perched faunas, *Geology*, 2, 479–482.

Johnson, J. G., Klapper, G., and Sandberg, C. A., 1985, Devonian eustatic fluctuations in Euramerica, *Geol. Soc. Amer. Bull.*, **96,** 567–587.

Kauffman, E. G., 1973, Cretaceous Bivalvia, in Hallam, A., ed., *Atlas of Palaeobiogeography*, Amsterdam, Elsevier, pp. 353–383.

Kauffman, E. G., 1978, Evolutionary rates and patterns among Cretaceous Bivalvia, *Philos. Trans. Roy. Soc. Lond.*, 284B, 277–304.

Kauffman, E. G., 1979, The ecology and biogeography of the Cretaceous–Tertiary extinction event, in Christensen, W. K., and Birkelund, T., eds., *Cretaceous-Tertiary Boundary Events Symposium*, Vol. 2, University of Copenhagen, pp. 29–37.

Kauffman, E. G., 1984, The fabric of Cretaceous marine extinctions, in Berggren, W. A., and Van Couvering, J. A., eds., *Catastrophes and Earth History*, Princeton, Princeton University Press, pp. 151–246.

Keigwin, L. D., Jr., 1980, Paleoceanographic change in the Pacific at the Eocene-Oligocene boundary, *Nature*, **287,** 722–725.

Keith, M. L., 1982, Violent volcanism, stagnant oceans and some inferences regarding petroleum, strata-bound ores and mass extinction, *Geochim. Cosmochim. Acta*, 46, 2631–2637.

Keller, G., 1983a, Paleoclimatic analyses of middle Eocene through Oligocene planktic foraminiferal faunas. *Palaeogeogr., Palaeoclimatol., Palaeoecol.*, 43, 73–94.

Keller, G., 1983b, Biochronology and paleoclimatic implications of middle Eocene to Oligocene planktic foraminiferal faunas, *Mar. Micropaleont.*, 7, 463–486.

Keller, G., D'Hondt, S., and Vallier, T. L., 1983, Multiple microtektite horizons in Upper Eocene marine sediments: No evidence for mass extinctions, *Science*, 221, 150–152.

Kennedy, W. J., 1977, Ammonite evolution, in Hallam, A., ed., *Patterns of Evolution*, Amsterdam, Elsevier, pp. 251–304.

Kennett, J. P., 1983, Paleo-oceanography: Global ocean evolution, *Rev. Geophys. Space Phys.*, 21, 1258–1274.

Kitchell, J. A., and Clark, D. L., 1982, Late Cretaceous-Paleogene paleogeography and paleocirculation: Evidence of north polar upwelling, *Palaeogeogr., Palaeoclimatol., Palaeoecol.*, 40, 135–165.

Kitchell, J. A., and Pena, D., 1984, Periodicity of extinctions in the geologic past: Deterministic versus stochastic explanations, *Science*, **226,** 689–691.

Knoll, A. H., 1984, Patterns of extinction in the fossil record of vascular plants, in Nitecki, M. H., ed., *Extinction*, Chicago, University of Chicago Press, pp. 21–68.

Kollmann, H. A., 1976–1982, Gastropoden aus den Losensteiner Schichten der Umgebung von Losenstein (Oberösterreich), I–IV, *Ann. Naturhist. Mus. Wien*, 80, 163–206; 81, 173–201; 82, 11–51, 84, 13–56.

Kollman, H. A., 1979, Distribution patterns and evolution of gastropods around the Cretaceous/Tertiary boundary, in Christensen, W. K., and Birkelund, T., eds., *Cretaceous-Tertiary Boundary Events Symposium*, Vol. 2, University of Copenhagen, pp. 83–87.

Kucha, H., 1981, Precious metal alloys and organic matter in the Zechstein copper deposits, Poland, *TMPM Tschermaks Min. Petr. Mitt.*, 28, 1–16.

Kyte, F. T., 1983, The Cretaceous-Tertiary boundary at 2 North Pacific sites, *Geol. Soc. Am. Abstr.*, 15, 620–621.

Kyte, F. T., Zhou, Z., and Wasson, J. T., 1980, Siderophile-enriched sediments from the Cretaceous–Tertiary boundary, *Nature*, 288, 651–656.

Kyte, F. T., Zhou, Z., and Wasson, J. T., 1981, High noble metal concentrations in a late Pliocene sediment, *Nature*, 292, 417–420.

Leggett, J. K., 1978, Eustacy and pelagic regimes in the Iapetus Ocean during the Ordovician and Silurian, *Earth Planet. Sci. Lett.*, **41,** 163–169.

Leggett, J. K., 1980, British lower Palaeozoic black shales and their palaeooceanographic significance, *J. Geol. Soc. Lond.*, 137, 139–156.

Lemcke, K., 1975, Mögliche Folgen des Eischlags von Grossmeteoriten ins Weltmeer, *N. Jb. Geol. Paläont. Mh.*, 1975, 719–726.

Lenz, A. C., 1976, Late Ordovician–Early Silurian glaciation and the Ordovician–Silurian boundary in the western Canadian Cordillera, *Geology*, 4, 313–317.

Levinton, J. S., 1974, Trophic group and evolution in bivalve molluscs, *Palaeontology*, 17, 579–585.

Lewis, J. S., Watkins, G. H., Hartman, H., and Prinn, R. G., 1982, Chemical consequences of major impact events on Earth, *Geological Society of America Special Paper 190*, 215–221.

Lockley, M. G., 1984, Faunas from a volcaniclastic debris flow from the Welsh Basin: A synthesis of palaeoecological and volcanological observations, in Bruton, D. L., ed., *Aspects of the Ordovician System*, Oslo, Universitetsforlaget, pp. 195–201.

Luck, J. M., and Turekian, K. T., 1983, Osmium-187/osmium-186 in manganese nodules at the Cretaceous–Tertiary boundary, *Science*, 222, 613–615.

MacArthur, R. H., and Wilson, E. O., 1967, *The Theory of Island Biogeography*, Princeton, Princeton University Press, 203 pp.

Mahoney, J., Macdougall, J. D., Lugmair, G. W., Murali, A. V., Sankar Das, M., and Gopalan, K., 1982, Origin of the Deccan Trap flows at Mahabaleshwar inferred from Nd and Sr isotopic and chemical evidence, *Earth Planet. Sci. Lett.*, **60**, 47–60.

McCoy, E. D., and Connor, E. F., 1980, Latitudinal gradients in the species diversity of North American mammals, *Evolution*, 34, 193–203.

McGhee, G. R., Jr., 1982, The Frasnian–Famennian extinction event: A preliminary analysis of Appalachian marine ecosystems, Geological Society of America Special Paper 190, pp. 491–500.

McGhee, G. R., Jr., Gilmore, J. S., Orth, C. J., and Olsen, E., 1984, No geochemical evidence for an asteroidal impact at Late Devonian mass extinction horizon, *Nature*, 308, 629–631.

McKerrow, W. S., 1979, Ordovician and Silurian changes in sea level, *J. Geol. Soc. Lond.*, 136, 137–146.

McLaren, D. J., 1970, Time, life and boundaries, *J. Paleontol.*, 44, 801–815.

McLaren, D. J., 1982, Frasnian–Famennian extinctions, Geological Society of America Special Paper 190, pp. 477–484.

McLaren, D. J., 1983, Bolides and biostratigraphy, *Geol. Soc. Am. Bull.*, 94, 313–324.

McLean, D., 1981, A test of terminal Mesozoic "catastrophe," *Earth Planet. Sci. Lett.*, 53, 103–108.

Martinell, J., and Hoffman, A., 1983, Species duration patterns on the Pliocene gastropod fauna of Emporda (Northeast Spain), *N. Jb. Geol. Paläont. Mh.*, 1983, 698–704.

Matsumoto, T., 1980, Inter-regional correlation of transgressions and regressions in the Cretaceous Period, *Cretac. Res.*, 1, 359–373.

Montanari, A., Hay, R. L., Alvarez, W., Asaro, F., Michel, H. V., Alvarez, L. W., and Smit, J., 1983, Spheroids at the Cretaceous–Tertiary boundary are altered impact droplets of basaltic composition, *Geology*, 11, 668–671.

Mörner, N.-A., 1978. Low sea levels, droughts, and mammalian extinctions, *Nature*, 271, 738–739.

Mörner, N-A., 1984, Low sea levels, droughts, and mammalian extinctions, in Berggren, W. A., and Van Couvering, J. A., eds., *Catastrophes and Earth History*, Princeton, Princeton University Press, pp. 387–393.

Nakazawa, K., and Runnegar, B., 1973, The Permian–Triassic boundary: A crisis for bivalves?, in Logan, A., and Hills, L. V., eds., *The Permian and Triassic Systems and Their Mutual Boundary*, Can. Soc. Petrol. Geol. Mem. 2, pp. 608–621.

Napier, W. M., and Clube, S. V. M., 1979, A theory of terrestrial catastrophism, *Nature*, 282, 455–459.

Neuman, R. B., 1972, Brachiopods of Early Ordovician volcanic islands, *Proc. 24th Int. Geol. Congr.*, 7, 297–302.

Neuman, R. B., 1976, Early Ordovician (late Arenig) brachiopods from Virgin Arm, New World Island, Newfoundland, *Bull. Geol. Surv. Can.*, 261, 11–61.

Newell, N. D., 1967, Revolutions in the history of life, *Geological Society of America Special Paper 89*, pp. 63–91.

Newell, N. D., 1971, An outline history of tropical reefs, *Am. Mus. Novit.*, 2465, 37 pp.

Newell, N. D., 1982, Mass extinctions: Illusion or realities?, *Geological Society of America Special Paper 190*, pp. 257–263.

Norris, G., 1982, Spore-pollen evidence for early Oligocene high-latitude cool climatic episode in northern Canada, *Nature*, 297, 387–389.

Officer, C. B., and Drake, C. L., 1983, The Cretaceous–Tertiary transition, *Science*, 219, 1383–1390.

Officer, C. B., and Drake, C. L., 1985, Terminal Cretaceous environmental events, *Science*, **227**, 1161–1167.

O'Keefe, J. A., 1980, The terminal Eocene event: Formation of a ring system around the Earth?, *Nature*, **285**, 309–311.

O'Keefe, J. D., and Ahrens, T. J., 1982, The interaction of the Cretaceous/Tertiary extinction bolide with the atmosphere, ocean and solid Earth, *Geological Society of America Special Paper 190*, pp. 103–120.

Olsen, P. E., and Galton, P. M., 1977, Triassic–Jurassic tetrapod extinctions: Are they real?, *Science*, 197, 983–986.

Olson, E. C., 1982, Extinctions of Permian and Triassic nonmarine vertebrates, *Geological Society of America Special Paper 190*, pp. 501–511.

Orth, C. J., Gilmore, J. S., Knight, J. D., Pillmore, C. L., Tschudy, R. H., and Fassett, J. E., 1981, An iridium anomaly at the palynological Cretaceous–Tertiary boundary in northern New Mexico, *Science*, 214, 1341–1343.

Orth, C. J., Gilmore, J. S., Knight, J. D., Pillmore, C. L., Tschudy, R. H., and Fassett, J. E., 1982, Iridium abundance measurements across the Cretaceous/Tertiary boundary in the San Juan and Raton Basins of northern New Mexico, *Geological Society of America Special Paper 190*, pp. 423–433.

Orth, C. J., Knight, J. D., Quintana, L. R., Gilmore, J. S., and Palmer, A. R., 1984, A search for iridium abundance anomalies at two late Cambrian biomere boundaries in western Utah, *Science*, 222, 163–165.

Palmer, A. R., 1965, Biomere, a new kind of biostratigraphic unit, *J. Paleontol.*, 39, 149–153.

Palmer, A. R., 1979, Biomere boundaries reexamined, *Alcheringa*, 3, 33–41.

Paul, C. R. C., 1982, The adequacy of the fossil record, in Joysey, K. A., and Friday, A. E., eds., *Problems of Phylogenetic Reconstruction*, Syst. Assoc. Spec. Vol. 21, London, Academic Press, pp. 75–117.

Perch-Nielsen, K., Ulleberg, K., Evensen, J. E., 1979, Comments on "The terminal Cretaceous event: A geologic problem with an oceanographic solution" (Gartner and Keany, 1978), in Christensen, W. K., and Birkelund, T., eds., *Cretaceous–Tertiary Boundary Events Symposium*, Vol 2, University of Copenhagen, pp. 106–111.

Perch-Nielsen, K., McKenzie, J., and He, Q., 1982, Biostratigraphy and isotope stratigraphy and the "catastrophic" extinction of calcareous nannoplankton at the Cretaceous/Tertiary boundary, *Geological Society of America Special Paper 190*, pp. 353–371.

Percival, S. F., Jr., and Fischer, A. G., 1977, Changes in calcareous nannoplankton in the Cretaceous–Tertiary biotic crisis at Zumaya, Spain, *Evol. Theory*, 2, 1–35.

Pitrat, C. W., 1973, Vertebrates and the Permo-Triassic extinction, *Palaeogeogr., Palaeoclimatol., Palaeoecol.*, 14, 249–264.

Playford, P. E., McLaren, D. J., Orth, C. J., Gilmore, J. S., and Goodfellow, W. D., 1984, Iridium anomaly in the Upper Devonian of the Canning Basin, Western Australia, *Science,* **226**, 1161–1167.

Pollack, J. B., Toon, O. B., Ackerman, T. P., McKay, C. P., and Turco, R. P., 1983, Environmental effects of an impact-generated dust cloud: Implications for the Cretaceous-Tertiary extinctions, *Science,* 219, 287–289.

Preston, F. W., 1962, The canonical distribution of commonness and rarity, *Ecology,* 43, 185–215, 410–432.

Quinn, J. F., 1983, Mass extinctions in the fossil record, *Science,* 219, 1239–1240.

Rampino, M. R., and Reynolds, R. C., 1983, Clay mineralogy of the Cretaceous-Tertiary boundary clay, *Science,* 219, 495–498.

Rampino, M. R., and Stothers, R. B., 1984, Terrestrial mass extinctions, cometary impacts and the Sun's motion perpendicular to the galactic plane, *Nature,* 308, 709–712.

Raup, D. M., 1978, Cohort analysis of generic survivorship, *Paleobiology,* 4, 1–15.

Raup, D. M., 1979, Size of the Permo-Triassic bottleneck and its evolutionary implications, *Science,* 206, 217–218.

Raup, D. M., 1981, Extinction: Bad genes or bad luck? *Acta Geol. Hispanica,* 16, 25–33.

Raup, D. M., and Sepkoski, J. J., Jr., 1982, Mass extinctions in the marine fossil record, *Science,* 215, 1501–1503.

Raup, D. M., and Sepkoski, J. J., Jr., 1984, Periodicities of extinctions in the geologic past, *Proc. Natl. Acad. Sci. U.S.A.,* 81, 801–805.

Reid, G. C., McAfee, J. R., and Crutzen, P. J., 1978, Effects of intense stratospheric ionisation events, *Nature,* 275, 489–492.

Rosenkrantz, A., 1960, Danian Mollusca from Denmark, *Rept. 21st. Int. Geol. Congr.,* 5, 193–198.

Rowell, A. J., and Brady, M. J., 1976, Brachiopods and biomeres, *Brigham Young Univ., Geol. Stud.,* 23, 165–180.

Rucklidge, J. C., de Gasparis, S., and Norris, G., 1982, Stratigraphic applications of accelerator mass spectrometry using ISOTRACE, *Proc. 3rd N. Am. Paleontol. Conv.,* 2, 455–460.

Russell, D. A., 1975, Reptilian diversity and the Cretaceous–Tertiary transition in North America, *Geol. Assoc. Can. Spec. Paper,* 13, 119–136.

Russell, D. A., 1979, The enigma of the extinction of the dinosaurs, *Ann. Rev. Earth Planet. Sci.,* 7, 163–182.

Russell, D. A., 1984, The gradual decline of the dinosaurs: Fact or fallacy?, *Nature,* 307, 360–361.

Russell, D. A., and Tucker, W., 1971, Supernovae and the extinction of the dinosaurs, *Nature,* 229, 553–554.

Russell, L. S., 1966, The changing environment of the dinosaurs in North America, *Adv. Sci.,* 23, 197–204.

Saul, L. R., 1983, *Turritella* Zonation across the Cretaceous-Tertiary boundary, California, *Univ. Calif. Pub. Geol. Sci.,* **125,** 165 pp.

Savage, D. E., and Russell, D. E., 1983, *Mammalian Paleofaunas of the World,* Reading, Massachusetts, Addison-Wesley, 432 pp.

Scheckler, S. E., 1984, Floral changes in the Frasnian and Famennian, *Geol. Soc. Am. Abstr.,* 16, 62 (Abstract).

Scheltema, R. S., 1977, Dispersal of marine invertebrate organisms: Paleobiogeographic and biostratigraphic implications, in Kauffman, E. G., and Hazel, J. E., eds., *Concepts and*

Methods of Biostratigraphy, Stroudsburg, Pennsylvania, Dowden, Hutchinson & Ross, pp. 73–108.

Scheltema, R. S., 1978, On the relationship between dispersal of pelagic veliger larvae and the evolution of marine prosobranch gastropods, in Battaglia, B., and Beardmore, J. A., eds., *Marine Organisms: Genetics Ecology and Evolution*, New York, Plenum, pp. 303–322.

Scheltema, R. S., and Williams I., 1983, Long-distance dispersal of planktonic larvae and the biogeography and evolution of some Polynesian and western Pacific mollusks, *Bull. Mar. Sci.*, 33, 545–565.

Schindewolf, O. H., 1963, Neokastrophismus?, *Deutsch Geol. Gesell. Zeitschr.*, 114, 430–445.

Schlager, W., 1981, The paradox of drowned reefs and carbonate platforms, *Geol. Soc. Am. Bull.*, 92, 197–211.

Schlanger, S. O., and Cita, M. B., eds., 1982, *Nature and Origin of Cretaceous Carbon-Rich Facies*, London, Academic Press.

Schopf, J. M., and Askin, R. A., 1980, Permian and Triassic floral biostratigraphic zones of southern land masses, in Dilcher, D. L., and Taylor, T. N., eds., *Biostratigraphy of Fossil Plants*, Stroudsburg, Pennsylvania, Dowden, Hutchinson & Ross, pp. 119–152.

Schopf, T. J. M., 1974, Permo-Triassic extinction: Relation to sea-floor spreading, *J. Geol.*, 82, 129–143.

Schopf, T. J. M., 1979, The role of biogeographic provinces in regulating marine faunal diversity through geologic time, in Gray J., and Boucot, A. J., eds., *Historical Biogeography, Plate Tectonics, and the Changing Environment*, Corvallis, Oregon, Oregon State University Press, pp. 449–457.

Schopf, T. J. M., 1982, Extinction of the dinosaurs: A 1982 understanding, *Geological Society of America Special Paper 190*, pp. 415–422.

Seilacher, A., 1984, Late Precambrian and Early Cambrian Metazoa: Preservational or real extinctions?, in Holland, H. D., and Trendall, A. F., eds., *Patterns of Change in Earth Evolution*, Berlin, Springer-Verlag, pp. 159–168.

Sepkoski, J. J., Jr., 1981, A factor analytic description of the Phanerozoic marine fossil record, *Paleobiology*, 7, 36–53.

Sepkoski, J. J., Jr., 1982a, Mass extinctions in the Phanerozoic oceans: A review, *Geological Society of America Special Paper 190*, pp. 283–289.

Sepkoski, J. J., Jr., 1982b, A compendium of fossil marine families, *Milwaukee Publ. Mus. Contrib. Biol. Geol.*, 51, 125 pp.

Sepkoski, J. J., Jr., 1984, A kinetic model of Phanerozoic taxonomic diversity. III. Post-Paleozoic families and mass extinctions, *Paleobiology*, **10**, 246–267.

Sepkoski, J. J., Jr., and Miller, A. I., 1984 (1985), Evolutionary faunas and the distribution of Paleozoic marine communities in space and time, in Valentine, J. W., ed., *Phanerozoic Diversity Patterns: Profiles in Macroevolution*, Princeton, Princeton University Press.

Sheehan, P. M., 1973, The relation of Late Ordovician glaciation to the Ordovician-Silurian changeover in North America brachiopod faunas, *Lethaia*, 6, 147–154.

Sheehan, P. M., 1979, Swedish Late Ordivician marine benthic assemblages and their bearing on brachiopod zoogeography, in Gray, J., and Boucot, A. J., eds., *Historical Biogeography, Plate Tectonics, and the Changing Environment*, Corvallis, Oregon, Oregon State University Press, pp. 61–73.

Sheehan, P. M., 1982, Brachiopod macroevolution at the Ordovician-Silurian boundary, *Proc. 3rd N. Am. Paleont. Conv.*, 2, 477–481.

Signor, P. W., III, and Lipps, J. H., 1982, Sampling bias, gradual extinction patterns, and catastrophes in the fossil record, *Geological Society of America Special Paper 190*, pp. 291–296.

Simberloff, D. S., 1972, Models in biogeography, in Schopf, T. J. M., ed., *Models in Paleobiology*, San Francisco, Freeman, Cooper, pp. 160–191.

Simberloff, D. S., 1974, Permo-Triassic extinctions: Effects of area on biotic equilibrium, *J. Geol.*, 82, 267–274.

Skevington, D., 1974, Controls influencing the composition and distribution of Ordovician graptolite faunal provinces, in Rickards, R. B., Jackson, D. E., and Hughes, C. P., eds., *Graptolite studies in honour of O. M. Bulman*, Special Papers in Palaeontology, 13, pp. 59–73.

Sloan, R. E., 1976, The ecology of dinosaur extinction, in Churcher, C. S., ed., *Athlon: Essays on Palaeontology in Honour of Loris Shano Russell*, Toronto, Royal Ontario Museum, pp. 134–154.

Smit, J., and Hertogen, J., 1980, An extraterrestrial event at the Cretaceous-Tertiary boundary, *Nature*, 285, 198–200.

Smit, J., and Klaver, G., 1981, Sanidine spherules at the Cretaceous-Tertiary boundary indicate a large impact event, *Nature*, 292, 47–49.

Smit, J., and Kyte, F. T., 1984, Siderophile-rich magnetic spheroids from the Cretaceous-Tertiary boundary in Umbria, Italy, *Nature*, 310, 403–405.

Smit, J., and Romein, A. J. T., 1985, A sequence of events across the Cretaceous-Tertiary Boundary, *Earth Planet. Sci. Lett.*, **74,** 155–170.

Smit, J., and ten Kate, W. G. H. Z., 1982, Trace element patterns at the Cretaceous-Tertiary boundary: Consequence of a large impact, *Cretac. Res.*, 3, 307–332.

Smit, J., and van der Kaars, S., 1984, Terminal Cretaceous extinctions in the Hell Creek area, Montana: Compatible with catastrophic extinction, *Science*, 223, 1177–1179.

Snyder, S. W., Muller, C., and Miller, K. G., 1984, Eocene-Oligocene boundary: Biostratigraphic recognition and gradual paleoceanographic change at DSDP Site 549, *Geology*, 12, 112–115.

Sohl, N. F., 1960, Archeogastropoda, Mesogastropoda and stratigraphy of the Ripley, Owl Creek, and Prairie Bluff Formations, *U.S. Geol. Surv. Prof. Paper*, 331-A, 1–151.

Sohl, N. F., 1964, Neogastropoda, Opisthobranchia and Basommatophora from the Ripley, Owl Creek, and Prairie Bluff Formation, *U.S. Geol. Surv. Prof. Paper*, 331-B, 153–333.

Sohl, N. F., 1971, North American Cretaceous biotic provinces delineated by gastropods, *Proc. N. Am. Paleontol. Conv.*, L, 1610–1638.

Sonnenfeld, P., 1978, Effects of a variable sun at the beginning of the Cenozoic Era, *Climatic Change*, 1, 355–382.

Stanley, G. D., Jr., 1981, Early history of scleractinian corals and its geological consequences, *Geology*, 9, 507–511.

Stanley, S. M., 1979, *Macroevolution, Pattern and Process*, San Francisco, Freeman, 332 pp.

Stanley, S. M., 1982a, Species selection involving alternative character states: An approach to macroevolutionary analysis, *Proc. 3rd N. Am. Paleontol. Conv.*, **2**, 505–510.

Stanley, S. M., 1982b, Glacial refrigeration and Neogene regional mass extinction of marine bivalves, in Gallitelli, E. M., ed., *Paleontology, Essential of Historical Geology*, Modena, Italy, S.T.E.M. Mucchi, pp. 179–191.

Stanley, S. M., 1984, Marine mass extinctions: A dominant role for temperature, in Nitecki, M. H., ed., *Extinctions*, Chicago, University of Chicago Press, pp. 69–117.

Stanley, S. M., and Campbell, L. D., 1981, Neogene mass extinction of western Atlantic molluscs, *Nature*, **293**, 457–459.

Steckler, M., 1984, Changes in sea level, in Holland, H. D., and Trendall, A. F., eds., *Patterns of Change in Earth Evolution*, Berlin, Springer-Verlag, pp. 103–121.

Stevens, C. H., 1977, Was development of brackish oceans a factor in Permian extinctions?, *Geol. Soc. Amer. Bull.*, **88,** 133–138.

Strathmann, R. R., 1978, The evolution and loss of feeding larval stages of marine invertebrates, *Evolution*, **32**, 894–906.

Strathmann, R. R., and Slatkin, M., 1983, The improbability of animal phyla with few species, *Paleobiology*, **9**, 97–106.

Sun, Y., Chai, Z., Ma, S., Mao, X., Xu, D., Zhang, Q., Yang, Z., Sheng, J., Chen, C., Rui, L., Liang, X., Zhao, J., and Hi, J., 1984, The discovery of iridium anomaly in the Permian-Triassic boundary clay in Changxing, Zhijing, China, and its significance, in Tu, G., ed., *Developments in Geoscience*, Beijing, Science Press, pp. 235–245.

Surlyk, F., and Johansen, M. B., 1984, End-Cretaceous brachiopod extinctions in the Chalk of Denmark, *Science*, **223**, 1174–1177.

Swain, T., 1976, Angiosperm-reptile coevolution, in d'A. Bellairs, A., and Cox, C. B., eds., *Morphology and Biology of Reptiles*, Linn. Soc. Symp. Ser. 3, pp. 107–122.

Tappan, H., 1968, Primary production, isotopes, extinctions and the atmosphere, *Palaeogeogr., Palaeoclimatol., Palaeoecol.*, **4**, 187–210.

Tappan, H., 1980, *The Paleobiology of Plant Protists*, San Francisco, W. H. Freeman, 1028 pp.

Tappan, H., 1982, Extinction or survival: Selectivity and causes of Phanerozoic crises, *Geological Society of America Special Paper 190*, pp. 265–276.

Tappan, H., and Loeblich, A. R., Jr., 1971, Geobiologic implications of phytoplankton evolution and time-space distribution, *Geological Society of America Special Paper, 127*, pp. 247–340.

Tappan, H., and Loeblich, A. R., Jr., 1973, Evolution of the oceanic plankton, *Earth-Sci. Rev.*, **9**, 207–240.

Thaddeus, P., and Chanan, G. A., 1985, Cometary impacts, molecular clouds and the motion of the Sun perpendicular to the galactic plane, *Nature*, **314**, 73–75.

Thierstein, H. R., 1981, Late Cretaceous nannoplankton and the change at the Cretaceous–Tertiary boundary, *Soc. Econ. Paleontol. Mineral. Spec. Pub.*, **32**, 355–394.

Thierstein, H. R., 1982, Terminal Cretaceous plankton extinctions: A critical assessment, *Geological Society of America Special Paper 190*, pp. 385–399.

Thierstein, H. R., and Berger, W. H., 1978, Injection events in ocean history, *Nature*, **276**, 461–466.

Thompson, S. L., and Barron, E. J., 1981, Comparison of Cretaceous and present Earth albedos: Implications for the causes of paleoclimates, *J. Geol.*, **89**, 143–167.

Toon, O. B., 1984, Sudden changes in atmospheric composition and climate, in Holland, H. D., and Trendall, A. F., eds., *Patterns of Change in Earth History*, Berlin: Springer-Verlag, pp. 41–61.

Toon, O. B., Pollack, J. B., Ackerman, T. P., Turco, R. P., McKay, C. P., and Liu, M. S., 1982, Evolution of an impact-generated dust cloud and its effects on the atmosphere, *Geological Society of America Special Paper 190*, pp. 187–200.

Urey, H. C., 1973, Cometary collisions and geological periods, *Nature*, **242**, 32–33.

Vail, P. R., Mitchum, R. M., and Thompson, S., 1977, Global cycles of relative changes of sea level, *Am. Assoc. Petrol. Geol. Mem.*, **26**, 83–97.

Valentine, J. W., 1972, Conceptual models of ecosystem evolution, in Schopf, T. J. M., ed., *Models in Paleobiology*, San Francisco, Freeman, Cooper, pp. 192–215.

Valentine, J. W., 1973, *Evolutionary Paleoecology of the Marine Biosphere*, Englewood Cliffs, New Jersey, Prentice-Hall, 511 pp.

Valentine, J. W., 1983, Seasonality: Effects in marine benthic communities, in Tevesz, M. J. S., and McCall, P. L., eds., *Biotic Interactions in Recent and Fossil Benthic Communities*, New York, Plenum Press, pp. 121–156.

Valentine, J. W., and Jablonski, D., 1983, Larval adaptations and patterns of brachiopod diversity in space and time, *Evolution*, **37**, 1052–1061.

Valentine, J. W., and Moores, E. M., 1973, Provinciality and diversity across the Permian–Triassic boundary, in Logan, A., and Hills, L. V., eds., *The Permian and Triassic Systems and Their Mutual Boundary*, Can. Soc. Petrol. Geol. Mem. 2, pp. 759–766.

Valentine, J. W., Foin, T. C., and Peart, D., 1978, A provincial model of Phanerozoic marine diversity, *Paleobiology*, **4**, 55–66.

Van Couvering, J. A., Aubrey, M.-P., Berggren, W. A., Bujak, J. P., Naeser, C. W., and Wieser, T., 1981, The Terminal Eocene Event and the Polish connection, *Palaeogeogr., Palaeoclimatol., Palaeoecol.*, **36**, 321–362.

Van Valen, L., 1973, A new evolutionary law, *Evol. Theory*, **1**, 1–30.

Van Valen, L., 1984, Catastrophes, expectations, and the evidence, *Paleobiology*, **10**, 121–137.

Van Valen, L., and Sloan, R. E., 1977, Ecology and the extinction of the dinosaurs, *Evol. Theory*, **2**, 37–64.

Vidal, G., and Knoll, A. H., 1982, Radiations and extinctions of plankton in the late Proterozoic and early Cambrian, *Nature*, **297**, 57–60.

Voigt, E., 1981, Critical remarks on the discussion concerning the Cretaceous-Tertiary boundary, *Newsl. Stratigr.*, **10**, 92–114.

Waddington, C. J., 1967, Paleomagnetic field reversals and cosmic radiation, *Science*, **158**, 913–915.

Ward, P. D., and Signor, P. W., III, 1983, Evolutionary tempo in Jurassic and Cretaceous ammonites, *Paleobiology*, **9**, 183–198.

Waterhouse, J. B., 1973, The Permian-Triassic boundary in New Zealand and New Caledonia and its relationship to world climatic changes and extinction of Permian life, in Logan, A., and Hills, L. V., eds., *The Permian and Triassic Systems and their Mutual Boundary*, Can. Soc. Petrol. Geol. Mem. 2, pp. 445–464.

Waterhouse, J. B., and Bonham-Carter, G. F., 1975, Global distribution and character of Permian biomes based on brachiopod assemblages, *Can. J. Earth Sci.*, **12**, 1085–1146.

Waterhouse, J. B., and Bonham-Carter, G. F., 1976, Range, proportionate representation, and demise of brachiopod families through the Permian Period, *Geol. Mag.*, **113**, 401–428.

Watts, A. B., 1982, Tectonic subsidence, flexure and global changes of sea level, *Nature*, **297**, 469–474.

Watts, N. L., Lapre, J. F., van Schijndel-Goester, F. S., and Ford, A., 1980, Upper Cretaceous and Lower Tertiary chalks of the Albuskjell area, North Sea: Deposition in a slope and base-of-slope environment, *Geology*, **8**, 217–221.

Weismann, P. R., 1985, Periodic impactors at geological boundary events: Comets or asteroids?, *Nature*, **314**, 517–518.

Wensink, H., 1973, Newer paleomagnetic results of the Deccan Traps, India, *Tectonophysics*, **17**, 41–59.

Wezel, F. C., Vannucci, S., and Vannucci, R., 1981, Decouverte de divers niveaux riches en iridium dans la "Scaglia rossa" et la "Scaglia bianca" de l'Apennin d'Ombrie-Marches (Italie), *C.R. Acad. Sci. Paris, Sér. II*, **293**, 837–844.

Whitmire, D. P., and Jackson, A. A., IV, 1984, Are periodic mass extinctions driven by a distant solar companion?, *Nature*, **308**, 713–715.

Wiedmann, J., 1973, Evolution or revolution of ammonoids at Mesozoic system boundaries, *Biol. Rev.*, **48**, 159–194.

Wolbach, W. S., Lewis, R. S., and Anders, E., 1985, Cretaceous extinctions: Evidence for wildfires and search for meteorite material, *Science*, **230**, 167–170.

Worrest, R. C., 1982, Review of literature concerning the impact of UV-B radiation upon marine organisms, in Calkins, J., ed., *The Role of Solar Ultraviolet Radiation in Marine Ecosystems*, New York, Plenum, pp. 429–457.

Worsley, T. R., 1974, The Cretaceous–Tertiary boundary event in the ocean, *Soc. Econ. Paleontol. Mineral. Spec. Pub.*, **20**, 94–125.

Xu, D., Ma, S., Chain, Z., Mao, X., Sun, Y., Zhang, Q., and Yang, Z., 1985, Abundance variation in iridium and trace elements at the Permian/Triassic boundary at Shangsi in China, *Nature,* **314**, 154–156.

Zoller, W. H., Parrington, J. R., and Phelankotra, J. M., 1983, Iridium enrichment in airborne particles from Kilauea volcano: January 1983, *Science,* **222**, 1118–1121.

11

SURVIVAL DURING BIOTIC CRISES: THE PROPERTIES AND EVOLUTIONARY SIGNIFICANCE OF REFUGES

GEERAT J. VERMEIJ

Department of Zoology
University of Maryland
College Park, Maryland

INTRODUCTION

Extinction occurs when some environmental challenge exceeds the adaptive capacities of individuals in a population and when there is no "safe" place to which individuals can move or become restricted. Any factor that can cause mortality to individuals can potentially bring about local or global extinction of a population. Relative to prevailing conditions, the extinction-causing environmental challenge is likely to be unusual in its intensity, frequency, or nature. All these circumstances conspire to render the causes of extinction difficult to identify. Moreover, the factor responsible for the final disappearance of a population may be much less dramatic and quite different from that responsible for a previous precipitous decline in population size. The inability of a population to expand after such a decline may make the population vulnerable even to rather minor hazards whose nature may never be known.

Given these intrinsic difficulties, the study of extinction would benefit from a shift in research emphasis. Instead of searching for causes of extinction, we might learn more about extinction by studying populations that survive episodes of biological impoverishment. If we understood the ecological, geographical, and populational characteristics that permit taxa to survive biotic crises, we might gain insight into a problem that has received little attention from evolutionists: what are the evolutionary consequences of extinction?

The characteristics that permit persistence are not necessarily the direct opposites of the causes of extinction, but they provide hints about the types of populations and habitats most likely to serve as sources of diversification and colonization. The identification of refuges is a first step in establishing the properties of habitats and populations that promote persistence in a crisis. This is the objective of this chapter. Although a quantitative and exhaustive survey of refuges is far beyond the scope of this chapter, comparisons between present and past distributions will be used to identify some of the more important marine refuges and suggest reasons why these areas and not others permit populations to persist. Finally, that some types of refuges are important to evolutionary events after crises whereas other refuges do little more than shelter the obsolete will be considered.

RECOGNITION OF REFUGES

Crucial to the recognition of refuges is a knowledge of the geographical or ecological distribution of a taxon before and after an observed episode of extinction. Habitat or geographical restriction is a necessary prerequisite for a hypothesis that a given area has served as a refuge. Moreover, several taxa should show similar ecological or geographical contraction to the same refuge. Ideally, it must be shown that taxa that underwent a distributional contraction occupied the refuge continuously during the period of contraction. In other words, reinvasion of a refuge from still another unidentified sanctuary must be ruled out. The fragmentary nature of the fossil record makes this a difficult criterion to satisfy. There is disagreement, for example, between workers who believe that tropical Eastern Pacific corals are remnants of a larger tropical American fauna, some of whose members also persisted in the Indo-West-Pacific (Heck and McCoy, 1978) and workers who think that these corals have reinvaded the Eastern Pacific from the Indo-West-Pacific after becoming extinct in the Americas after the Pliocene (Porter, 1972; Glynn et al., 1972; Dana, 1975; Vermeij, 1978; Glynn, 1982).

The persistence of some populations while others are becoming extinct does not necessarily imply the existence of a refuge. If a population inhabits an environment or region where the extinction-causing agency had little impact, the population would persist even though there was no contraction in distribution. This type of persistence may be very important in subsequent evolution, but we shall not deal with it here.

There is at present very little quantitative information on the importance or duration of refuges. The ecological and biogeographical data required for the recognition of refuges are contained in the primary taxonomic literature. Examples of range restriction therefore come to mind easily, but estimates of the proportion of a given fossil fauna that has become confined to a given refuge are difficult to make except for the few well-known groups for which reliable compilations exist. In attempting to review patterns of ecological and geographical restriction, I have come to the realization that faunistic compilations of Recent as well as fossil biotas are essential and that most of our present information on refuges is still at the anecdotal level.

GEOGRAPHIC REFUGES

There are at least three reasons that a region or habitat serves as a refuge: (1) large-scale environmental changes that resulted in extinctions elsewhere did not occur or were of smaller magnitude; (2) because of shifting climatic belts or other ecological conditions, the refuge may acquire characteristics compatible with the adaptations of species that have been pushed out of or have migrated from another region; and (3) either because of large extent or high productivity, a refuge may be appropriate for the reexpansion of populations that have been decimated in that area and that would otherwise have become extinct. Of course, there may be more than one reason that a given habitat or region is a refuge. The discussion below will clarify these points.

Consider the tropical Indo-West-Pacific region, for example. Today this is the region with the highest local and global diversity of shallow-water marine species, many of which extend from the Red Sea and southeastern Africa to Hawaii and southeastern Polynesia. This region as a whole, and its continental shores in particular, served as a refuge throughout the Cenozoic. Of 27 genera of reef-building scleractinian corals listed as living in the tropical Eastern Pacific during the Eocene by Heck and McCoy (1978), 8 (29%) are found today only in the Indo-West-Pacific. Most or all of these 8 genera represent instances of geographical restriction to the Indo-West-Pacific. Of 10 Pliocene genera in the Eastern Pacific, 3 (30%) are restricted in the Recent fauna to the Indo-West-Pacific (calculation from data given by Heck and McCoy, 1978). Many other groups show a similar pattern of restriction. They include the gastropod family Campanilidae (Houbrick, 1981b), the gastropod genus *Columbarium* (Harasewych, 1983), the pelecypod genus *Fimbria* (Woodring, 1982), and the crab *Scylla* (Vermeij, 1977). All these taxa became extinct in the Americas after the Eocene but persist in the modern fauna of the Indian and Western Pacific Oceans. Of 62 gastropod subgenera (as compiled by Vermeij and Petuch, 1985) that became extinct in tropical America after the Pliocene, 11 (18%) still survive as relicts in the Indo-West-Pacific. Lunulite ectoprocts, which lived on unconsolidated bottoms throughout the tropics until at least Pliocene time, are now mostly restricted to the Indo-West-Pacific region (Cook and Chimonides, 1983).

Contraction of ranges within the Indo-West-Pacific to the shores of continents and high islands since the Pleistocene have been documented for several gastropod species. Members of the genus *Clavocerithium* are known from the Neogene of Indonesia and the Pleistocene of Guam but today are confined to New Guinea

(Houbrick, 1975, 1978). The related *Rhinoclavis vertagus* is known from the Pleistocene of the Marshall Islands but today is restricted to the shores of continents and high islands well to the west of the Marshalls (Houbrick, 1978). Other recent restrictions to continental or high-island shores have been documented for the potamidid mangrove-associated genera *Cerithidea* and *Terebralia* and for the reef-associated genus *Gourmya* (Vermeij, 1978; Houbrick, 1981a). Of 62 Pleistocene gastropod species from Guam, 14 (22%) have become locally extinct, being found today only on islands and continents to the west and south of Guam (Vermeij, 1984).

Although the evidence is not yet compelling, data on Pleistocene sea surface temperatures suggest that the Indo-West-Pacific region underwent less fluctuation in temperature than did the tropical Atlantic (for a review see Vermeij, 1978). If, as many believe, temperature fluctuations were responsible for much of the biological impoverishment in the Atlantic during the Pliocene and Pleistocene (Petuch, 1981, 1982; Stanley and Campbell, 1981; Stanley, 1982), the persistence of the Indo-West-Pacific fauna may be due partly to the relatively constant temperature regime in that region.

The pattern of geographical restriction within the Indo-West-Pacific further suggests that persistence is associated with nutrient-rich conditions. The shores of continents and high islands are typically richer in terrestrially derived nutrients than are the shores of atolls (Bakus, 1964; Marsh, 1977). Moreover, there appears to be seasonal upwelling of nutrients in many parts of the Philippines, Indonesia, New Guinea, New Caledonia, the northwestern Indian Ocean, and elsewhere along continental coasts of the Indian and Western Pacific Oceans. Nutrient-rich conditions may protect small populations that were decimated by some catastrophe from complete extinction. Experience in the temperate zone (Lewis et al., 1982) has shown that small or sparse populations near the distributional limits of a species often disappear because of their failure to reproduce or recruit successfully. In areas of high productivity, the chances of successful recruitment may be substantially enhanced. Birkeland (1982) offers strong support for his hypothesis that large quantities of terrestrial runoff containing nitrates, phosphates, and other nutrients permit the mass survival of planktotrophic larvae that would otherwise have starved. He formulated the hypothesis to account for massive population explosions of the coral-eating crown-of-thorns sea-star *Acanthaster planci* in the tropical Pacific. After examining weather records from places with known *Acanthaster* outbreaks, Birkeland discovered that the latter occur about 3 yr after unusually heavy rains that are in turn preceded by prolonged droughts. The heavy rains transport accumulated nutrients into coastal waters. These nutrients stimulate a phytoplankton bloom, which in turn provides food for planktotrophic zooplankters like the larvae of *Acanthaster* and many other reef-associated invertebrates. Larvae take about 3 yr to grow large enough to be noticed as pests on coral reefs. Birkeland's hypothesis spawned several predictions about the timing and location of population outbursts of *Acanthaster*, which were verified subsequently. No outburst is known, for example, from atolls. White (1976, 1978) independently reached similar conclusions about the role of high-nutrient inputs to explain population explosions of locusts.

A great deal still needs to be learned about how populations with different types of larvae respond to fluctuations in nutrient concentrations. Are species with non-planktotrophic larvae as dependent on high nutrient concentrations for rapid population growth as are species with planktotrophic larvae? How do high nutrient concentrations affect individual growth and fecundity? These and other questions must be answered before the importance of upwelling and terrestrial runoff to the persistence and reexpansion capacities of populations can be judged adequately.

Despite our lack of knowledge, some additional biogeographic evidence supports the hypothesis that regions with a high concentration of nutrients have served as important refuges. The tropical Eastern Pacific is characterized by widespread upwelling and by extensive terrestrial runoff from numerous short rivers. Many barnacles, pelecypods, gastropods, and crabs are notably larger than their Western Atlantic counterparts and therefore may grow faster (Vermeij, 1978). Rates of colonization of clean panels by larvae of sessile organisms are much higher in Pacific Panama than on the Atlantic coast (Birkeland, 1977).

The tropical Eastern Pacific has been a major refuge for marine species. This has been demonstrated best for shelled mollusks (Woodring, 1966; Vermeij, 1978), the only animals for which quantitative information is available. During and after the Pliocene, the tropical Western Atlantic fauna suffered a substantial impoverishment, with about 36% of gastropod subgenera becoming locally extinct. Of the locally extinct gastropods, 57% still persist in the Eastern Pacific; these taxa comprise some 20% of the Pliocene gastropod fauna of the southern Caribbean (Vermeij and Petuch, 1985). By contrast, only 15% of Pliocene Eastern Pacific subgenera became extinct in that region, and of the extinct groups, only 18% persist in the Caribbean region; that is, less than 3% of the Pliocene Eastern Pacific fauna became restricted to the Western Atlantic. Other instances of restrictions to the Eastern Pacific from a Neogene amphi-American distribution have been documented for the mangrove genus *Pelliseria* (Graham, 1977), the echinoids *Toxopneustes* and *Metalia* (Chesher, 1972), the ectoproct family Cupuladriidae (Cook and Chimonides, 1983), and barnacles of the *Balanus concavus, B. nubilis*, and *B. pacificus* species groups (Newman and Ross, 1976; Newman, 1979b). (It is not known whether *Toxopneustes* persisted in the Eastern Pacific or whether it recolonized that region from the Indo-West-Pacific along with many other reef-associated species.) No restriction to the Western Atlantic is known in these groups.

The pattern of post-Pliocene extinction in corals would seem to provide a cogent counterexample to the generalization above, but in fact it provides additional support for the view that nutrient availability is important to the persistence of populations. Of 10 Pliocene genera of reef-building scleractinian corals listed by Heck and McCoy (1978) from the Eastern Pacific, 2 (20%) have become restricted to the Western Atlantic. So far as is known, no reef-building coral genus living in the Western Atlantic during the Pliocene has become restricted to the Eastern Pacific. Birkeland (1977) noted that zooxanthella-bearing scleractinians under conditions of upwelling do not compete well with barnacles, hydroids, and other animals lacking these symbiotic algal cells. The pattern of coral restrictions to the Western Atlantic and to the Indo-West-Pacific (see above) may reflect this competitive relationship.

It reminds us to be cautious about generalizations and that the conditions that favor the persistence of some (or even many) populations are inimical to others.

Geographical restriction of mollusks within the Western Atlantic further supports the hypothesis that high nutrient concentrations permit persistence. The north and east coasts of South America have served as a refuge for more than a dozen subgenera of gastropods and for many pelecypods (Vermeij and Petuch, 1985; Petuch, 1976, 1982). The Caribbean coast of eastern Colombia and Venezuela is characterized by strong upwelling (Antonius, 1980, and references therein), and the region from eastern Venezuela to central Brazil has many large rivers emptying into the Atlantic.

The status of tropical West Africa as a refuge remains a matter of speculation because of the poorly known Late Tertiary fossil record there. Upwelling occurs along most of the West African coast, and many large sediment-laden rivers empty into the Eastern Atlantic between Senegal and Angola. That the tropical Eastern Atlantic is a refuge for taxa with a former distribution encompassing the Eastern Atlantic, Western Atlantic, and Eastern Pacific is suggested by the disjunct distribution of several genera and species groups that today are found only in the Eastern Atlantic and Eastern Pacific. Groups with this distribution are the gastropods *Purpurellus, Zonaria,* and *Sveltia* (Vokes, 1964; Foin, 1976; Petit, 1976); the stomatopod genus *Coronida*; and species groups within the stomatopod genera *Squilla* and *Eurysquilla* (Manning, 1977; Reaka and Manning, 1980). Western Atlantic fossils are known for each of the gastropod genera. There are several subgeneric pairs that indicate very close affinities between West African taxa on the one hand and taxa that have become restricted to the Eastern Pacific on the other. Examples include the arcid pelecypods *Senilia* (West Africa) and *Grandiarca* (Eastern Pacific) (Reinhart, 1935; Olsson, 1961) and the potamidid gastropods *Tympanotonus* (West Africa) and *Rhinocoryne* (Eastern Pacific). A few genera are circumtropical except for the Western Atlantic. Although definitive evidence of these genera having lived in the Western Atlantic is lacking, it is possible that this region was occupied by them during the Late Tertiary. Examples known to me include the hermit crab *Trizopagurus* (Forest, 1952), the stomatopod *Pseudosquillopsis* (Reaka and Manning, 1980), the crab *Daldorfia* (Vermeij, 1977), and the gastropod *Lentigo* (Abbott, 1960). The alternative explanation for this distributional pattern, that Indo-West-Pacific representatives reached West Africa during warm intervals of the Pleistocene, cannot be ruled out, but it seems unlikely in the cases of *Trizopagurus* and *Pseudosquillopsis*.

West Africa has apparently also served as a refuge for taxa that were eliminated from the Mediterranean during the Pleistocene. Although no West African fossils are known from the Neogene, taxa that during the Pliocene were common in the Mediterranean area are known today only from tropical West Africa. Well-documented examples include the helmet-shell (cassid) genera *Cassis* and *Cypraecassis* (Abbott, 1968) and at least five lineages of terebrid gastropods (Bouchet, 1981).

Many temperate coasts that trend north to south were also refuges during the Pleistocene, when sea levels rose and fell and climatic belts moved south and north with the coming and going of glaciers. Despite the post-Pliocene biological impoverishment in the Atlantic, for example, the warm-temperate to subtropical waters between North Carolina and Florida clearly served as a refuge for many taxa that during the Pliocene extended north into Virginia and Maryland (Vermeij, in preparation). Similar latitudinal movements have been documented in the northeast Pacific (MacNeil, 1965; Strauch, 1972; Marincovich, 1977; Nelson, 1978; and many other references). Latitudinal shifts were not possible in all instances, however. The rocky-shore biota of pre-Pleistocene New England probably perished because there is no rocky shore south of the southernmost limit of Pleistocene glaciers on the east coast of North America (Vermeij, 1978).

The warm-temperate faunas of New Zealand and Australia contain many relict taxa descended from forms common in Western Europe during the Eocene and Oligocene. Well-known examples include trigonniid pelecypods (see Newell and Boyd, 1975), the cerithiacean gastropods *Campanile* and *Diastoma* (Houbrick, 1981b,c), the muricid gastropod *Subpterynotus*, and bernayine cypraeid gastropods (Foin, 1976). The ophiuroid genus *Ophiocrossota*, known from the Eocene of the northeast Pacific, is today found only in Australia (Blake, 1975). Many ancient taxa that have become restricted to the Pacific and Indian Oceans include in their current range the temperate coasts of Australia and New Zealand. Examples include the pelecypod genus *Cucullaea* (Habe, 1964), the limpet *Cellana*, and the ectoproct family Lunulitidae (Cook and Chimonides, 1983). To my knowledge, no quantitative estimates of the importance of these temperate shores as refuges are available.

Detailed studies of Late Tertiary mollusks in the northern hemisphere temperate zone are beginning to show that the northern Pacific has served as a refuge for many taxa that lived in the North Atlantic during the Pleistocene. Documented examples include the hiatellid pelecypod *Panomya trapezoides* (Strauch, 1972); the astartid pelecypod *Tridonta alaskensis* (Janssen, 1981); the tellinid pelecypods *Macoma obliqua, M. logeni,* and *M. crassula* (Coan, 1969, 1971); and the slipper limpet *Crepidula adunca* (Hoagland, 1977). It is curious that most or all of these species originated in the North Pacific.

ECOLOGIC REFUGES

Several years ago I suggested that species belonging to extinction-resistant lineages are typically found in environments where the variety and rate of biological activity are chronically or periodically limited by the scarcity of energy (light and heat) or essential nutrients (water, ions, oxygen). In the sea, these stressful environments can be described as deep, dark, and cold (Vermeij, 1978). The fossil record shows that the association between persistence and life in a stressful environment is often the result of ecological restriction, with members in physically less demanding environments often becoming extinct.

Probably the best-known and most satisfactorily documented ecological refuge is deep water. Jablonski et al. (1983) have shown that faunas originate in shallow near-shore environments and slowly come to occupy exclusively deeper-water and offshore positions with the passage of time as new faunas evolve (see also Vermeij, 1978). This pattern was exemplified by brachiopods during the Frasnian–Famennian and Permo-Triassic extinction events (Copper, 1977; Steele-Petrovic, 1979), stalked crinoids after the Triassic (Meyer and Macurda, 1977), rugosan corals during the Frasnian–Famennian event (Pedder, 1982), limpetlike monoplacophoran mollusks after the Devonian, parallelodontid pelecypods after the Cretaceous (Morton, 1982), abyssochrysid gastropods after the Mesozoic (Houbrick, 1979), pholadomyid anomalodesmatan pelecypods after the Cretaceous (Morton, 1981), glypheid decapod crustaceans after the Mesozoic (George and Main, 1968; Schram, 1982), and doubtless numerous other animals. Many striking primitive animals are known only from deep water in the modern ocean and represent probable instances of habitat restriction. Examples include the rift limpet *Neomphalus*, which is possibly the descendant of coiled Paleozoic and Mesozoic euomphalacean gastropods (McLean, 1981); the shelled cephalopod *Nautilus*; and the lepadomorph barnacle *Neolepas*, whose level of organization is like that of Paleozoic and Triassic forms (Newman, 1979a).

Competition or predation are usually thought to be the causes of ecological restriction of these groups to deep water. Although this explanation seems highly plausible, independent evidence of the biological shortcomings of relicts at the time of their restriction is rarely available. To be sure, deep-water relicts today are likely to be outcompeted or eaten when placed with shallow-water forms (although even this has rarely been demonstrated), but the biological incompetence of relicts today does not necessarily extend to the ancestors of these relicts at the time that ecological restriction was in progress.

Undersurfaces of ledges, marine caves, and other mobility-inhibiting and shaded shallow-water environments have been widely recognized as places of refuge for groups that once occupied open surfaces. Taxa that have undergone this kind of ecologic contraction include phyletically diverse sponges with rigid skeletons of silica or calcium carbonate (Jackson et al., 1971; Vacelet, 1979, 1981; Reid, 1968; Hartman, 1979), brachiopods (Jackson et al., 1971; Steele-Petrovic, 1979), various groups of Paleozoic Crustacea (Schram, 1977, 1982), the ectoproct *Stomatopora* (Palmer, 1982), and various serpulid polychaetes (Palmer, 1982). Most of these contractions occurred during the Mesozoic. Again, biological factors have been invoked to explain these contractions. Fast-growing zooxanthella-bearing scleractinian corals and other post-Paleozoic reef-associated animals with algal symbionts perhaps outcompeted the slower-growing sponges, brachiopods, and other forms mentioned above that presumably lacked algal symbionts (Jackson et al., 1971). Caution must be exercised, however, in inferring properties of ancient organisms from those of clearly relictual modern representatives.

Polar areas may also function as refuges for many marine groups. Brachiopods, hexactinellid sponges, and a number of pelecypod genera (e.g., *Thracia, Limopsis,* and *Astarte*) now have a characteristically high-latitude distribution, whereas during and before the Jurassic, they were often or chiefly tropical (Hallam, 1976; Vermeij, 1978; Jablonski and Valentine, 1981). The Frasnian–Famennian extinction event in the Late Devonian and the crisis at the end of the Cretaceous had a much greater impact on warm-water taxa than on those living in cold waters at high latitudes (Copper, 1977; Kollmann, 1979; Taylor et al., 1980; Hsu et al., 1982; Birkelund and Hakansson, 1982; Pedder, 1982). In fact, this may be a characteristic of all the major extinction events (Sepkoski, 1982; Sheehan, 1982).

As already mentioned, selection imposed by biological agents in stressful environments may be less intense than that in physically less rigorous habitats (Vermeij, 1978, 1982). Skeletal breakage and other forms of predation, for example, elicit more numerous and more specialized adaptations in tropical species than in high-latitude forms and also increase in evolutionary importance from fresh to salt water and from deep to shallow levels below the sediment surface. The rate of herbivory declines linearly with water depth even in the photic zone (Hay, 1981a,b; Hay et al., 1983). If these ecological and geographical patterns in the evolutionary importance of biological agents of selection existed in the past, then many species unable to adapt to improvements in coexisting predators and competitors would have been displaced or restricted to environments where selection due to these agents was weaker.

If these considerations were important during episodes of extinction, then island-like habitats that are composed of few species should be good refuges for relict taxa. This indeed seems to be so for the terrestrial biotas of islands. The relatively small tetrapod faunas of South America, Australia, and Madagascar show reduced levels of adaptation to running (Bakker, 1980) and other characteristics fashioned by biological agencies of selection compared to the faunas of large continents like Africa (Janzen, 1976); the island faunas are composed chiefly of primitive orders that had a broader distribution in the Tertiary or Cretaceous (Fooden, 1972; Jardine and McKenzie, 1972). Lizards on islands show less tail damage than do related continental forms and may therefore be under weaker regimes of selection from predators (Rand, 1954). In the sea, however, the status of oceanic islands as refuges is much less certain. Although the biotas of small islands are usually less diverse than those of nearby large islands or continents, there is no compelling evidence that the intensity of biologically imposed selection is less on islands (Vermeij, 1978; Vermeij et al., 1980). More importantly, no molluscan examples of relict marine species that are now confined to oceanic islands are known at present; in fact, as already pointed out, quite the reverse seems to be the case: relict taxa appear to have continental distributions. The only empirical case that has been made for oceanic islands being refuges is for the primitive coronuloid balanomorph barnacle genus *Tesseropora* (Newman and Ross, 1977). When it originated during the Oligocene, this genus occurred in Europe, but today it is restricted to volcanic oceanic islands in the Pacific and Atlantic (see Edwards and Lubbock, 1983, for additional records). Faster-growing and perhaps more predation-resistant advanced coronuloids of the genus *Tetraclita*, which are abundant and widespread on many tropical and subtropical shores of large islands and continents, are absent where *Tesseropora* is found.

To say that relicts often persist in habitats where selection due to biotic agents is weak does not imply that biotic agents played a significant direct role in extinction. In order to bring about the extinction of a species, predators or competitors should have a success rate of near 100% in killing their victims when encounters take place. At least for predators, this is rarely achieved; most predator–prey interactions are characterized by a success rate well below 80% (Vermeij, 1982). Only on small oceanic islands, whose terrestrial biotas have typically evolved in the presence of very few predators and competitors, is extinction due to biotic agents likely. Many endemic species on oceanic islands have become extinct in historic time because of habitat alteration and the introduction by man of competitors and predators that evolved in continental biotas. In most other instances, species that are inferior in competitive ability or antipredatory capacity can migrate or become restricted to regions or habitats where these vulnerable species are free from their enemies. These habitats may be little affected by the physical catastrophes that bring about extinction in biotically more demanding habitats. Long-term shielding of the sun by clouds, for example, will have a great impact on organisms in well-lit habitats in shallow water but will have only indirect consequences for organisms in deep water, polar regions, or caves. Persistence of relicts in these habitats results from lack of exposure to catastrophe and not from the rarity of biotic challenges.

Several stressful environments have not been refuges in that species with a previously broader range did not become restricted to them. Deep layers of sediments, for example, contain species with long geological life spans (Vermeij, 1978), but no species are known that have become confined to this habitat after an initially broader ecological distribution. The high intertidal zone is another such environment, but in this case the fossil record is meager and inadequate. Stanley and Newman (1980) argued, for example, that chthamaloid barnacles have become restricted to high intertidal habitats as a result of competition with balanoid barnacles and perhaps because of predation (Paine, 1981). Unfortunately, the preserved record of chthamaloids is exceedingly sparse, so that verification of this plausible hypothesis is impossible on presently available fossil evidence.

REFUGES AND POSTCRISIS EVOLUTION

Persistence of a taxon during an episode of biological impoverishment does not guarantee that the taxon will play a significant role in evolution after the crisis. In fact, most refuges represent evolutionary backwaters—"old folks' homes"—that receive and protect many species but export very few. The chief reason for the limited role of most refuges seems to be that the environmental conditions in the refuges often elicit, either through adaptation or by physiological necessity, life history characteristics associated with a low population growth rate. Under conditions of low temperature, for example, growth rates of individual ectothermic animals are low (Clarke, 1980). Increase in body size is more apt to be affected by cell enlargement, resulting in a high DNA content, than by cell division (Grime and Mowforth, 1982) because enlargement is less temperature sensitive than is division. Typically, species in chronically stressful environments have few large offspring with long individual life spans (Valentine and Jablonski, 1983). This also seems to be true for the relicts in temperate New Zealand and Australia. These reproductive and populational characteristics make dispersal to new habitats or regions unlikely, except perhaps at high latitudes where transport of large non-planktonic propagules on seaweeds or ice can result in substantial dispersal. Even if individuals from populations with a low intrinsic growth rate should disperse to less stressful habitats from time to time, they would be at a numerical disadvantage to species with faster individual growth rates and higher fecundities during times when the environment was being repopulated. This is the essence of Valentine and Jablonski's (1983) argument about the inability of brachiopods to reestablish numerical supremacy after the crisis at the end of the Permian Period.

The association between poor dispersibility and persistence in refuges would seem to contradict the well-established generalization that geographically widespread species with widely dispersing (usually planktotrophic) larval stages are stratigraphically long-lived (Jackson, 1974; Shuto, 1974; Scheltema, 1977, 1978; Hansen, 1978, 1980, 1982, 1983; Jablonski, 1980, 1982; Jablonski and Valentine, 1981). Wide dispersal and long stratigraphical persistence are characteristic chiefly of shallow-water taxa (Jackson, 1974; Jablonski, 1980). Populations may be decimated or even become locally extinct, but recolonization and subsequent population expansion are so rapid that, from the geological perspective, any one area seems to be occupied continuously. Areas that act as temporary refuges for these populations are probably productive or at least have conditions that do not interfere with a high growth rate or with the production of numerous offspring. Only the potentially rapidly expanding populations in productive refuges are likely to play a substantial role in postcrisis evolution. Not only does large population size provide some buffering against that population's extinction but the populations that can expand out of refuges also influence the type and intensity of selection in newly constituted communities outside the refuge. It is well known, for example, that populations that become established early will have at least a temporary numerical competitive advantage over, and potentially a large selective impact on, those that come later, particularly if adults of the first species interfere with the settlement or growth of members of later species (Birkeland, 1974; Woodin, 1976). The key to reestablishment in biologically decimated areas may therefore not be the same as the recipe for persistence in established assemblages.

If we are ever to understand how extinction affects the subsequent course of evolution, we need to study refuges and their potential to contribute to the biotas of decimated nonrefuge areas. As seems clear, we know very little about this topic. Moreover, we need to understand whether selection immediately after an extinction event differs from that in subsequent assemblages and, if so, how. This requires a careful analysis of the architecture, pattern of relative abundance, and signs of biological interactions in assemblages that appear at various levels above the one marked by the disappearance of many taxa.

REFERENCES

Abbott, R. T., 1960, The genus *Strombus* in the Indo-Pacific, *Indo-Pacific Mollusca*, **1**, 33–146.

Abbott, R. T., 1968, The helmet shells of the world (Cassidae), *Indo-Pacific Mollusca*, **2**, 2–202.

Antonius, A., 1980, Occurrence and distribution of stony corals in the Gulf of Cariaco, Venezuela, *Int. Rev. Ges. Hydrobiol.*, **65**, 321–338.

Bakker, R. T., 1980, Dinosaur heresy–dinosaur renaissance: Why we need endothermic archosaurs for a comprehensive theory of bioenergetic evolution, in Thomas, R. D. K., and

Olson, E. C., eds., *A Cold Look at the Warm-Blooded Dinosaurs*, AAAS Selected Symposium, 28, Boulder, Westview Press, pp. 351–462.

Bakus, G. J., 1964, *The Effects of Fish-Grazing on Invertebrate Evolution in Shallow Tropical Waters*, Los Angeles, Allan Hancock Foundation Publication, **27**, pp. 1–29.

Birkeland, C., 1974, Interactions between a sea pen and seven of its predators, *Ecol. Monogr.*, **44**, 211–232.

Birkeland, C., 1977, The importance of rate of biomass accumulation in early successional stages of benthic communities to the survival of coral recruits, *Proc. Third Int. Coral Reef Symp.*, **1**, 15–21.

Birkeland, C., 1982, Terrestrial runoff as a cause of outbreaks of *Acanthaster planci* (Echinodermata: Asteroidea), *Mar. Biol.*, **69**, 175–185.

Birkelund, T., and Hakansson, E., 1982, The terminal Cretaceous extinction in boreal shelf seas: A multicausal event, Geological Society of America Special Paper 190, pp. 373–384.

Blake, D. B., 1975, A new West American Miocene species of the modern Australian ophiuroid *Ophiocrossota*, *J. Paleontol.*, **49**, 501–507.

Bouchet, P., 1981, Evolution of larval development in Eastern Atlantic Terebridae (Gastropoda), Neogene to Recent, *Malacologia*, **21**, 363–369.

Chesher, R. H., 1972, The status of knowledge of Panamanian echinoids, 1971, with comments on other echinoderms, *Bull. Biol. Soc. Washington*, **2**, 139–158.

Clarke, A., 1980, A reappraisal of the concept of metabolic cold adaptation in polar marine invertebrates, *Biol. J. Linn. Soc.*, **14**, 77–92.

Coan, E. V., 1969, Recognition of an Eastern Pacific *Macoma* in the Coralline Crag of England and its biogeographic significance, *Veliger*, **11**, 277–279.

Coan, E. V., 1971, The northwest American Tellinidae, *Veliger*, 14 (Suppl.), 1–63.

Cook, P. L., and Chimonides, P. J., 1983, A short history of the lunulite bryozoa, *Bull. Mar. Sci.*, **33**, 566–581.

Copper, P., 1977, Paleolatitudes in the Devonian of Brazil and the Frasnian-Famennian mass extinction, *Palaeogeogr., Palaeoclimatol., Palaeoecol.*, **21**, 165–207.

Dana, T. F., 1975, Development of contemporary Eastern Pacific coral reefs, *Mar. Biol.*, **33**, 355–374.

Edwards, A., and Lubbock, R., 1983, Marine zoogeography of St. Paul's Rocks, *J. Biogeogr.*, **10**, 65–72.

Foin, T. C., 1976, Plate tectonics and the biogeography of the Cypraeidae (Mollusca: Gastropoda), *J. Biogeogr.*, **3**, 19–34.

Fooden, J., 1972, Breakup of Pangaea and isolation of relict mammals in Australia, South America, and Madagascar, *Science*, **175**, 894–898.

Forest, J., 1952, Contribution a la révision des crustacés Paguridae, I. Le genre *Trizopagurus*, *Mem. Mus. Natn. Hist. Nat. Paris (A, Zool.)*, **5**, 1–40.

George, R. W., and Main, A. R., 1968, The evolution of spiny lobsters (Palinuridae): A study of evolution in the marine environment, *Evolution*, **22**, 803–820.

Glynn, P. W., 1982, Coral communities and their modifications relative to past and prospective Central American seaways, *Adv. Mar. Biol.*, **19**, 91–132.

Glynn, P. W., Stewart, H. H., and McCosker, J. E., 1972, Pacific coral reefs of Panama: Structure, distribution and predators, *Geol. Rundschau*, **61**, 481–519.

Graham, A., 1977, New records of *Pelliseria* (Theaceae/Pelliseriaceae) in the Tertiary of the Caribbean, *Biotropica*, **9**, 48–52.

Grime, J. P., and Mowforth, M. A., 1982, Variations in genome size: An ecological interpretation, *Nature*, **299**, 151–153.

Habe, T., 1964, Notes on the genus *Cucullaea* (Mollusca), *Bull. Nat. Sci. Mus.*, **7**, 259–261.

Hallam, A., 1976, Stratigraphic distribution and ecology of European Jurassic bivalves, *Lethaia*, **9**, 245–259.

Hansen, T. A., 1978, Larval dispersal and species longevity in Lower Tertiary gastropods, *Science*, **199**, 885–887.

Hansen, T. A., 1980, Influence of larval dispersal and geographic distribution on species longevity in neogastropods, *Paleobiology*, **6**, 193–207.

Hansen, T. A., 1982, Modes of larval development in Early Tertiary neogastropods, *Paleobiology*, **8**, 367–377.

Hansen, T. A., 1983, Modes of larval development and rates of speciation in Early Tertiary neogastropods, *Science*, **220**, 501–502.

Harasewych, M. G., 1983, A review of the Columbariinae (Gastropoda: Turbinellidae) of the Western Atlantic with notes on the anatomy and systematic relationships of the subfamily, *Nemouria*, **27**, 1–42.

Hartman, W. D., 1979, A new sclerosponge from the Bahamas and its relationship to Mesozoic stromatoporoids, *Colloq. Int. Centre Nat. Rech. Sci.*, **291**, 467–474.

Hay, M. E., 1981a, The functional morphology of turf-forming seaweeds: Persistence in stressful marine habitats, *Ecology*, **62**, 739–750.

Hay, M. E., 1981b, Herbivory, algal distribution, and the maintenance of between-habitat diversity on a tropical fringing reef, *Am. Nat.*, **118**, 520–540.

Hay, M. E., Colburn, T., and Downing, D., 1983, Spatial and temporal patterns in herbivory on a Caribbean fringing reef: The effects on plant distribution, *Oecologia (Berlin)*, **58**, 299–308.

Heck, K. L. Jr., and McCoy, E. D., 1978, Long-distance dispersal and the reef-building corals of the Eastern Pacific, *Mar. Biol.*, **48**, 349–356.

Hoagland, K. E., 1977, Systematic review of fossil and Recent *Crepidula* and discussion of evolution of the Calyptraeidae, *Malacologia*, **16**, 353–420.

Houbrick, R. S., 1975, *Clavocerithium (Indocerithium) taeniatum*, a little-known and unusual cerithiid from New Guinea, *Nautilus*, **89**, 99–105.

Houbrick, R. S., 1978, The family Cerithiidae in the Indo-Pacific. Part I: The genera *Rhinoclavis*, *Pseudovertagus* and *Clavocerithium*, *Monogr. Mar. Mollusca*, **1**, 1–130.

Houbrick, R. S., 1979, Classification and systematic relationships of the Abyssochrysidae, a relict family of bathyal snails (Prosobranchia: Gastropoda), *Smithsonian Contrib. Zool.*, **290**, 1–21.

Houbrick, R. S., 1981a, Anatomy and systematics of *Gourmya gourmyi* (Prosobranchia: Cerithiidae), a Tethyan relict from the Southwest Pacific, *Nautilus*, 95, 2–11.

Houbrick, R. S., 1981b, Anatomy of *Diastoma melanioides* (Reeve, 1849) with remarks on the systematic position of the family Diastomatidae (Prosobranchia: Gastropoda), *Proc. Biol. Soc. Washington*, **94**, 598–621.

Houbrick, R. S., 1981c, Anatomy, biology and systematics of *Campanile symbolicum* with reference to adaptive radiation of the Cerithiacea (Gastropoda: Prosobranchia), *Malacologia*, **21**, 263–289.

Hsu, K. J., He, Q., McKenzie, J. A., Weissert, H., Perch-Nielsen, K., Oberhänsli, H., Kelts, K., LaBrecque, J., Tauxe, L., Krähenbühl, U., Percival, S. F., Wright, R., Karpoff, A. M., Petersen, N., Tucker, P., Poore, R. Z., Gombos, A. M., Pisciotto, K., Carman, Jr., M. F., and Schreiber, E., 1982, Mass mortality and its environmental and evolutionary consequences, *Science*, **216**, 249–256.

Jablonski, D., 1980, Apparent versus real biotic effects of transgression and regression, *Paleobiology*, **6**, 397–407.

Jablonski, D., 1982, Evolutionary rates and modes in Late Cretaceous gastropods: Role of larval ecology, *Third N. Am. Paleontol. Conv. Proc.*, **1**, 257–262.

Jablonski, D., and Valentine, J. W., 1981, Onshore-offshore gradients in Recent Eastern Pacific shelf faunas and their paleobiogeographic significance, in Scudder, G. G. E., and Reveal, J. L., eds., *Evolution Today*, p. 441–453.

Jablonski, D., Sepkoski, J. J., Jr., Bottjer, D. J., and Sheehan, P. M., 1983, Onshore-offshore patterns in the evolution of Phanerozoic shelf communities, *Science*, **222**, 1123–1125.

Jackson, J. B. C., 1974, Biogeographic consequences of eurytopy and stenotopy among marine bivalves and their evolutionary significance, *Am. Nat.*, **108**, 541–560.

Jackson, J. B. C., Goreau, T. F., and Hartman, W. D., 1971, Recent brachiopod-coralline sponge communities and their paleoecological significance, *Science*, **173**, 623–625.

Janssen, A. W., 1981, *Tridonta zelandica* Janssen and van der Slik, 1974, a junior synonym of *Astarte alaskensis* Dall, 1903, *Basteria*, **45**, 85–86.

Janzen, D. H., 1976, The depression of reptile biomass by large herbivores, *Am. Nat.*, **110**, 371–400.

Jardine, N., and McKenzie, D., 1972, Continental drift and the dispersal and evolution of organisms, *Nature*, **235**, 20–24.

Kollmann, H. A., 1979, Distribution patterns and evolution of gastropods around the Cretaceous–Tertiary boundary, in Kegel Christensen, W., and Birkelund, T., eds., *Cretaceous–Tertiary Boundary Events, II. Proceedings*, University of Copenhagen Press, pp. 83–87.

Lewis, J. R., Bowman, R. S., Kendall, M. A., and Williamson, P., 1982, Some geographical components in population dynamics: Possibilities and realities in some littoral species, *Neth. J. Sea Res.*, **16**, 18–28.

MacNeil, F. S., 1965, Evolution and distribution of the genus *Mya*, and Tertiary migrations of Mollusca, *U.S. Geol. Surv. Prof. Paper*, 483G, G1–G51.

Manning, R. B., 1977, A monograph of the West African stomatopod Crustacea, *Atlantide Rep. Danish Exped. Coasts Trop. W. Afr. 1945–46*, **12**, 25–181.

Marincovich, L. Jr., 1977, Cenozoic Natacidae (Mollusca: Gastropoda) of the Northeastern Pacific, *Bull. Am. Paleontol.*, **70**, 169–494.

Marsh, J. A. Jr., 1977, Terrestrial inputs of nitrogen and phosphorus on fringing reefs of Guam, *Proc. Third Int. Coral Reef Symp.*, **1**, 331–336.

McLean, J. H., 1981, The Galapagos rift limpet *Neomphalus*: Relevance to understanding the evolution of a major Paleozoic-Mesozoic radiation, *Malacologia*, **21**, 291–336.

Meyer, D. L., and Macurda, D. B., 1977, Adaptive radiation of the comatulid crinoids, *Paleobiology*, **3**, 74–82.

Morton, B., 1981, The Anomalodesmata, *Malacologia*, **21**, 35–60.

Morton, B., 1982, The functional morphology of *Bathyarca pectunculoides* (Bivalvia: Arcacea) from a deep Norwegian fjord with a discussion of the mantle margin in the Arcoida, *Sarsia*, **67**, 269–282.

Nelson, C. M., 1978, *Neptunea* (Gastropoda: Buccinacea) in the Neogene of the North Pacific and adjacent Bering Sea, *Veliger*, **21**, 203–215.

Newell, N. D., and Boyd, D. W., 1975, Parallel evolution in early trigoniacean bivalves, *Bull. Am. Mus. Nat. Hist.*, **154**, 53–162.

Newman, W. A., 1979a, A new scalpellid (Cirripedia); a Mesozoic relic living near an abyssal hydrothermal spring, *Trans. San Diego Soc. Nat. Hist.*, **19**, 153–167.

Newman, W. A., 1979b, Californian transition zone: Significance of short-range endemics, in Gray, J., and Boucot, A. J., eds., *Historical Biogeography, Plate Tectonics, and the Changing Environment*, Corvallis, Oregon State University Press, pp. 399–416.

Newman, W. A., and Ross, A., 1976, Revision of the balanomorph barnacles; including a catalog of the species, *Mem. San Diego Soc. Nat. Hist.*, **9**, 1–108.

Newman, W. A., and Ross, A., 1977, A living *Tesseropora* (Cirripedia: Balanomorpha) from Bermuda and the Azores: First records from the Atlantic since the Oligocene, *Trans. San Diego Soc. Nat. Hist.*, **18**, 207–216.

Olsson, A. A., 1961, *Mollusks of the Tropical Eastern Pacific, Particularly from the Southern Half of the Panamic-Pacific Faunal Province (Panama to Peru)*, Ithaca, New York, Paleontological Research Institute, 574 pp.

Paine, R. T., 1981, Barnacle ecology: Is competition important?, *Paleobiology*, **7**, 553–560.

Palmer, T. J., 1982, Cambrian to Cretaceous changes in hardground communities, *Lethaia*, **15**, 309–323.

Pedder, A. E. H., 1982, The rugose coral record across the Frasnian-Famennian boundary, Geological Society of America Special Paper 190, pp. 485–489.

Petit, R. E., 1976, Notes on the Cancellariidae (Mollusca: Gastropoda)—III, *Tulane Stud. Geol. Paleontol.*, **12**, 33–43.

Petuch, E. J., 1976, An unusual molluscan assemblage from Venezuela, *Veliger*, **18**, 322–325.

Petuch, E. J., 1981, A relict Neogene caenogastropod fauna from northern South America, *Malacologia*, **20**, 307–347.

Petuch, E. J., 1982, Geographical heterochrony: Contemporaneous coexistence of Neogene and Recent molluscan faunas in the Americas, *Palaeogeogr., Palaeoclimatol., Palaeoecol.*, **37**, 277–312.

Porter, J. W., 1972, Ecology and species diversity of coral reefs on opposite sides of the isthmus of Panama, *Bull. Biol. Soc. Wash.*, **2**, 89–116.

Rand, A. S., 1954, Variation and predator pressure in an island and a mainland population of lizards, *Copeia*, 1954, 26–262.

Reaka, M. L., and Manning, R. B., 1980, The distributional ecology and zoogeographical relationships of stomatopod Crustacea from Pacific Costa Rica, *Smithsonian Contrib. Mar. Sci.*, **7**, 1–29.

Reid, R. E. H., 1968, Bathymetric distributions of Calcarea and Hexactinellida in the present and the past, *Geol. Mag.*, **105**, 546–559.

Reinhart, P. W., 1935, Classification of the pelecypod family Arcidae, *Bull. Mus. Roy. Hist. Nat. Belg.*, **11**, 1–68.

Scheltema, R. S., 1977, Dispersal of marine invertebrate organisms: Paleobiogeographic and biostratigraphic implications, in Kauffman, E. G., and and Hazel, J. E., eds., *Concepts and Methods of Biostratigraphy*, Stroudsburg, Dowden, Hutchinson, and Ross, pp. 73–108.

Scheltema, R. S., 1978, On the relationship between dispersal of pelagic veliger larvae and the evolution of marine prosobranch gastropods, in Battaglia, B., and and Beardmore, J. A., eds., *Marine Organisms*, New York, Plenum, pp. 303–322.

Schram, F. R., 1977, Paleozoogeography of Late Paleozoic and Triassic Malacostraca, *Syst. Zool.*, **26**, 367–379.

Schram, F. R., 1982, The fossil record and the evolution of Crustacea, in Abele, L. G., ed., *The Biology of Crustacea*, Vol. 1, *Systematics, the Fossil Record, and Biogeography*, New York, Academic Press, pp. 93–147.

Sepkoski, J. J. Jr., 1982, Mass extinctions in the Phanerozoic oceans: A review, Geological Society of America Special Paper 190, pp. 283–289.

Sheehan, P. M., 1982, Brachiopod macroevolution at the Ordovician-Silurian boundary, *Third N. Am. Paleontol. Conv. Prod.*, **2**, 477–481.

Shuto, T., 1974, Larval ecology of prosobranch gastropods and its bearing on biogeography and paleontology, *Lethaia*, **7**, 239–256.

Stanley, S. M., 1982, Species selection involving alternative character states: An approach to macroevolutionary analysis, *Third N. Am. Paleontol. Conv. Proc.*, **2**, 505–510.

Stanley, S. M., and Campbell, L. D., 1981, Neogene mass extinction of Western Atlantic molluscs, *Nature*, **293**, 457–459.

Stanley, S. M., and Newman, W. A., 1980, Competitive exclusion in evolutionary time: The case of the acorn barnacles, *Paleobiology*, **6**, 173–183.

Steele-Petrovic, H. M., 1979, The physiological differences between articulate brachiopods and filter-feeding bivalves as a factor in the evolution of marine level-bottom communities, *Paleontology*, **22**, 101–134.

Strauch, F., 1972, Phylogenese, Adaptation und Migration einiger nordischer mariner Molluskengenera (*Neptunea, Panomya, Cyrtodaria*, und *Mya*), *Senckenb. Naturf. Ges. Abh.*, 531, 1–211.

Taylor, J. D., Morris, N. J., and Taylor, C. N., 1980, Food specialization and the evolution of predatory prosobranch gastropods, *Paleontology*, **23**, 375–409.

Vacelet, J., 1979, Description et affinités d'une éponge spinctozoaire actuelle, *Colloq. Int. Centre Nat. Rech. Sci.*, **291**, 483–493.

Vacelet, J., 1981, Eponges hypercalcifiées ("pharetronides", "sclerosponges") des cavités des récifs coralliens de Nouvelle-Caledonie, *Bull. Mus. Nat. Hist. Nat. Paris (4)*, **3**, 313–351.

Valentine, J. W., and Jablonski, D., 1983, Larval adaptations and patterns of brachiopod diversity in space and time, *Evolution*, **37**, 1052–1061.

Vermeij, G. J., 1977, Patterns in crab claw size: The geography of crushing, *Syst. Zool.*, **26**, 138–151.

Vermeij, G. J., 1978, Biogeography and adaptation: Patterns of marine life, Cambridge, Harvard University Press, 332 pp.

Vermeij, G. J., 1982, Unsuccessful predation and evolution, *Am. Nat.*, **120**, 701–720.

Vermeij, G. J., 1986, The biology of human-caused extinctions of species, in Norton, B. G., and Shue, H., eds., *The Preservation of Species*, in press.

Vermeij, G. J., and Petuch, E. J., 1985, Differential extinction in tropical American molluscs: Endemism, architecture, and the Panama land bridge, *Malacologia*, **28**, in press.

Vermeij, G. J., Zipser, E., and Dudley, E. C., 1980, Predation in time and space: Peeling and drilling in terebrid gastropods, *Paleobiology*, **6**, 352–364.

Vokes, E. H., 1964, Supraspecific groups in the subfamilies Muricinae and Tritonaliinae (Gastropoda: Muricidae), *Malacologia*, **2**, 1–41.

White, T. C. R., 1976, Weather, food and plagues of locusts, *Oecologia (Berlin)*, **22**, 119–134.

White, T. C. R., 1978, The importance of a relative shortage of food in animal ecology, *Oecologia (Berlin)*, **33**, 71–86.

Woodin, S. A., 1976, Adult-larval interactions in dense infaunal assemblages: Patterns of abundance, *J. Mar. Res.*, **34**, 25–41.

Woodring, W. P., 1966, The Panama land bridge as a sea barrier, *Am. Philos. Soc. Proc.*, **110**, 425–433.

Woodring, W. P., 1982, Geology and paleontology of Canal Zone and adjoining parts of Panama. Description of Tertiary mollusks (Pelecypods: Propeamussiidae to Cuspidariidae; additions to families covered in P 306-E; additions to gastropods; cephalopods), *U.S. Geol. Surv. Prof. Pap.*, **306F**, 541–759.

12

AREA-BASED EXTINCTION MODELS IN CONSERVATION

WILLIAM J. BOECKLEN
Department of Biological Sciences and Department of Mathematics
Northern Arizona University
Flagstaff, Arizona

DANIEL SIMBERLOFF
Department of Biological Sciences
Florida State University
Tallahassee, Florida

INTRODUCTION

The equilibrium theory of island biogeography (MacArthur and Wilson, 1963, 1967; Simberloff, 1974a) has been proposed as a nomothetic explanation for mass extinctions in geological time (Webb, 1969, Schopf, 1974; Simberloff, 1974b; Flessa, 1975; Sepkoski, 1976; Flessa and Sepkoski, 1978). The theory views taxonomic diversity as a dynamic balance between origination (speciation or immigration) and extinction. Extinction rates are assumed to be inversely related to population sizes, which, in turn, are positively related to area. Thus, when area is reduced (as during shallow sea regressions), population sizes decrease, extinction probabilities increase, and number of species (genera, families, etc.) declines until a new equilibrium is reached.

Support for this equilibrium model of geological phenomena is based primarily on a rough concordance between episodes of mass extinction and periods of reduction in habitable area (Chamberlain, 1898; Moore, 1954; Newell, 1967; Schopf,

1974) and on several demonstrations of significant diversity–area relationships spanning geological lengths of time. Webb (1969) argues that the number of genera of North American land mammals during the late Cenozoic was in dynamic equilibrium. Schopf (1974) suggests that the number of marine invertebrate families at the Permo–Triassic boundary constituted a dynamic equilibrium mediated by shallow-sea regressions and transgressions; Simberloff (1974b) demonstrates a family–area relationship for Schopf's data. Significant diversity–area relationships exist for genera of North and South American mammals in the late Cenozoic (Flessa, 1975) and for marine invertebrate species in the Phanerozoic (Sepkoski, 1976; Flessa and Sepkoski, 1978). The equilibrium model as applied to geological events says little about the proximate causes of particular mass extinctions, and one may question whether an area change per se or a concomitant change (such as a climatic one), whether caused by the area change or not, is directly responsible for a particular extinction (e.g., Jablonski, 1985). Nevertheless, the model may serve as a useful heuristic device for depicting aggregate extinction pressures (Webb, 1969).

The equilibrium theory of island biogeography has also been proposed as a nomothetic principle underlying the optimal design of nature reserves to minimize extinctions in ecological time. Willis first suggested this principle in a series of lectures beginning in 1971, in which he analogized a network of refuges to an archipelago of islands (Willis, 1984). Although Willis feels that early writings, notably by Diamond (1973), failed to credit his inspiration properly, his publication (Wilson and Willis, 1975) included "rules of design of natural preserves, based on current biogeographic theory" that are often repeated (e.g., Terborgh, 1974, 1975; Diamond, 1975; Diamond and May, 1976). The most frequently cited contention is that species–area relationships dictate that a group of small refuges will preserve fewer species than will a single large one of equal total area. Subsequent, related approaches include faunal collapse models (e.g., Soulé et al., 1979) and calculations of relaxation times (e.g., Diamond, 1972). These models begin with the calculation from species–area curves of the expected number of species in a refuge of given area. Rates of extinctions resulting from refuge insularization are then estimated. It is hypothesized that larger reserves will not only have larger equilibrium numbers of species but will also have lower extinction rates following refuge insularization. Thus, groups of small refuges will have shorter relaxation times to their new equilibria than will a single large refuge of the same total area, and a faunal collapse will ensue.

Recommendations based on the species–area relationship have been widely accepted by conservation planners and biologists in spite of early, strong criticism by Simberloff and Abele (1976, 1982). The International Union for Conservation of Nature, the major umbrella group of conservation organizations from around the world, in its summary statement of conservation measures necessary to stem an impending tide of extinctions (I.U.C.N., 1980), states that refuge design criteria and management practices should be in accord with the equilibrium theory of island biogeography, as outlined by Wilson and Willis (1975). Simberloff (1985) documents how the I.U.C.N. document has become the justification for policy statements by other organizations wishing to aid conservation efforts. Faaborg (1979) and

Samson (1980) have designated the elaboration of refuge design principles the "crowning achievement" of island biogeographic theory.

Here we criticize the application of the theory to conservation and argue that faunal collapse and relaxation models are unlikely to be faithful representations of the course of extinction.

SPECIES–AREA RELATIONSHIPS

The species–area relationship is a weak foundation for predicting extinctions in ecological time because the mechanism by which area per se is said to generate extinction is increased turnover on smaller sites, as described above. However, the very existence in most insular systems of substantial turnover as envisaged by the equilibrium theory is largely unsubstantiated (Simberloff, 1974a; Gilbert, 1980). Whatever the cause of observed species–area relationships in nature, the existence of the relationship alone generates ambiguous predictions about whether one large or several small sites would conserve the most species (Simberloff and Abele, 1976; Higgs, 1981; Higgs and Usher, 1980). And the relationship has little predictive power (Boecklen and Gotelli, 1984).

Equilibrium theory predicts a dynamic balance of immigrations and extinctions such that species composition changes but species richness remains quite constant (Figure 1*a*). Extinction rates are lower on large islands and immigration rates are lower on distant islands, so equilibrium species richness increases with area (Figure 1*b*) but decreases with isolation (Figure 1*c*). Gilbert suggests three necessary, but not sufficient, conditions for acceptance of the equilibrium model in any particular system: species turnover, a species–area relationship, and a distance effect.

Evidence for substantial species turnover is scarce and equivocal (Simberloff,

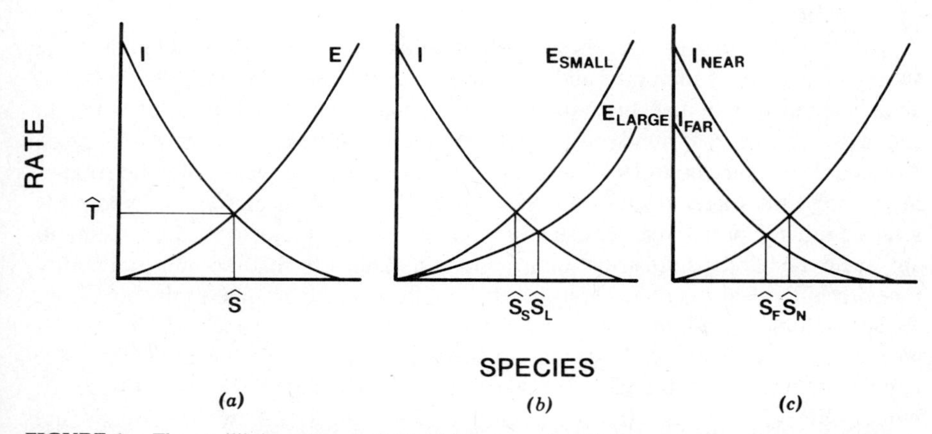

FIGURE 1. The equilibrium theory of island biogeography (*a*). Increasing extinction rates E and decreasing immigration rates I result in an equilibrium number of species S_{eq} maintained by a turnover rate T. The theory predicts both an area effect (*b*) and a distance effect (*c*).

1976a, 1983; Abbott, 1980; Gilbert, 1980; McCoy, 1982). Part of the problem is that changes in lists of species present need not indicate extinction; transient movement of individuals within populations would also produce such changes (Smith, 1975). For example, Simberloff and Wilson (1969, 1970) present experimental evidence that number of arthropod species on mangrove islets is in approximate dynamic equilibrium, but it is likely that a substantial fraction of the turnover represents transient movement of individuals rather than extinction events (Simberloff, 1976a). Similarly, Lynch and Johnson (1974) argued that much of the turnover reported by Diamond (1969) for birds of the California Channel Islands is simply transient movement. Jones and Diamond (1976) support Diamond's original interpretation, though there is still little clear indication of what constitutes the breeding population for many of these species. Whitcomb et al. (1977) report a 15–25% turnover rate over 30 yr for the avifauna of eastern forest fragments, but McCoy (1982), reanalyzing these data, excluded transient and anomalous species and found no evidence of extinction.

There also appear to be a number of oceanic islands in which the number of bird species remains constant but composition does not change—there is no extinction (references in Simberloff, 1983). Finally, extinction rate is beclouded by infrequent censuses, so that estimates of actual extinction rate must be very imprecise (Simberloff, 1969, Diamond and May, 1977). In sum, Gilbert (1980) reviews 25 studies purporting to demonstrate equilibrium turnover but finds that, of these, only Simberloff and Wilson (1969, 1970) and Jones and Diamond (1976) can reasonably claim that dynamic equilibrium remains a tenable hypothesis.

The demonstration of significant species–area relationships has often been adduced to support the dynamic equilibrium model. However, the relationship alone cannot be construed as strong support because the same type of relationship is predicted by other models (Simberloff, 1974a). In particular, the passive sampling hypothesis (Connor and McCoy, 1979), the random placement hypothesis (Coleman et al., 1982), and the habitat heterogeneity hypothesis (Williams, 1964) generates the same prediction.

The passive sampling hypothesis is simply that any site or island contains some subset of a large community and so can be viewed as "sampling" the species–abundance distribution. If the number of individuals in each site is proportional to site area, the expected number of species should decrease as area decreases, since any sample is expected to lack individuals of at least some species, and the smaller the sample, the more species will be missing. The random placement hypothesis seems similar. For a given number of individuals divided into species according to any species–abundance distribution, one may hypothesize that the null probability for each individual to colonize any particular site i is proportional to the area A_i of the site so long as the individuals colonize independently of one another. This model also predicts a monotonically increasing species–area curve. Coleman et al. (1982) argue that the random placement hypothesis is an appropriate null hypothesis against which observed species–area relationships should be tested, while Connor and McCoy (1979) make the same claim for the passive sampling hypothesis. The underlying motivation for these claims is that the equilibrium hypothesis and the habitat diversity hypothesis both posit some sort of determinate relationship, either between the habitat and the presence of particular species or between the species themselves, and the equilibrium theory in addition posits extinctions. Though exact predictions generated by either the passive sampling hypothesis or the random placement hypothesis require one to specify the species–abundance distribution, and it is quite conceivable that the latter was determined by, for example, species interactions or the relative amounts of different habitats, nevertheless, these two hypotheses require fewer assumptions about forces underlying the observations. However, such null hypotheses have rarely been tested. It is interesting that Coleman et al. (1982) cannot reject the random placement hypothesis for birds on islands in Pymatuning Lake. Distances in this system are so small that the islands may well be viewed as samples from a larger community.

The habitat diversity hypothesis for the species–area relationship is simply that as area increases, the number of habitats increases, and each habitat has its own set of more or less characteristic species. Empirical support for this effect is strong and rests primarily on multiple regressions between species richness as dependent variable and area and various measures of habitat heterogeneity as independent variables. Among the measures employed have been number of soil types (Johnson and Simberloff, 1974), elevation (Hamilton and Armstrong, 1965), numbers of habitats (Watson, 1964), and numbers of plant species (Power, 1972). These measures are all assumed by the respective authors to represent meaningful habitat diversity for the taxa under consideration, and in every instance the regression was significantly improved by adding such a measure as an independent variable after area. In some cases, such a measure of habitat diversity predicts species richness better than area does. It is noteworthy that Williams (1964) and many earlier authors who found monotonic relationships between species richness and area assumed that area was simply a surrogate for habitat diversity, which they could not easily measure directly. Probably any naturalist would agree that the habitat diversity effect contributes to the species–area relationship, though there would be many that would argue that the other three potential contributors play little or no role in most instances.

Whatever the basis for the species–area relationship, the recommendation that a single large refuge is preferable to several small ones of equal total area is not a unique consequence of the relationship (Simberloff and Abele, 1976, 1982; Higgs and Usher, 1980; Margules et al., 1982; Kindlmann, 1983). Simberloff and Abele (1976) manipulated the most widely cited form of the species–area relationship, $S = cA^z$ (where S is species richness, A is area, and c and z and fitted constants). They found that two (or more) small refuges may contain more or fewer species than a single large one, depending on the degree of biotic similarity between the small refuges. This similarity, in turn, depends on the gradient of colonizing abilities of the species in the species pool. If this gradient is so strong that the same species are represented on all the small refuges, then the single large refuge will contain more species. If, on the other hand, the species are equally adept colonists, the archipelago of small refuges will contain more species. Since each community has a unique gradient of colonizing abilities, no generalization is possible.

Simberloff and Abele (1982) found that for a variety of studies in which species lists of large islands or sites are compared to combined species lists of groups of small islands or sites, one of two results is usually obtained. Either the group of small sites contained more species or else there was no clear difference in species richness between one large site and a group of smaller ones of equal total area. They interpret this result as consistent with the hypothesis that habitat diversity is the key determinant of species richness: on average, a group of separate sites would have greater habitat diversity than would a single large site of equal total area. Kitchener et al. (1980) similarly argue that a group of small sites very far apart but carefully chosen to contain a variety of habitats maintain lizard species richness so high that it would take a single contiguous site several hundred times larger to have equal species richness. Western and Ssemakula (1981) also argue that for African savanna ungulates small refuges containing widely diverse habitats chosen to be optimal for particular species of interest can play a crucial role in conservation that could not have been predicted from species–area relationships alone.

In sum, determining an optimal design strategy is an empirical rather than theoretical matter. Connor and Simberloff (1978) and Simberloff (1978) suggest a method to approximate the colonizing gradient if species distributions are known over an entire archipelago, while Higgs and Usher (1980) and Higgs (1981) present an empirical index that will indicate the optimal design.

Even when the theoretical basis for the species–area relationship is uncertain, the relationship could be of use to conservationists if it generates precise and reliable predictions. Boecklen and Gotelli (1984) statistically analyze several recent conservation applications of the relationship and conclude that it has little predictive power. Area alone typically explains only half the variation in species richness (cf. Connor and McCoy, 1979). Furthermore, the species–area models themselves are poorly estimated; slope and intercept estimates are heavily influenced by single observations. Consequently, point estimates extracted from these models often vary over several orders of magnitude following casewise deletion of single observations. For example, Kitchener et al. (1980) calculate the species–area relationship for lizards in Western Australia and use this model to estimate the minimum area required to preserve all 45 species. The slope estimate varies from 4.16 to 5.76, and the intercept estimate varies from 0.96 to 5.20 following deletion of single observations. The point estimate thus varies between areas equivalent to Texas and to the entire African continent. In addition, species–area relationships generate imprecise estimates: 95% prediction and inverse prediction intervals routinely span two or more orders of magnitude. Boecklen and Gotelli (1984) conclude that the species–area relationship should be subordinate to autecologic considerations in conservation policy.

MINIMUM VIABLE POPULATION SIZES

The scenario sketched above suggests that in some instances groups of small refuges might be preferable to single large ones of equal total area. However, Simberloff and Abele (1976, 1982) observe that there must be some minimum area such that rapid extinction of one or more species is assured when the small refuges fall below this size. This is because each species has a minimum viable population size, and isolated populations below this size are unlikely to persist. Shaffer (1981) and Soulé (1983) review literature on this notion and describe five sorts of phenomena that place very small populations at great risk of extinction.

First is a group of factors that Shaffer terms *demographic stochasticity*. In a number of demographic traits, variation is proportionally larger the smaller the population size. For example, the probability that all offspring in some generation will be male (ensuring extinction) is much greater in smaller populations. MacArthur and Wilson (1967) and Richter-Dyn and Goel (1972) suggest that for certain types of demographic stochasticity the relationship between population size and probability of extinction is not linear: there is a threshold C such that populations below C will be quickly extinguished while populations greater than C will persist almost indefinitely.

A second sort of danger to very small populations is posed by a force that Shaffer (1981) terms *environmental stochasticity*. Here he refers to the normal range of variation in physical factors, like rainfall, and interacting populations, like predators. If a population is already very small, unusual values of one or more of these factors could cause extinction, but a large population is unlikely to be eliminated by such means. Natural catastrophes are a third class of danger to very small populations. A hurricane or major fire, for example, could threaten the very existence of a small population. So could an epizootic disease.

Soulé (1983) suggests that dysfunction of social behavior constitutes a fourth force that endangers small populations. For example, many animal species form large breeding aggregations (Wynne-Edwards, 1962), and the characteristic behavior associated with these aggregations may require a threshold number of individuals.

Both Shaffer (1981) and Soulé (1983) emphasize a fifth phenomenon—inbreeding depression—as the major long-term threat to small populations. There are two aspects of inbreeding depression. First, debilitating diseases caused by recessive alleles of one or a few genes increase greatly in frequency as population size decreases. A second and even more important phenomenon is loss of heterosis. The more genes for which an individual is heterozygous, the more vigorous and fertile that individual is likely to be. This is true even if there is no debilitating disease present among less heterozygous individuals. Since inbreeding decreases average heterozygosity, and decreases it faster the smaller the population size (Hartl, 1981), some loss of vigor and fertility is a common outcome of inbreeding.

Ralls and Ballou (1983) note that such effects are subtle, so there continues to be skepticism among wildlife biologists about the importance of inbreeding depression. Nevertheless, detailed consideration of vigor and fertility in zoo animals with extensive pedigree records confirms this effect. For plants, there is less evidence that inbreeding is so major a problem, but there is good evidence that in at least some species inbred populations with lower heterozygosity have lower fertility and also decreases in other traits associated with fitness (e.g., Schaal and Levin, 1976; Clegg and Brown, 1983). Exactly how small a population must be before inbreeding depression becomes a major force, perhaps facilitating extinction, is not well known. Franklin (1980) argues that an effective population size of about 50 would suffice to stem severe inbreeding. Since effective population size is often much less than the number of individuals in a population, depending on the breeding system and sex ratio as well as other factors (Franklin, 1980; Hartl, 1981), prevention of inbreeding depression could require many more than 50 individuals.

Franklin adds, however, that even if a population surpassed this inbreeding depression threshold of $N_e = 50$, it might still be so small that it would lose alleles by genetic drift. This loss, in turn, could mean that there would be insufficient genetic variability to allow the population to respond to the environmental changes that may well occur in the course of a few centuries or millennia. Franklin (1980) feels that an effective population size of perhaps 500 would be required for this purpose, while Berry (1971) believes that a population with $N_e = 50$ would retain enough genetic variability for future evolution. Because there are no data that directly measure a population's ability to adapt to future environmental changes, analogous to the way zoo inbreeding records or plant breeding records directly assess inbreeding depression, arguments about the threat of extinction from this direction are unlikely to be very cogent. The models will necessarily rest on unmeasured parameters.

Direct determination of minimum viable population sizes from any combination of the above five forces has rarely been attempted. Consequently, one rarely knows the minimum area that could maintain a population of a particular species. Shaffer's (1978) study of the grizzly bear (*Ursus arctos*) is one direct attempt to estimate both the minimum viable population size and the minimum area required to maintain a population of this size. By contrast, most attempts to estimate the minimum area required for a species are indirect. They consist of scanning sets of sites to see what the smallest site is that a given species occupies. Smaller sites are then hypothesized to be incapable of supporting a population of that species. Simberloff and Gotelli (1984) show that this approach consistently overestimates the minimum necessary area and suggest a computer simulation for any particular set of data that indicates whether any of the species in the set are likely to require an area larger than the smallest area in the set.

However difficult it may be to estimate minimum viable population sizes, estimating the area required to maintain a population of that size is considerably more difficult. The reason is the same one that renders the species–area relationship a poor predictor of species richness. Namely, each species has a particular set of habitat requirements. Though these may be very subtle and difficult to elucidate, they are absolutely crucial to a prediction of whether a particular site will suffice to maintain a particular species. A desert, for example, even one as large as the Sahara, will never support a bird of montane forest. Arguments about minimum areas and about whether one large or several small refuges would be preferable are often of trifling consequence, and prevention of extinction must rest first and foremost on intimate knowledge of a species biology, including habitat requirements.

There may also be inextricable ecological links between a species of interest and one or more other species, so that we may often have to conserve groups of species as a unit if we wish to conserve any one of them. In this instance, the minimum area for the group is the largest minimum area for any of its component species (Simberloff, 1985). That different species will have different minimum areas for long-term conservation is obvious from the most cursory consideration of their different ecologies. Raptors, for example, will surely require enormous areas, while some plants and insects may persist indefinitely in very small sites, so long as these have the requisite habitats and are protected (Schonewald-Cox, 1983).

It is clear, then, that there must be some lower limit to the area of an effective refuge, but it is equally clear that there are many other forces that are at least as likely to cause extinction if refuges are poorly designed.

FAUNAL COLLAPSE AND RELAXATION MODELS

Faunal collapse models (e.g., Soulé et al., 1979) and relaxation models (e.g., Diamond, 1972) attempt to forecast species extinctions in nature reserves resulting from reserve insularization. The forces causing the extinctions are not specified but are assumed to be all of those that place smaller populations at greater risk. The expected species loss is based on comparisons of continental and insular species–area curves. Preston (1960) observes that when progressively larger areas are sampled from continuous regions like continents, species–area relationships typically have larger intercepts and lower slopes than when isolated areas like islands are sampled (Figure 2). Preston suggests that nonisolated areas have an underlying truncated log-normal distribution of individuals into species that has a larger species-to-individuals ratio than does the complete log-normal distribution characteristic of isolated areas. Consequently, small samples from continuous areas have disproportionately more species, depressing the slope of the species–area relationship. MacArthur and Wilson (1967) suggest that transients in nonisolated areas inflate species counts and produce the different curves that Preston (1960) noted for islands and nonisolated areas.

Faunal collapse and relaxation models predict species loss as refuges change in status from nonisolated areas to islands (Figure 2). Most refuges currently are

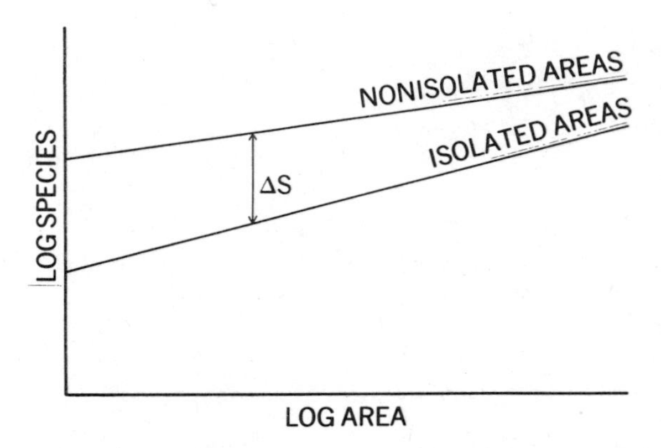

FIGURE 2. Species–area relationships characteristic of nonisolated and isolated areas. Species loss, ΔS, is assumed to occur when reserves change in status from nonisolated to isolated areas.

contiguous with areas of similar habitat and should have species richness characteristic of nonisolated areas. However, as more and more of the contiguous areas become uninhabitable, refuges become more insular and extinctions should cause species richness to "relax" to new equilibria appropriate for islands. This insularization of the landscape is occurring in natural habitats of all sorts (Burgess and Sharpe, 1981).

The relaxation scenario necessitates species–area relationships with slopes that are intrinsic properties of isolated and nonisolated areas. Connor and McCoy (1979) review 100 species–area relationships and cite numerous examples of continental and insular relationships with slopes outside the ranges suggested by Preston (1960) and MacArthur and Wilson (1967). Moreover, they suggest that the apparent regularity in slope estimates is more parsimoniously explained as an artifact of the regression procedure. Consequently, they are skeptical of equilibrium interpretations of species–area relationships and suggest that slope estimates are best viewed as "fitted constants devoid of biological meaning." Sugihara (1981) reaffirms the equilibrium interpretation of the slope values, while Connor et al. (1983) continue to demur.

Various assumptions regarding extinction rate following refuge isolation have

Table 1. Relaxation and Faunal Collapse Models

Author	Model	Extinction Rate	Species Loss	Coefficient Estimate
		Relaxation Models		
Diamond 1972, 1975; Diamond and May, 1976; Case, 1975; Miller, 1978		$dS/dt = -kS$	$S(t) = (S_0 - S_{eq})e^{-t_r/t} + S_{eq}$	$t_r = -t/\ln\{[S(t) - S_{eq}]/(S_0 - S_{eq})\}$
		Faunal Collapse Models		
Terborgh, 1975	1	$dS/dt = -k$	$S(t) = S_0 - kt$	$k = -[S(t) - S_0]/t$
Terborgh, 1975; Soulé et al., 1979; Wilcox 1980	2	$dS/dt = -kS$	$S(t) = S_0 e^{-kt}$	$k = -\ln[S(t) - S_0]/t$
Terborgh, 1974, 1975; Soulé et al., 1979; Wilcox, 1980	3	$dS/dt = -kS^2$	$S(t) = S_0/(1 + kS_0 t)$	$k = [1/S(t) - 1/S_0]/t$
Soulé et al., 1979; Wilcox, 1980	4	$dS/dt = -kS^3$	$S(t) = [S_0^2/2S_0^2kt + 1)]^{1/2}$	$k = [1/S(t)^2 - 1/S_0^2]/2t$
Soulé et al., 1979; Wilcox, 1980	5	$dS/dt = -kS^4$	$S(t) = [S_0^3/(3S_0^3kt + 1)]^{1/3}$	$k = [1/S(t)^3 - 1/S_0^3]/3t$

led to a plethora of models describing species loss through time (Table 1). However, the choice of an extinction rate is arbitrary, as relevant empirical data are very few (e.g., Simberloff, 1976b; Rey, 1981). Nevertheless, five distinct extinction rates have been proposed, all with the general form

$$\frac{dS}{dt} = -kS^n \tag{1}$$

where k, the extinction coefficient, is a constant and n takes integer values from 0 through 4. The extinction coefficient is assumed to be taxon specific (Wilcox, 1980), invariant over time, and inversely related to refuge area. Diamond (1972, 1975) suggests that extinction rates are proportional to the number of species present, $dS/dt = -kS$. Rates of this form characterize exponential decay models, such as

$$S(t) = S_0 e^{-kt} \tag{2}$$

where $S(t)$ is the number of species at time t and S_0 is the initial species richness. However, Diamond proposes that species loss follows the relaxation model

$$S(t) = (S_0 - S_{eq})e^{-t_r/t} + S_{eq} \tag{3}$$

where S_{eq} is the postrelaxation equilibrium number of species and t_r is the relaxation time. The relaxation time is defined as the interval required for the relaxation process to be 36.8% complete. Relaxation is then 90% complete after 2.303 relaxation times.

Terborgh (1974) suggests that extinction rates are proportional to the square of species richness, $dS/dt = -kS^2$. He argues that the square of species richness reflects the importance of interspecific interactions in the extinction process. As competitors become extinct, extinction rates for surviving species decrease. The number of species surviving at time t is given by the faunal collapse model,

$$S(t) = \frac{S_0}{1 + kS_0 t} \tag{4}$$

where k is the extinction coefficient. Soulé et al. (1979) add extinction rates proportional to the third and fourth power of species richness, while Terborgh (1975) proposes a constant extinction rate.

The five extinction rates yield faunal collapse models that estimate very different amounts of species loss through time. For example, Figure 3 presents the faunal collapse of the Nairobi National Park as predicted by the five faunal collapse models. After 500 yr of refuge isolation, the models estimate that 55 species (model 1) to 11 species (model 5) will remain in the Nairobi Park. As the choice of an extinction rate is quite arbitrary (i.e., no extinctions have been observed), so are these estimates arbitrary. Thus, these models will have little practical applicability to conservation practice until extinction rates can be determined precisely for specific situations.

Western and Ssemakula (1981) used a species–area relationship derived from observations in East African grasslands and predicted much lower rates of extinction for the Nairobi National Park and other African parks than did Soulé et al. (1979). They also argued that close attention to habitat requirements of the target species

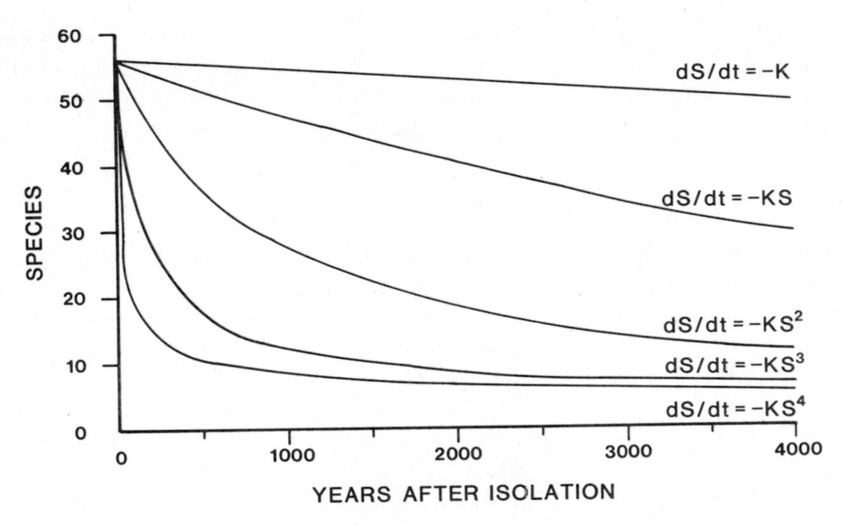

FIGURE 3. The faunal collapse of the Nairobi National Park as predicted by the faunal collapse models in Table 1 (data from Soulé et al., 1979).

is crucial, so that small refuges with exactly the right habitat may preserve species that will be extinguished on much larger refuges. East (1983) notes that, in addition to the fact that Soulé et al. (1979) used data from islands in the Malay Archipelago while Western and Ssemakula (1981) used data from African grasslands to predict the fate of the Nairobi National Park, the two approaches differ in their view of how large minimum viable population sizes are likely to be. Soulé et al. believe them to be large, while Western and Ssemakula feel that small isolated populations can persist. As stated above, direct measurement of minimum viable population sizes is virtually nonexistent, so it is difficult assess the merits of these two views. As East (1983) notes, probably all sides in this debate would agree that intensive management and protection will be required if extinctions are to be avoided in isolated refuges.

Relaxation models differ from faunal collapse models in several respects. Relaxation models seem mathematically more realistic; they predict that species richness will eventually relax to the equilibrium number of species appropriate for isolated areas, according to equation (3). Faunal collapse models predict either negative species richness (model 1) or complete extinction (models 2–5) as t becomes infinitely large. Relaxation models require estimates of t_r, S_0, and S_{eq}, while faunal collapse models require estimates of k and S_0. It is impossible to calculate relaxation times if postrelaxation equilibrium species richness exceeds present species richness. The extinction coefficient can be calculated for any situation. Relaxation times are assumed to vary positively with area, extinction coefficients negatively.

Principal support for faunal collapse and relaxation models is based on comparisons of the biotas of landbridge and oceanic islands. Landbridge islands are separated from continental areas by shallow seas and probably were connected to the mainland during the Pleistocene when sea levels dropped by about 100–200 m. As continental areas, landbridge islands are assumed to have had species–area relationships characteristic of nonisolated areas. However, when sea levels rose, landbridge islands were supersaturated with species and have since been relaxing to new equilibria appropriate for isolated areas. Oceanic islands are assumed not to have changed equilibrium states since the Pleistocene and are thus assumed to represent the ultimate equilibrium condition for landbridge islands. It is also assumed that differences in climate, habitat heterogeneity, and resource availability between landbridge and oceanic islands have little effect on species richness.

Species numbers on landbridge islands intermediate between those in mainland areas and on oceanic islands have been adduced as evidence for the relaxation process. Diamond (1972) shows that landbridge islands in the New Guinea satellite island group typically have more species of birds for a given area than do oceanic islands. Terborgh (1974) demonstrates that avifaunas of landbridge islands in the West Indies are smaller than those of comparable mainland areas but larger than those of comparable oceanic islands. Case (1975) and Wilcox (1980) claim a similar pattern for the herpetofaunas of islands in the Gulf of California.

Comparisons of species richness in mainland areas, on landbridge islands, and on oceanic islands provide estimates of relaxation times, while comparisons of mainland areas and landbridge islands provide estimates of extinction coefficients. For example, Terborgh (1974) compares the avifaunas of mainland areas in South America and landbridge islands in the West Indies to calculate the extinction coefficient for Trinidad, a landbridge island. Terborgh assumes that the extinction rate is proportional to the square of the number of species and that sea levels rose 10,000 yr ago. Trinidad currently has 236 species, and Terborgh estimates that Trinidad had 380 species when it was connected to the mainland. The extinction coefficient for Trinidad is given by

$$k = \frac{-S_p^{-1} - S_0^{-1}}{t} = \frac{1/236 - 1/380}{10000} = -1.61 \times 10^{-7} \tag{5}$$

where S_p is the present number of species. Terborgh estimates extinction coefficients for four other landbridge islands and demonstrates a significant inverse relationship between extinction coefficient and island area. Extinction coefficients have also been estimated for mammals on landbridge islands on the Sunda Shelf and in nature refuges in East Africa (Soulé et al., 1979; Wilcox, 1980). Relaxation times have been estimated for birds on the New Guinea satellite islands (Diamond, 1972, 1975), lizards on islands in the Gulf of California (Case, 1975; Wilcox, 1978), and mammals in the Mkomazi Game Reserve in East Africa (Miller, 1978).

Faunal collapse models predict that extinction coefficients monotonically decrease with area. Soulé et al. (1979) fit negative linear regressions to four sets of extinction coefficients and areas from seven landbridge islands in the Malay Archipelago. Three of these are significant at the 5% level. Soulé et al. (1979) use these four models to estimate extinction coefficients for 19 nature reserves in East Africa. However, the data suggest a curvilinear relationship (Figure 4). In each case, a

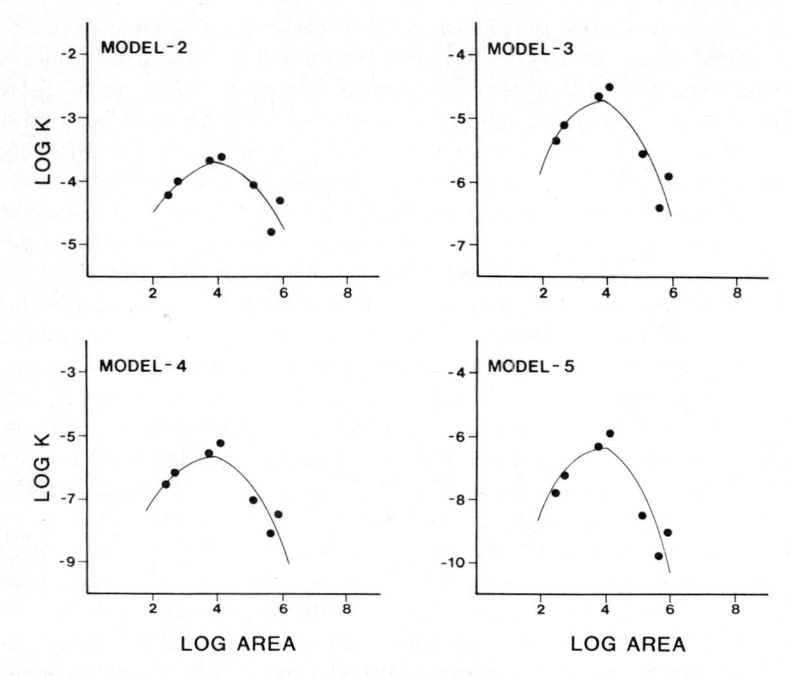

FIGURE 4. Extinction coefficient–area relationships for landbridge islands in the Malay archipelago. In each case, a quadratic regression was a significant improvement over a linear regression at $p<.025$ significance level (data from Soulé et al., 1979).

quadratic regression is a significant improvement over the linear models. The observation that intermediate areas have larger extinction coefficients than do small areas appears to contradict equilibrium theory.

Critics of faunal collapse models and relaxation models (Simberloff and Abele, 1976; Abele and Connor, 1979; Faeth and Connor, 1979; Kitchener et al., 1980; Abbott, 1980) contend that estimates of relaxation times and extinction coefficients are based on unreasonable assumptions regarding the numbers of species initially present on landbridge islands. Diamond (1972) assumes that each landbridge island in the New Guinea satellite island group had the full complement of the New Guinea avifauna, 325 species, even though these islands range over three orders of magnitude in area. Case (1975) also assumes that each landbridge island in the Gulf of California initially contained the 15 lizard species currently found on the mainland. These islands also range over three orders of magnitude in area. Terborgh and Winter (1980) assume that each of three forest reserves in Amazonia, which range over two orders of magnitude in area, initially contained 304 species of birds. With these initial conditions, the existence of a species–area relationship for these landbridge islands or forest reserves, whether the result of a dynamic balance of immigrations and extinctions, habitat heterogeneity, random placement, or passive sampling, guarantees that estimates of relaxation times vary positively with area and that extinction coefficients vary negatively.

Estimating initial species richness is problematic even if one assumes that landbridge islands originally had species–area relationships characteristic of mainland areas. Variance about species–area relationships is typically large (Haas, 1975; Boecklen and Gotelli, 1984). Slope and intercept estimates vary among archipelagos (Connor and McCoy, 1979) and may change historically as habitats change (Abbott and Grant, 1976; Abele and Connor, 1979; Picton, 1979; Usher, 1979). In addition, the construction of the initial species–area relationship often lacks statistical rigor. For example, Soulé et al. (1979) construct the species–area relationships for the entire Sunda Shelf based on one sample observation: they select a z value characteristic of mainland areas and solve the equation $S = cA^z$ for c. They then use this equation to estimate species numbers originally present on the landbridge islands of the Sunda Shelf and to calculate extinction coefficients for these islands. They find a significant inverse relationship between island areas and extinction coefficients and then use this relationship to estimate extinction coefficients for 19 nature reserves in East Africa. Wilcox (1980) adds another sample observation that does little to improve statistical rigor. Miller (1978) estimates the postrelaxation equilibrium number of species and the relaxation time for the Mkomazi Game Reserve from a species–area relationship constructed from no data whatever. He arbitrarily selects values for both z and c and solves the appropriate equations for S_{eq} and t_r. Terborgh (1974) estimates initial species richnesses for landbridge islands in the West Indies from "interpolation" of mainland South America species counts and areas, but we have been unable to duplicate these estimates using least-squares techniques on every published functional form of the species–area relationship.

Terborgh's (1974) estimates of original species richness on West Indian landbridge islands may also be suspect on historical grounds. He assumes that the current mainland species–area relationship has not changed since the Pleistocene and is identical to the species–area relationship for the landbridge islands when they were connected to the mainland. Haffer (1969), Vuilleumier (1971), and Simpson and Haffer (1978) suggest that distributions of South American birds have not been static since the Pleistocene, and, owing to Pleistocene climatic changes, much of this avifauna was restricted to several small forest refugia. These refugia do not include the landbridge islands, and it is unlikely that forest species persisted there. Moreover, Pregill and Olson (1981) argue that much of the avifauna of West Indian oceanic islands became extinct coincidentally with Pleistocene climatic changes.

Distance effects may also confound comparisons of landbridge and oceanic islands biotas and influence calculations of relaxation times and extinction coefficients. The dynamic equilibrium theory predicts an inverse relationship between species richness and distance. Oceanic islands are typically further from the mainland than are landbridge islands. Thus, the observation that landbridge islands are supersaturated with species relative to oceanic islands may be an artifact of distance. Case (1975) estimates postrelaxation equilibrium lizard species richness for landbridge islands in the Gulf of California from a multiple regression relating species richness to area, elevation, plant species richness, and elevation based on data from 16 oceanic islands. However, this equation consistently underestimates lizard species richness for near islands. A significant negative relationship exists between the

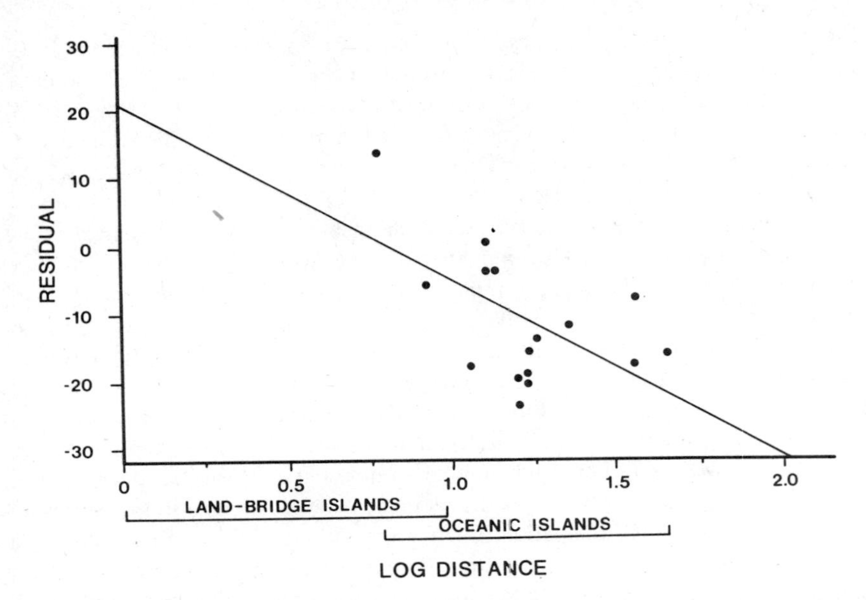

FIGURE 5. Unstandardized residual–distance relationship for Case's (1975) multiple regression analysis of lizard species richness on landbridge islands in the Gulf of California. The relationship is R = 18.7–24.7 log distance ($F_{1,14}$ = 6.93, p<.025).

unstandardized residuals from this equation and the logarithm of island distance (Figure 5). Because of this pattern of residuals, landbridge islands will contain on average 0.68 more species than would be estimated from the multiple regression equation. This excess closely agrees with the observed average difference, 0.91 species. Case's (1975) demonstration of long relaxation times for the landbridge islands may be largely artifactual.

PREDICTIVE POWER AND IMPLICATIONS FOR REFUGE DESIGN

Variation in estimates of extinction coefficients and relaxation times produces variation in projected species richness through time. There are two sources of variation in these estimates: variance in estimating initial or postrelaxation species richness and variance in the extinction coefficient–area (EC–A) or relaxation times–area (RT–A) relationship. The 95% prediction intervals for Terborgh's (1974) estimates (assuming a log species–log area model) of original species richness on landbridge islands in the West Indies typically span 500 species (Figure 6). Consequently, 95% prediction intervals for the extinction coefficients encompass a twofold range (Table 2). The 95% prediction intervals for Case's (1975) estimates of the postrelaxation lizard species richness for landbridge islands in the Gulf of California typically span seven species. The lower bounds of these intervals indicate that landbridge islands are supersaturated with species, while the upper bounds suggest

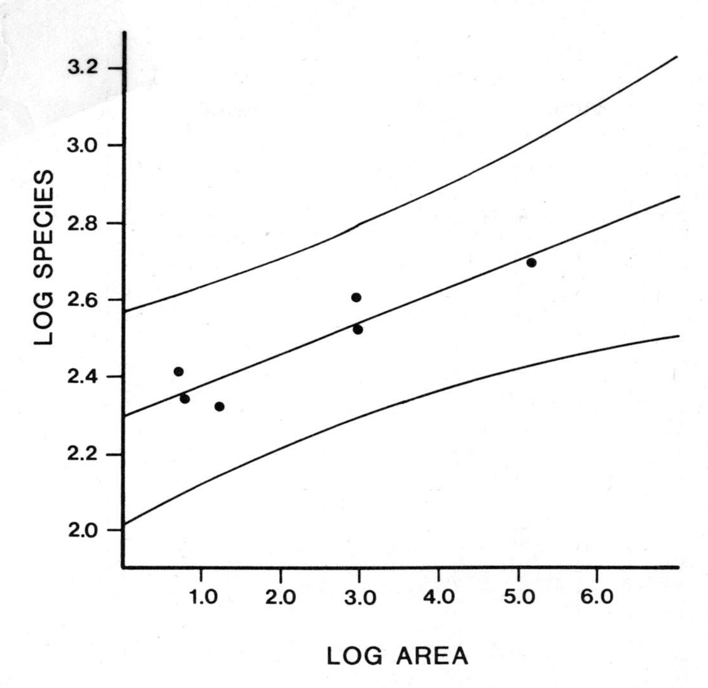

FIGURE 6. Species–area relationship for mainland South American avifaunas. The curved lines are the 95% simultaneous prediction interval (data from Terborgh, 1974).

that they are depauperate. The upper bounds of the 95% prediction intervals for relaxation times are twice as large as the point estimates; the lower bounds are uncalculable. The variation in estimating initial and postrelaxation species numbers results in variation in estimates of extinction coefficients and relaxation times that is intolerable for specific recommendations.

Variation in RT–*A* and EC–*A* relationships is typically large. Diamond (1972) calculates relaxation times for eight landbridge islands near New Guinea. The 95% simultaneous prediction band for this relationship spans nearly an order of magnitude (Figure 7*a*). Case (1975) calculates relaxation times for five landbridge islands in the Gulf of California. The 95% simultaneous prediction band for this relationship is more than an order of magnitude wide (Figure 7*b*). Variation in these relationships results in imprecise point estimates. For example, Terborgh (1974) calculates extinction coefficients for five landbridge islands in the West Indies, constructs an EC–*A* relationship, and estimates the extinction coefficient for Barro Colorado Island from this relationship. The 95% prediction interval for Terborgh's point estimate spans three orders of magnitude (Table 3). Soulé et al. (1979) construct EC–*A* relationships for four faunal collapse models (models 2–5, Table 1) based on data from seven landbridge islands in the Malay Archipelago and use these relationships to estimate extinction coefficients for 19 nature reserves in East Africa. The 95% prediction intervals for the extinction coefficients of model 2 routinely span three

Table 2. Estimates of Pleistocene Avifauna Richness, $\hat{S}_0$, and Extinction Coefficients, $\hat{k}$, for Land Bridge Islands in the West Indies and Estimates of Postrelaxation Lizard Species Richness, $\hat{S}_{eq}$, and Relation Times, $\hat{t}_r$, for Land Bridge Islands in the Gulf of California[a]

West Indies[b]	Area (km^2)	Species	$\hat{S}_0$		$\hat{k}$	
Trinidad	4828	380	391	(218, 702)	1.68×10^{-7}	$(3.50 \times 10^{-8}, 2.81 \times 10^{-7})$
Margarita	1150	320	349	(199, 614)	1.06×10^{-6}	$(8.49 \times 10^{-7}, 1.19 \times 10^{-6})$
Coiba	453	250	323	(184, 567)	9.72×10^{-7}	$(7.39 \times 10^{-7}, 1.19 \times 10^{-6})$
Tobago	300	300	313	(179, 546)	9.00×10^{-7}	$(6.61 \times 10^{-7}, 1.04 \times 10^{-6})$
Rey	249	255	308	(176, 538)	1.85×10^{-6}	$(1.61 \times 10^{-6}, 1.99 \times 10^{-6})$
California[a]	**Area (km^2)**	**Species**	**$\hat{S}_{eq}$**		**$\hat{t}_r$**	
Tiberon	1196.0	11	8	(3, 13)	11,802	(—, 24,663)
San Marcos	31.5	9	7	(4, 10)	7,213	(—, 12,683)
Coronados	8.5	8	7	(4, 11)	4,809	(—, 9,885)
San Jose	194.0	10	7	(3, 11)	10,195	(—, 18,553)
San Francisco	2.6	7	6	(2, 9)	4,551	(—, 10,466)
Portida Sur and Espirito Santo	99.0	10	7	(3, 10)	10,195	(—, 18,553)

[a]The values in parentheses are 95% prediction intervals.
[b]From Terborgh, 1974.
[c]From Case, 1975.

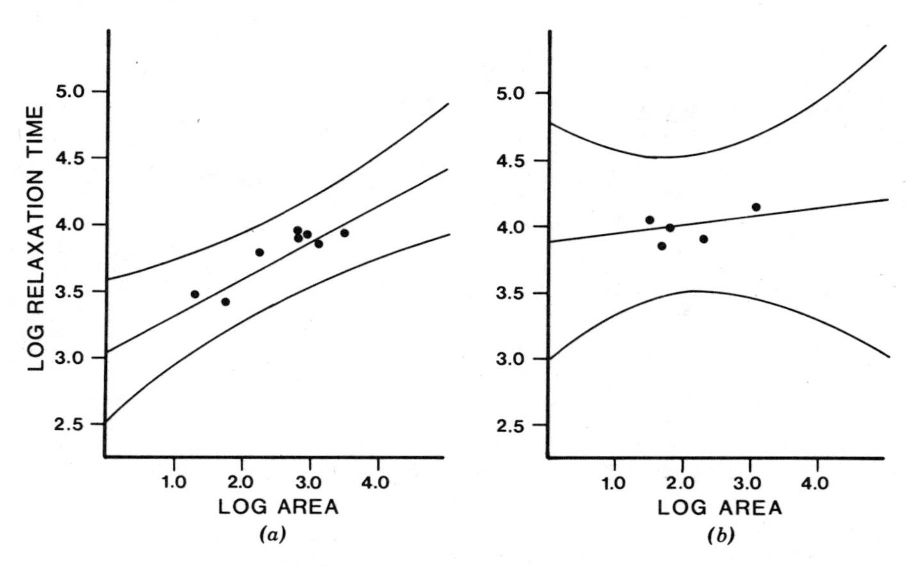

FIGURE 7. Relaxation time–area relationships for Diamond's (1972) analysis of bird species richness on landbridge islands in the New Guinea satellite island group (*a*) and for Case's (1975) analysis of lizard species richness on landbridge islands in the Gulf of California (*b*). The curved lines are 95% simultaneous prediction intervals.

orders of magnitude, and they are even wider for the other models. Substituting the lower and upper bounds of the 95% prediction intervals for the point estimates into the appropriate faunal collapse model gives a 95% prediction interval for projected species loss. The faunal collapse models of Soulé et al. (1979) typically predict with a 95% level of confidence that 0.2%–100% of the original species will be extinguished in these reserves within 5000 yr of insularization (Figure 8).

The parameter estimates of RT–*A* and EC–*A* relationships are sensitive to particular observations. For example, the slope estimate for Terborgh's (1974) EC–*A* relationship varies from -0.20 to -0.74 following systematic deletion of single observations. The intercept estimate varies from -3.98 to -5.44 (Figure 9*a*). Consequently, Terborgh's point estimate, the extinction coefficient for Barro Colorado Island, varies from 2.06×10^{-6} to 1.28×10^{-5}. The slope estimate for Soulé et al.'s (1979) EC–*A* model 3 varies from -0.19 to -0.46, and the intercept estimate varies from -3.26 to -4.42 (Figure 9*b*). As a result, the point estimate of the extinction coefficient for the Nairobi National Park varies from 1.56×10^{-5} to 6.13×10^{-5}.

The predictive power of faunal collapse models and relaxation models is far too low to warrant specific conservation recommendations. However, even imprecise models may illustrate general principles. For example, an imprecise species–area relationship illustrates that species richness generally increases with area. Whether faunal collapse and relaxation models illustrate that extinction rates generally decrease with increasing area is questionable owing to the shortcomings discussed above.

Table 3. Estimates of Extinction Coefficients, $\hat{k}$, *for Barro Colorado Island*[a] *and Some East African Savanna Reserves*[a]

	Model	Area (km^2)	$\hat{k}$
Barro Colorado Island[b]	3	17	8.56×10^{-6} $(1.87 \times 10^{-7}, 3.91 \times 10^{-4})$
East African Savanna Reserves[c]			
Tsavo National Park, Kenya	2	20,808	7.86×10^{-5} $(3.20 \times 10^{-6}, 1.93 \times 10^{-3})$
	3		3.64×10^{-6} $(2.35 \times 10^{-8}, 5.64 \times 10^{-4})$
	4		2.42×10^{-7} $(1.79 \times 10^{-10}, 3.26 \times 10^{-4})$
	5		1.46×10^{-8} $(7.49 \times 10^{-13}, 2.85 \times 10^{-4})$
Serengeti National Park, Tanzania	2	14,504	8.27×10^{-5} $(3.45 \times 10^{-6}, 1.98 \times 10^{-3})$
	3		4.07×10^{-6} $(3.16 \times 10^{-8}, 5.24 \times 10^{-4})$
	4		2.89×10^{-7} $(2.14 \times 10^{-10}, 3.90 \times 10^{-4})$
	5		1.86×10^{-8} $(9.54 \times 10^{-13}, 3.63 \times 10^{-4})$
Ruaha National Park, Tanzania	2	12,950	8.40×10^{-5} $(3.50 \times 10^{-6}, 2.02 \times 10^{-3})$
	3		4.22×10^{-6} $(3.28 \times 10^{-8}, 5.44 \times 10^{-4})$
	4		3.05×10^{-7} $(2.26 \times 10^{-10}, 4.11 \times 10^{-4})$
	5		2.01×10^{-8} $(1.03 \times 10^{-12}, 3.92 \times 10^{-4})$
Ngorongoro Conservation Area, Tanzania	2	6,475	9.26×10^{-5} $(3.77 \times 10^{-6}, 2.27 \times 10^{-3})$
	3		5.23×10^{-6} $(3.97 \times 10^{-8}, 6.89 \times 10^{-4})$
	4		4.29×10^{-7} $(3.04 \times 10^{-10}, 6.06 \times 10^{-4})$
	5		3.22×10^{-8} $(1.54 \times 10^{-12}, 6.73 \times 10^{-4})$
Murchison Falls National Park, Uganda	2	4,040	9.89×10^{-5} $(3.94 \times 10^{-6}, 2.48 \times 10^{-3})$
	3		6.05×10^{-6} $(4.38 \times 10^{-8}, 8.35 \times 10^{-4})$
	4		5.41×10^{-7} $(3.66 \times 10^{-10}, 8.00 \times 10^{-4})$
	5		4.44×10^{-8} $(1.89 \times 10^{-12}, 1.04 \times 10^{-3})$
Amboseli Game Reserve, Kenya	2	3,261	1.02×10^{-4} $(3.97 \times 10^{-6}, 2.62 \times 10^{-3})$
	3		6.47×10^{-6} $(4.58 \times 10^{-8}, 9.14 \times 10^{-4})$
	4		6.00×10^{-7} $(3.87 \times 10^{-10}, 9.29 \times 10^{-4})$
	5		5.14×10^{-8} $(2.19 \times 10^{-12}, 1.20 \times 10^{-3})$

Marasabit National Park, Kenya	2	2,072	1.09×10^{-4}	$(4.05 \times 10^{-6}, 2.93 \times 10^{-3})$
	3		7.45×10^{-6}	$(4.92 \times 10^{-8}, 1.13 \times 10^{-3})$
	4		7.50×10^{-7}	$(4.42 \times 10^{-10}, 1.27 \times 10^{-3})$
	5		7.00×10^{-8}	$(2.60 \times 10^{-12}, 1.88 \times 10^{-3})$
Queen Elizabeth National Park, Uganda	2	1,986	1.09×10^{-4}	$(4.05 \times 10^{-6}, 2.93 \times 10^{-3})$
	3		7.54×10^{-6}	$(4.98 \times 10^{-8}, 1.14 \times 10^{-3})$
	4		7.66×10^{-7}	$(4.51 \times 10^{-10}, 1.30 \times 10^{-3})$
	5		7.20×10^{-8}	$(2.68 \times 10^{-12}, 1.94 \times 10^{-3})$
Masai Mara Game Reserve, Kenya	2	1,813	1.11×10^{-4}	$(4.12 \times 10^{-6}, 2.99 \times 10^{-3})$
	3		7.76×10^{-6}	$(5.01 \times 10^{-8}, 1.20 \times 10^{-3})$
	4		8.01×10^{-7}	$(4.61 \times 10^{-10}, 1.39 \times 10^{-3})$
	5		7.66×10^{-8}	$(2.72 \times 10^{-12}, 2.16 \times 10^{-3})$
Midepo Valley National Park, Uganda	2	1,259	1.16×10^{-4}	$(4.17 \times 10^{-6}, 3.27 \times 10^{-3})$
	3		8.69×10^{-6}	$(5.36 \times 10^{-8}, 1.41 \times 10^{-3})$
	4		9.57×10^{-7}	$(5.02 \times 10^{-10}, 1.83 \times 10^{-3})$
	5		9.82×10^{-8}	$(3.11 \times 10^{-12}, 3.11 \times 10^{-3})$
Mikumi National Park, Tanzania	2	1,165	1.18×10^{-4}	$(4.19 \times 10^{-6}, 3.33 \times 10^{-3})$
	3		8.90×10^{-6}	$(5.36 \times 10^{-8}, 1.48 \times 10^{-3})$
	4		9.94×10^{-7}	$(5.10 \times 10^{-10}, 1.94 \times 10^{-3})$
	5		1.03×10^{-7}	$(3.18 \times 10^{-12}, 3.33 \times 10^{-3})$
Meru Game Reserve, Kenya	2	1,021	1.20×10^{-4}	$(4.16 \times 10^{-6}, 3.46 \times 10^{-3})$
	3		9.27×10^{-6}	$(5.46 \times 10^{-8}, 1.57 \times 10^{-3})$
	4		1.06×10^{-6}	$(5.31 \times 10^{-10}, 2.11 \times 10^{-3})$
	5		1.13×10^{-7}	$(3.33 \times 10^{-12}, 3.83 \times 10^{-3})$
Aberdares National Park, Kenya	2	590	1.29×10^{-4}	$(4.17 \times 10^{-6}, 3.99 \times 10^{-3})$
	3		1.10×10^{-5}	$(5.77 \times 10^{-8}, 2.10 \times 10^{-3})$
	4		1.39×10^{-6}	$(5.79 \times 10^{-10}, 3.33 \times 10^{-3})$
	5		1.64×10^{-7}	$(3.84 \times 10^{-12}, 7.00 \times 10^{-3})$

Table 3. (*Continued*)

	Model	Area (km^2)	$\hat{k}$	
Mt. Kenya National Park	2	588	1.30×10^{-4}	$(4.21 \times 10^{-6}, 4.02 \times 10^{-3})$
	3		1.10×10^{-5}	$(5.77 \times 10^{-8}, 2.10 \times 10^{-3})$
	4		1.39×10^{-6}	$(5.79 \times 10^{-10}, 3.33 \times 10^{-3})$
	5		1.65×10^{-7}	$(3.78 \times 10^{-12}, 7.20 \times 10^{-3})$
Lake Manyara National Park, Tanzania	2	319	1.41×10^{-4}	$(4.16 \times 10^{-6}, 4.78 \times 10^{-3})$
	3		1.33×10^{-5}	$(6.08 \times 10^{-8}, 2.91 \times 10^{-3})$
	4		1.88×10^{-6}	$(6.37 \times 10^{-10}, 5.55 \times 10^{-3})$
	5		2.50×10^{-7}	$(4.25 \times 10^{-12}, 1.50 \times 10^{-2})$
Samburu-Isiolo Game Reserve, Kenya	2	298	1.42×10^{-4}	$(4.10 \times 10^{-6}, 4.92 \times 10^{-3})$
	3		1.36×10^{-5}	$(6.07 \times 10^{-8}, 3.04 \times 10^{-3})$
	4		1.94×10^{-6}	$(6.42 \times 10^{-10}, 5.86 \times 10^{-3})$
	5		2.62×10^{-7}	$(4.35 \times 10^{-12}, 1.60 \times 10^{-2})$
Nairobi National Park, Kenya	2	114	1.63×10^{-4}	$(3.91 \times 10^{-6}, 1.00 \times 10^{-2})$
	3		1.83×10^{-5}	$(6.20 \times 10^{-8}, 5.40 \times 10^{-3})$
	4		3.11×10^{-6}	$(6.80 \times 10^{-10}, 1.42 \times 10^{-2})$
	5		5.03×10^{-7}	$(4.69 \times 10^{-12}, 5.39 \times 10^{-2})$
Ngurdoto Crater National Park, Tanzania	2	54	1.81×10^{-4}	$(3.70 \times 10^{-6}, 1.00 \times 10^{-2})$
	3		2.31×10^{-5}	$(6.22 \times 10^{-8}, 8.58 \times 10^{-3})$
	4		4.48×10^{-6}	$(6.78 \times 10^{-10}, 2.96 \times 10^{-2})$
	5		8.36×10^{-7}	$(4.81 \times 10^{-12}, 1.45 \times 10^{-1})$
Olorgessailie National Park Kenya	2	0.2	3.96×10^{-4}	$(1.73 \times 10^{-6}, 9.07 \times 10^{-2})$
	3		1.31×10^{-4}	$(3.29 \times 10^{-8}, 5.22 \times 10^{-1})$
	4		6.96×10^{-5}	$(3.13 \times 10^{-10}, 1.43 \times 10^{-1})$
	5		3.76×10^{-5}	$(1.80 \times 10^{-12}, 7.86 \times 10^{-2})$

[a]The values in parentheses are 95% prediction intervals.
[b]From Terborgh, 1974.
[c]From Soulé et al., 1979.

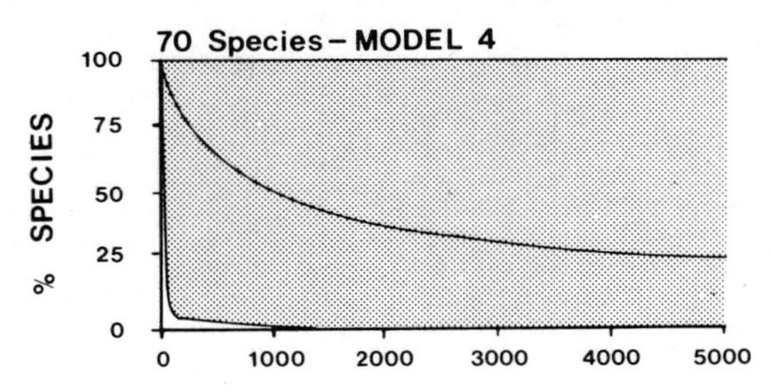

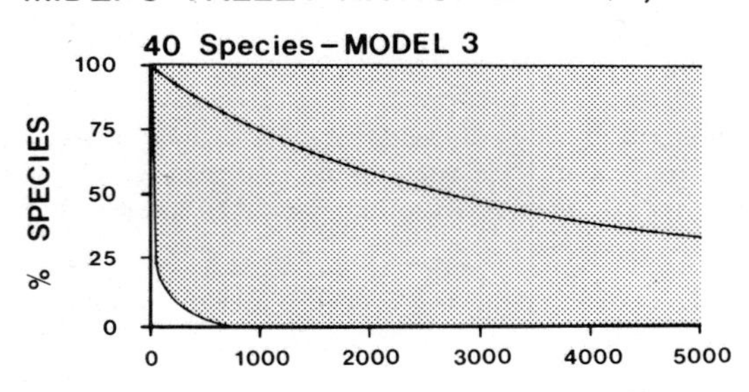

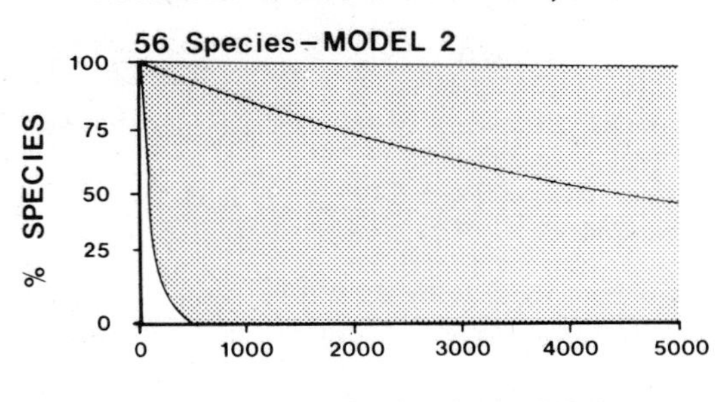

FIGURE 8. Faunal collapse of three East African savanna reserves. The stippled areas are 95% simultaneous prediction intervals (data from Soulé et al., 1979).

Even if they do, the pattern is relevant to conservation if and only if the models generate unique recommendations about refuge design.

The recommendation that a single large refuge is preferable to several small refuges of equal total area is not a unique consequence of faunal collapse and relaxation models. As with species–area relationships, the optimal design strategy will depend on the degree of faunal similarity among the small refuges. It will also depend on the slope of the EC–A or RT–A relationship. For example, the Nairobi National Park is 114 km^2 and currently contains 56 species of large mammals.

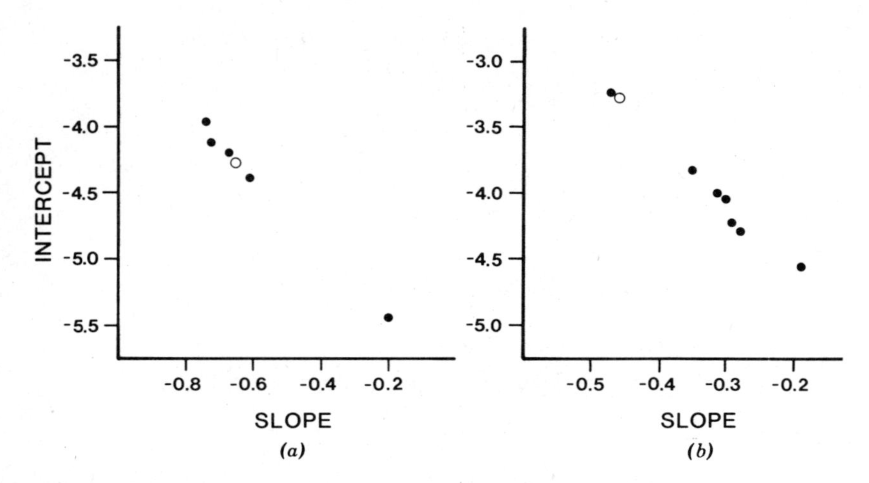

FIGURE 9. Slope and intercept estimates following casewise deletion of single observations for Terborgh's (1974) extinction coefficient–area relationship (*a*) and Soulé et al.'s (1979) extinction coefficient–area relationship for model 3 (*b*). The open circles are the parameter estimates using all data.

Soulé et al. (1979) estimate its extinction coefficient for model 2, 1.63×10^{-4}, from the equation

$$\log k = -0.14 \log A - 3.5 \tag{6}$$

The faunal collapse of the Nairobi National Park is given by

$$S(t) = 56e^{-kt} \tag{7}$$

The analogous expression for a network of refuges is (Boecklen and Bell, personal communication)

$$S(t) = \sum n_1[1 - (1 - e^{-kt})^i] \tag{8}$$

where n_i is the number of species found in i refuges and $k = 1.63 \times 10^{-4}$. This model predicts that the Nairobi refuge will contain only 25 species 5000 yr from now (Table 4).

Solving equation (6) for 57 km^2 gives the extinction coefficients for two small refuges each half the area of the Nairobi National Park. If the two small refuges contain all 56 species but share no species, then 23 species will remain 5000 yr after insularization. This is only two species less than the number remaining in the single large refuge. Moreover, less than a single species difference exists for the first 500 yr of isolation. However, if the two refuges have species in common, then more species may be preserved in the two small refuges than in the single large one. For example, if the two share 10 species, the number remaining after 5000 yr is given by equation (8) as 25. If the two small refuges share 15 species, then 26 species will remain after 5000 yr. Similar results also obtain for three small refuges, and these results are all probably conservative. Between-refuge migration would increase species survivorship in an archipelago.

Table 4. Effects of Reserve Fragmentation on the Faunal Collapse (Model 2) of the Nairobi National Park, Kenya[a]

Reserves	Species Shared	Years after Isolation					
		0	50	500	1000	3000	5000
1	—	56.00	55.55	51.62	47.58	34.35	24.79
2	0	56.00	55.50	51.19	46.80	32.68	22.82
	5	56.00	55.54	51.58	47.48	33.89	24.02
	10	56.00	55.59	51.98	48.71	35.11	25.23
	15	56.00	55.63	52.37	48.85	36.33	26.44
3	0,0	56.00	55.47	50.92	46.13	31.67	21.66
	4, 1	56.00	55.52	51.35	47.03	33.00	22.99
	8,2	56.00	55.56	51.76	47.79	34.34	24.32
	12, 3	56.00	55.61	52.18	48.53	35.67	25.66

[a]Data from Soulé et al., 1979. The three reserves share the first number of species in two reserves, the second number in three reserves.

The slope of the EC–A relationship determines the degree of faunal similarity necessary for an archipelago to preserve more species than a single refuge does. The steeper the slope, the more species must be shared within the archipelago. However, since the EC–A relationship will vary from situation to situation, as will faunal similarities, determination of the optimal design strategy for a particular application is an empirical rather than a theoretical matter.

CONCLUSIONS

We have seen, then, that the application of faunal collapse and relaxation models to conservation is beset with the same problems that render species–area relationships rather poor predictors of extinction. In particular, they are not useful in resolving the question of whether one large or several small refuges would be better in any particular instance.

First, the models themselves do not render unique predictions. One can get different results about the speed of species loss (or the number of species remaining after a fixed period) by using different parameters and/or models.

Second, to measure the parameters (or to verify the models) that would be necessary to generate unique predictions would at least require extensive observations of the sort that are virtually nonexistent in the biological literature. Some sorts of observations (e.g., numbers of species present on landbridge islands at various times in the past) can probably not be made at all.

Third, faunal collapse and relaxation models, like the species–area models that are their progenitors, take no account of habitat or of the biology of the species in any particular community. The abundant evidence that habitat and, occasionally, specific idiosyncratic species interactions are critical to the continued existence of all species should make one suspicious about extinction predictions that do not take these factors into account. In fact, just as species–area curves have wide confidence limits and vary greatly from system to system, so do we find that confidence limits around predictions generated from relaxation and faunal collapse models are enormous. They are so large, in fact, that it is difficult to conceive of how they could fruitfully be used to produce specific recommendations about refuge design.

Finally, in spite of the fact that it was quickly shown that species–area curves can tell us very little about how to design refuges, early recommendations based on such curves have insinuated themselves into the planning literature as apodictic principles. There is great danger that proposals based on relaxation and faunal collapse models will become similarly entrenched. For example, the concluding chapter (Schonewald-Cox, 1983) of a recent conference on managing animal and plant populations for conservation sponsored by the U.S. Man and the Biosphere Program states as fact the predictions of Soulé et al. (1979) on faunal collapse in African refuges and fails even to cite the very different conclusions of Western and Ssemakula (1981) on the same topic. Such selective and uncritical treatment of the literature is exactly the reason that the species–area relationship has been incorrectly accepted as a key element in conservation planning.

REFERENCES

Abbott, I., 1980, Theories dealing with the ecology of landbirds on islands, *Adv. Ecol. Res.*, **11**, 329–371.

Abbott, I., and Grant, P. R., 1976, Non-equilibrial bird faunas on islands, *Amer. Nat.*, **110**, 507–528.

Abele, L. G., and Connor, E. F., 1979, Application of island biogeography theory to refuge design: Making the right decision for the wrong reasons, in Linn, R. M., ed., *Proceedings of the First Conference on Scientific Research in the National Parks*, Vol. 1, Washington, U.S. Department of the Interior.

Berry, R. J., 1971, Conservation aspects of the genetic constitution of populations, in Duffy, E., and Watt, A. S., eds., *The Scientific Management of Animal and Plant Communities for Conservation*, Oxford, Blackwell, pp. 177–206.

Boecklen, W. J., and Gotelli, N. J., 1984, Island biogeographic theory and conservation practice: Species–area or specious–area relationships?, *Biol. Conserv.*, **29**, 63–80.

Burgess, R. L., and Sharpe, D. M., 1981, Introduction, in Burgess, R. L., and Sharpe, D. M., eds., *Forest Island Dynamics in Man-Dominated Landscapes*, New York, Springer-Verlag, pp. 1–5.

Case, T. J., 1975, Species numbers, density compensation, and colonizing ability of lizards on islands in the Gulf of California, *Ecology*, **56**, 3–18.

Chamberlain, T. C., 1898, A systematic source of evolution of provincial faunas, *J. Geol.*, **6**, 597–608.

Clegg, M. T., and Brown, A. H. D., 1983, The founding of plant populations, in Schonewald-Cox, C. M., Chambers, S. M., MacBryde, B., and Thomas, W. L., eds., *Genetics and Conservation*, Menlo Park, California, Benjamin/Cummings, pp. 216–228.

Coleman, B. D., Mares, M. A., Willig, M. R., and Hsieh, Y., 1982, Randomness, area, and species richness, *Ecology*, **63**, 1121–1133.

Connor, E. F., and McCoy, E. D., 1979, The statistics and biology of the species-area relationship, *Am. Nat.*, **113**, 791–833.

Connor, E. F., and Simberloff, D. S., 1978, Species number and compositional similarity of the Galapagos flora and avifauna, *Ecol. Monogr.*, **48**, 219–248.

Connor, E. F., McCoy, E. D., and Cosby, B. J., 1983, Model discrimination and expected slope values in species-area studies, *Am. Nat.*, **122**, 789–796.

Diamond, J. M., 1969, Avifaunal equilibria and species turnover on the Channel Islands of California, *Proc. Nat. Acad. Sci. U.S.A.*, **64**, 57–63.

Diamond, J. M., 1972, Biogeographic kinetics: Estimation of relaxation times for avifaunas of Southwest Pacific Islands, *Proc. Nat. Acad. Sci. U.S.A.*, **69**, 3199–3203.

Diamond, J. M., 1973, Distributional ecology of New Guinea birds, *Science*, **179**, 759–769.

Diamond, J. M., 1975, The island dilemma: Lessons of modern biogeographic studies for the design of natural reserves, *Biol. Conserv.*, **7**, 129–146.

Diamond, J. M., and May, R. M., 1976, Island biogeography and the design of natural reserves, in May, R. M., ed., *Theoretical Ecology*, Philadelphia, Saunders, pp. 163–186.

Diamond, J. M., and May, R. M., 1977, Species turnover rates on islands: Dependence on census interval, *Science*, **197**, 266–270.

East, R., 1983, Application of species-area curves to African savannah reserves, *Afr. J. Ecol.*, **21**, 123–128.

Faaborg, J., 1979, Qualitative patterns of avian extinction on neotropical land-bridge islands: Lessons for conservation, *J. Appl. Ecol.*, **16**, 99–107.

Faeth, S. H., and Connor, E. F., 1979, Supersaturated and relaxing island faunas: A critique of the species–age relationship, *J. Biogeogr.*, **6**, 311–316.

Flessa, K. W., 1975, Area, continental drift and mammalian diversity, *Paleobiology*, **1**, 189–194.

Flessa, K. W., and Sepkoski, J. J., Jr., 1978, On the relationship between Phanerozoic diversity and changes in habitable area, *Paleobiology*, **4**, 359–366.

Franklin, I. R., 1980, Evolutionary change in small populations, in Soule, M. E., and Wilcox, B. A., eds., *Conservation Biology: An Evolutionary–Ecological Perspective*, Sunderland, Massachusetts, Sinauer, pp. 135–149.

Gilbert, F. S., 1980, The equilibrium theory of island biogeography: Fact or fiction? *J. Biogeogr.*, **7**, 209–235.

Haas, P. H., 1975, Some comments on the use of the species–area curve, *Am. Nat.*, **109**, 371–373.

Haffer, J., 1969, Speciation in Amazonian forest birds, *Science*, **165**, 131–137.

Hamilton, T. H., and Armstrong, N. E., 1965, Environmental determination of insular variation in bird species abundance in the Gulf of Guinea, *Nature*, **207**, 148–151.

Hartl, D. L., 1981, *A Primer of Population Genetics*, Sunderland, Massachusetts, Sinauer.

Higgs, A. J., 1981, Island biogeography theory and nature reserve design, *J. Biogeogr.*, **8**, 117–124.

Higgs, A. J., and Usher, M. B., 1980, Should nature reserves be large or small?, *Nature*, **285**, 568–569.

I.U.C.N., 1980, *World Conservation Strategy*, Gland, Switzerland, International Union for Conservation of Nature and Natural Resources, United Nations Environmental Program, World Wildlife Fund.

Jablonski, D., 1985, Mass extinction, hypothesis testing, and the fossil record: Ecologic and biogeographic patterns as critical data, in Elliot, D. K., ed., *Dynamics of Extinction*, New York, Wiley.

Johnson, M. P., and Simberloff, D., 1974, Environmental determinants of island species numbers in the British Isles, *J. Biogeogr.*, **1**, 149–154.

Jones, H. L., and Diamond, J. M., 1976, Short-term-base studies of turnover in breeding bird populations on the California Channel Islands, *Condor*, **78**, 526–549.

Kindlmann, P., 1983, Do archipelagoes really preserve fewer species than one island of the same total area, *Oecologia*, **59**, 141–144.

Kitchener, D. J., Chapman, A., Dell, J., Muir, B. G., and Palmer, M., 1980, Lizard assemblage and reserve size and structure in the Western Australian wheatbelt: Some implications for conservation, *Biol. Conserv.*, **17**, 25–62.

Lynch, J. F., and Johnson, N. K., 1974, Turnover and equilibria in insular avifaunas with special reference to the California Channel Islands, *Condor*, **76**, 370–384.

MacArthur, R. H., and Wilson, E. O., 1963, An equilibrium theory of insular zoogeography, *Evolution*, **17**, 373–387.

MacArthur, R. H., and Wilson, E. O., 1967, *The Theory of Island Biogeography*, Princeton, New Jersey, Princeton University Press.

Margules, C., Higgs, A. J., and Rafe, R. W., 1982, Modern biogeographic theory: Are there any lessons for nature reserve design, *Biol. Conserv.*, **24**, 115–128.

McCoy, E. D., 1982, The application of island-biogeographic theory to forest tracts: Problems in the determination of turnover rates, *Biol. Conserv.*, **22**, 217–227.

Miller, R. I., 1978, Applying island biogeographic theory to an East African reserve, *Environ. Conserv.*, **5**, 191–195.

Moore, R. C., 1954, Evolution of Late Paleozoic invertebrates in response to major oscillations of shallow seas, *Bull. Mus. Comp. Zool.*, **112**, 259–286.

Newell, N. D., 1967, Revolutions in the history of life, Geological Society of America Special Paper 89, 63–91.

Picton, H. D., 1979, The application of insular biogeographic theory to the conservation of large mammals in the northern Rocky Mountains, *Biol. Conserv.*, **15**, 73–79.

Power, D. M., 1972, Numbers of bird species on the California islands, *Evolution*, **26**, 451–463.

Pregill, G. K., and Olson, S. L., 1981, Zoogeography of West Indian Vertebrates in relation to Pleistocene climatic cycles, *Ann. Rev. Ecol. Syst.*, **12**, 75–98.

Preston, F. W., 1960, Time and space and the variation of species, *Ecology*, **41**, 612–627.

Ralls, K., and Ballou, J., 1983, Extinction: Lessons from zoos, in Schonewald-Cox, C. M., Chambers, S. M., MacBryde, B., and Thomas, W. L., eds., *Genetics and Conservation*, Menlo Park, California, Benjamin/Cummings, pp. 164–184.

Rey, J. R., 1981, Ecological biogeography of arthropods on *Spartina* islands in northwest Florida, *Ecol. Monogr.*, **51**, 237–265.

Richter-Dyn, N., and Goel, N. S., 1972, On the extinction of a colonizing species, *Theor. Pop. Biol.*, **3**, 406–433.

Samson, F. B., 1980, Island biogeography and conservation of nongame birds, Transactions of North American Wildlife and Natural Resources Conference, 45th, pp. 245–251.

Schaal, B. A., and Levin, D. A., 1976, The demographic genetics of *Liatris cylindracea* Michx. (Compositae), *Am. Nat.*, **110**, 191–206.

Schonewald-Cox, C. M., 1983, Conclusions: Guidelines to management: A beginning attempt, in Schonewald-Cox, C. M., Chambers, S. M., MacBryde, B., and Thomas, W. L., eds., *Genetics and Conservation*, Menlo Park, California, Benjamin/Cummings, pp. 414–445.

Schopf, T. J. M., 1974, Permo-Triassic extinctions: Relation to sea-floor spreading, *J. Geol.*, **82**, 129–143.

Sepkoski, J. J., Jr., 1976, Species diversity in the Phanerozoic: Species-area effects, *Paleobiology*, **2**, 298–303.

Shaffer, M. L., 1978, Determining minimum viable population sizes: A case study of the grizzly bear (*Ursus arctos* L.) Ph.D. dissertation, Duke University.

Shaffer, M. L., 1981, Minimum population sizes for species conservation, *BioSci.*, **31**, 131–134.

Simberloff, D. S., 1969, Experimental zoogeography of islands: A model for insular colonisation, *Ecology*, **50**, 296–314.

Simberloff, D. S., 1974a, Equilibrium theory of island biogeography and ecology, *Ann. Rev. Ecol. Syst.*, **5**, 161–182.

Simberloff, D. S., 1974b, Permo-Triassic extinctions: Effects of area on biotic equilibrium, *J. Geol.*, **82**, 267–274.

Simberloff, D. S., 1976a, Species turnover and equilibrium island biogeography, *Science*, **194**, 572–578.

Simberloff, D., 1976b, Experimental zoogeography of islands: Effects of island size, *Ecology*, **57**, 629–648.

Simberloff, D. S., 1978, Using island biogeographic distributions to determine if colonisation is stochastic, *Am. Nat.*, **112**, 713–726.

Simberloff, D., 1983, Biogeographic models, species distributions, and community organization, in Sims, R. W., Price, J. H., and Whalley, P. E. S., eds., *Evolution, Time and Space: The Emergence of the Biosphere*, London, Academic Press, pp. 57–83.

Simberloff, D., 1985, Design of nature reserves, in Usher, M. B., ed., *Wildlife Conservation Evaluation*, London, Chapman and Hall, in press.

Simberloff, D. S., and Abele, L. G., 1976, Island biogeography theory and conservation practice, *Science*, **191**, 285–286.

Simberloff, D., and Abele, L. G., 1982, Refuge design and island biogeographic theory: Effects of fragmentation, *Am. Nat.*, **120**, 41–50.

Simberloff, D., and Gotelli, N., 1984, Effects of insularisation on plant species richness in the prairie–forest ecotone, *Biol. Conserv.*, **29**, 27–46.

Simberloff, D. S., and Wilson, E. O., 1969, Experimental zoogeography of islands: The colonisation of empty islands, *Ecology*, **50**, 278–296.

Simberloff, D. S., and Wilson, E. O., 1970, Experimental zoogeography of islands: A two-year record of colonisation, *Ecology*, **51**, 934–937.

Simpson, B. B., and Haffer, J., 1978, Speciation patterns in the Amazonian forest biota, *Ann. Rev. Ecol. Syst.*, **9**, 497–518.

Smith, F. E., 1975, Ecosystems and evolution, *Bull. Ecol. Soc. Am.*, **56**, 2.

Soulé, M. E., 1983, What do we really known about extinction?, in Schonewald-Cox, C. M., Chambers, S. M., MacBryde, B., and Thomas W. L., eds., *Genetics and Conservation*, Benjamin/Cummings, pp. 111–124.

Soulé, M. E., Wilcox, B. A., and Holtby, C., 1979, Benign neglect: A model of faunal collapse in the game reserves of East Africa, *Biol. Conserv.*, **15**, 259–272.

Sugihara, G., 1981, $S = CA^z$, $z = 1/4$: A reply to Connor and McCoy, *Am. Nat.*, **117**, 790–793.

Terborgh, J., 1974, Preservation of natural diversity: The problem of extinction prone species, *BioSci*, **24**, 715–722.

Terborgh, J., 1975, Faunal equilibria and the design of wildlife preserves, in Golley, F., and Medina, E., eds., *Tropical Ecological Systems: Trends in Terrestrial and Aquatic Research*, New York, Springer, pp. 369–380.

Terborgh, J., and Winter, B., 1980, Some causes of extinction, in Soulé, M. E., and Wilcox, B. A., eds., *Conservation biology: An Evolutionary-Ecological Perspective*, Sunderland, Massachusetts, Sinauer, pp. 119–133.

Usher, M. B., 1979, Changes in the species-area relation of higher plants on nature reserves, *J. Appl. Ecol.*, **16**, 213–215.

Vuilleumier, B. S., 1971, Pleistocene changes in the fauna and flora of South America, *Science*, **173**, 771–780.

Watson, G., 1964, Ecology and evolution of passerine birds on the islands of the Aegean Sea, Ph.D. dissertation, Yale University.

Webb, S. D., 1969, Extinction-origination equilibria in late Cenozoic land mammals of North America, *Evolution*, **23**, 688–702.

Western, D., and Ssemakula, J., 1981, The future of the savannah ecosystems: Ecological islands or faunal enclaves?, *Afr. J. Ecol.*, **19**, 7–19.

Whitcomb, B. L., Whitcomb, R. F., and Bystrak, D., 1977, Island biogeography and habitat islands of Eastern forest, III. Long-term turnover and effects of selective logging on the avifauna of forest fragments, *Am. Birds*, **31**, 17–23.

Wilcox, B. A., 1978, Supersaturated islands faunas: A species-age relationship for lizards on post-Pleistocene land-bridge islands, *Science*, **199**, 996–998.

Wilcox, B. A., 1980, Insular ecology and conservation, in Soulé, M. E., and Wilcox, B. A., eds., *Conservative Biology: An Evolutionary-Ecological Perspective*, Sunderland, Massachusetts, Sinauer, pp. 95–117.

Williams, C. B., 1964, Patterns in the balance of nature and related problems in quantitative ecology, New York, Academic Press.

Willis, E. O., 1984, Conservation, subdivision of reserves, and the antidismemberment hypothesis, *Oikos*, **42**, 396–398.

Wilson, E. O., and Willis, E. O., 1975, Applied biogeography, in Cody, M. L., and Diamond, J. M., eds, *Ecology and Evolution of Communities*, Cambridge, Massachusetts, Belknap Press, pp. 522–534.

Wynne-Edwards, V. C., 1962, *Animal Dispersion in Relation to Social Behavior*, New York, Hafner.

AUTHOR INDEX

SUBJECT INDEX